AF678371

Advance in Sedimentology and Coastal and Marine Geology

Advance in Sedimentology and Coastal and Marine Geology

Editor

Gemma Aiello

MDPI

Basel • Beijing • Wuhan • Barcelona • Belgrade • Novi Sad • Cluj • Manchester

Editor
Gemma Aiello
CNR ISMAR Sezione
Secondaria di Napoli
Napoli
Italy

Editorial Office
MDPI
St. Alban-Anlage 66
4052 Basel, Switzerland

This is a reprint of articles from the Special Issue published online in the open access journal *Journal of Marine Science and Engineering* (ISSN 2077-1312) (available at: https://www.mdpi.com/journal/jmse/special_issues/sedimentology_geology).

For citation purposes, cite each article independently as indicated on the article page online and as indicated below:

Lastname, A.A.; Lastname, B.B. Article Title. *Journal Name* **Year**, *Volume Number*, Page Range.

ISBN 978-3-7258-1087-1 (Hbk)
ISBN 978-3-7258-1088-8 (PDF)
doi.org/10.3390/books978-3-7258-1088-8

© 2024 by the authors. Articles in this book are Open Access and distributed under the Creative Commons Attribution (CC BY) license. The book as a whole is distributed by MDPI under the terms and conditions of the Creative Commons Attribution-NonCommercial-NoDerivs (CC BY-NC-ND) license.

Contents

About the Editor

Gemma Aiello

Dr. Gemma Aiello was born in Aversa (CE), Italy, on 24th October 1964. In 1989, she graduated in Geological Sciences at the University of Naples "Federico II". In 1993, she earned her Ph.D. in Sedimentary Geology at the University of Naples "Federico II", Department of Earth Sciences, Faculty of Geological Sciences. She completed a two-year postdoctoral fellowship at the University of Naples "Federico II", a CNR-CEE fellowship, and several contracts at the Research Institute "Geomare Sud", CNR, Naples, Italy. Since 1998, she has been a full-time researcher at the Italian CNR. Dr. Aiello has 25 years of experience in the fields of sedimentary geology, marine geology, and geophysics, participating in different research projects of the Italian National Research Council (CARG, Vector, Centri Regionali di Competenza). She has worked as a contract professor of sedimentology and stratigraphy at the Parthenope University of Naples, Italy, and as a teacher in formation courses of technicians in marine science and engineering in the same city.

Journal of
Marine Science and Engineering

MDPI

Editorial

Advances in Sedimentology and Coastal and Marine Geology

Gemma Aiello

Institute of Marine Sciences (ISMAR), National Research Council of Italy (CNR), 80133 Naples, Italy; gemma.aiello@cnr.it

Citation: Aiello, G. Advances in Sedimentology and Coastal and Marine Geology. *J. Mar. Sci. Eng.* **2023**, *11*, 329. https://doi.org/10.3390/jmse11020329

Received: 17 January 2023
Revised: 19 January 2023
Accepted: 19 January 2023
Published: 3 February 2023

Copyright: © 2023 by the author. Licensee MDPI, Basel, Switzerland. This article is an open access article distributed under the terms and conditions of the Creative Commons Attribution (CC BY) license (https://creativecommons.org/licenses/by/4.0/).

The Special Issue "Advances in Sedimentology and Coastal and Marine Geology" has collected significant research articles advancing the state of the art of the corresponding sub-disciplines. Beach rocks, hybrid event beds and quaternary marine sedimentation and shallow gas represent important topics contained within this Special Issue. Different areas have been studied, including both Mediterranean and extra-Mediterranean areas. In the Mediterranean areas, the Bay of Naples and the Cilento offshore (Southern Tyrrhenian Sea, Italy) have been studied, along with the Red Sea (Saudi Arabia). The Bay of Naples displays the physiographic domains of the Somma–Vesuvius volcanic complex, the Campi Flegrei volcanic complex, the Ischia and Capri islands, and the Sorrento Peninsula. The Cilento offshore, studied by two papers of the Special Issue, is the marine area surrounding the Cilento Promontory and represents a structural high resulting from the seaward prolongation of the Licosa Cape's high structure, bounded northwards and southwards by two half-graben basins: the Salerno Valley and the Policastro Gulf. The Red Sea, studied by two papers within the Special Issue, separates the coasts of Egypt, Sudan, and Eritrea to the west from those of Saudi Arabia and Yemen to the east. The Red Sea contains some of the world's warmest and saltiest seawater, and its name is derived from the colour changes observed in its waters. Normally, the Red Sea is an intense blue-green; occasionally; however, it is populated by extensive blooms of the algae *Trichodesmium erythraeum*, which, upon dying off, turn the sea a reddish-brown colour. In the extra-Mediterranean areas, Quanzhou Bay and the South China Sea (China), the Taiwan Strait (Taiwan), and north-western Borneo (Malaysia) have been examined. Quanzhou Bay is a semi-enclosed bay located in the southeast coast of China, with its mouth opening towards the Taiwan Straits. Two rivers, the Jinjiang and Luoyang, carry sediments to the bay and are connected to the Taiwan Straits. The sediments are also introduced to the bay from the open sea and the anthropogenic factors have a strong influence on the bay. The South China Sea is a marginal sea south of China, included in the Pacific Ocean, and encompassing an area from Singapore to the Taiwan Strait. The sea stretches in a southwest to northeast direction, whose southern border is 3 degrees. The Gulf of Thailand covers the western portion of the South China Sea. The South China Sea overlies a drowned continental shelf; during recent ice ages, the global sea level was hundreds of meters lower, and Borneo was part of the Asian mainland. The Taiwan Strait is an arm of the Pacific Ocean, 160 km wide at its narrowest point, lying between the coast of China's Fukien province and the island of Taiwan (Formosa). The strait extends from southwest to northeast between the South and East China seas, reaching a depth of about 70 m. Borneo is the third-largest island in the world and the largest in Asia. At the geographic centre of Maritime Southeast Asia, in relation to major Indonesian islands, it is located north of Java, west of Sulawesi, and east of Sumatra. The island is politically divided among three countries: Malaysia and Brunei in the north, and Indonesia to the south.

Beach rocks, which are the subject of two papers of this Special Issue, are carbonate-cemented sandstones, individuating on the shoreline of tropical and warm temperate beaches. Beach rocks represent a good potential indicator of sea-level position. Their location within the intertidal zone makes beach rocks potentially good indicators of former sea level, particularly if the tidal range is small. Most beach rocks are found in low-latitude locations, although there are particularly extensive outcrops around the Mediterranean Sea. In the studied areas, beach rocks mainly consist of calcareous biogenic remains (Red Sea, Arabian Gulf); alternatively, they consist of siliciclastic grains in near the tectonically active areas (Gulf of Aqaba).

Hybrid event beds are the deposits of such flows which do not resemble traditional end-member sediment gravity flow facies (debris flows or turbidites), as they result from a combination of turbulent, transitional and laminar transportation mechanisms, all as part of the same event. Hybrid event beds include transitional flow deposits, slurry beds and matrix-rich sandstones. Hybrid event beds comprise up to five (H1-H5) vertically stacked divisions, including a basal matrix-poor turbidite sandstone (H1), overlain by a banded sandstone (H2) with alternating matrix-rich and matrix-poor bands; this gives way to a matrix-rich sandstone (H3) which constitutes the 'linked debrite'. The matrix-rich sandstone of H3, typically contains mudstone clasts, mud chips, dispersed clay matrix, mica flakes, sand injections, pseudo nodules and organic matter. A couple of thin, well-structured sandstones/siltstones (H4) and a mudstone (H5) complete the typical sequence of hybrid event beds.

Aiello and Caccavale [1] have shown the depositional environments of the Cilento offshore based on marine geological data, including sedimentologic and seismo-stratigraphic data and seismic stratigraphy. Based on marine geological mapping, coupled with seismo-stratigraphic interpretation and confirmed by the sedimentological data, littoral, inner shelf and outer shelf environments compose the highstand system tract of the Late Quaternary depositional sequence. The seismo-stratigraphic results are basic for further developments of Aiello and Caccavale on shallow gas in the same area [2]. The first area, distinguished by acoustic blanking is located offshore from the Licosa Cape promontory at water depths ranging between 30 and 90 m; the second area, distinguished by shallow gas pockets, is located in the northern Cilento promontory from the seaward prolongation of the Paestum Plain to the Tresino Cape; the third area, distinguished by shallow gas pockets and by the seismic units impregnated by gas ("gassy sediments"), is located on the northern Cilento promontory, starting from the offshore prolongation of the Paestum Plain up to the Licosa Cape promontory, at water depths ranging between 10 and 60 m.

Aiello [3] has studied the seismo-stratigraphic setting of the eastern Bay of Naples, recognizing new seismic units and correlating them with buried tuff rings; these were compared with the Porto Miseno, Archiaverno and Averno and Astroni tuff rings onshore (Campi Flegrei) and a new seismic unit, recognized offshore the Somma–Vesuvius volcano, correlating with the fallout deposits representing the base of the AD 79 eruptive sequence based on isopach maps available in the literature. The impact of PDCs (pyroclastic density currents) on this area, both onshore and offshore, has been discussed, due to the dispersal, thickness and extent of the pyroclastic deposits, corresponding with seven plinian and sub-plinian eruptions.

Ghandour et al. [4] studied the lithofacies characteristics, petrographic, XRD, and stable isotope data of Al-Mejarma beach rocks, located in the Red Sea, Saudi Arabia, to reconstruct its geologic evolution. Beach rock was deposited during the Middle-Late Holocene as a shoreface-beach barrier in two stages, attesting a landward migrating sediment accumulation and a rapid marine cementation. The beach rock consists of massive, planar, and ripple cross-laminated sand with foresets trending parallel to the slightly oblique Red Sea coastline. Six vertically stacked lithofacies have been distinguished through the textural characteristics and the sedimentary structures. The obtained results are basic for the subsequent paper of Mannaa et al. [5], dealing with the use of beach rocks as palaeo-shoreline indicators.

Hybrid event beds in deep marine environments have been studied by Jamil et al. [6]. In particular, these authors have discussed the development of hybrid event beds in submarine lobes, analyzing proximal-to-distal, frontal-to-lateral relationships and evolution during lobe progradation. A geological survey was carried out on the Late Paleogene Crocker Fan to understand the relationship between the character of hybrid bed facies and lobe architecture. The obtained results have shown that the hybrid facies are well developed in proximity to medial lobes, whereas distal lobes are mainly parallel to cross-laminated clean-to-muddy hybrid facies. Deep marine environments usually include a fan lobe system and only locally hybrid event beds. Muddy sandstone and clay-rich hybrid event bed facies negatively affect the reservoir potential of sandy lobe intervals, hindering the pore network and connectivity for lateral and vertical migration of fluids from sandstone reservoirs.

Shan et al. [7] have analyzed the effects of the mineralogic composition of the clays on the dynamic properties and fabric of artificial marine clays. In particular, the marine clays found in the South China Sea deltas have been studied. Raw non-clay minerals (such as quartz, albite) and clay minerals (such as Na-montmorillonite and kaolinite) have been used to produce artificial marine clay, whose dynamic properties were studied through the mineralogic composition. Dynamic triaxial laboratory tests for artificial marine clay comprising various clay minerals have been carried out, showing that marine clay with a high montmorillonite content exhibited slower development of strain than the clay with a low montmorillonite content. The obtained results have shown that the clay minerals, especially montmorillonite, strongly control the dynamic properties of large strain.

Xiao et al. [8] detailed an integrated modeling of Quanzhou Bay (China), associating the geomorphologic evolution and the oceanographic parameters. Based on the topographic and hydrological data of the bay, a numerical model was used to simulate the hydrodynamics of the bay under the influence of human activities. The coupling relationship between human activities and the evolution of geomorphology and hydrodynamics in the urban bay at different stages of industrialization has been discussed. This study has demonstrated the anthropogenic impact on geomorphology and oceanography in highly industrialized settings.

Finally, He et al. [9] have studied the sedimentary evolution of the Western Taiwan Shoal Area during the Late Pleistocene. A new pollen analysis and major and trace element contents has been carried out on a gravity core recovered from the Taiwan Shoal (sand ridges), south of the Taiwan Strait. The variations in the pollen assemblage and concentration have highlighted the climate change in the Taiwan Shoal and the strength of the Zhe-Min Coastal Current. Five phases of the sedimentary evolution in the Taiwan Shoal area have been distinguished based on the analysis of pollen and major trace elements with the combination of the AMS ^{14}C dating results.

Funding: This research received no external funding.

Conflicts of Interest: The author declares no conflict of interest.

References

1. Aiello, G.; Caccavale, M. The Depositional Environments in the Cilento Offshore (Southern Tyrrhenian Sea, Italy) Based on Marine Geological Data. *J. Mar. Sci. Eng.* **2021**, *9*, 1083. [CrossRef]
2. Aiello, G.; Caccavale, M. New Seismoacoustic Data on Shallow Gas in Holocene Marine Shelf Sediments, Offshore from the Cilento Promontory (Southern Tyrrhenian Sea, Italy). *J. Mar. Sci. Eng.* **2022**, *10*, 1992. [CrossRef]
3. Aiello, G. Submarine Stratigraphy of the Eastern Bay of Naples: New Seismo-Stratigraphic Data and Implications for the Somma-Vesuvius and Campi Flegrei Volcanic Activity. Volcanic Activity. *J. Mar. Sci. Eng.* **2022**, *10*, 1520. [CrossRef]
4. Ghandour, I.M.; Al-Washmi, H.A.; Khan, A.A.; Mannaa, A.A.; Aljahdali, M.H.; Jones, B.G. Depositional Setting and Cementation Pattern of Al-Mejarma Beachrocks, Saudi Arabia: A Proxy for the Late Quaternary Red Sea Coastal Evolution. *J. Mar. Sci. Eng.* **2021**, *9*, 1012. [CrossRef]
5. Mannaa, A.A.; Haredy, R.A.; Ghandour, I.M. Beachrock as a Paleoshoreline Indicator: Example from Wadi Al-Hamd, South Al-Wajh, Saudi Arabia. *J. Mar. Sci. Eng.* **2021**, *9*, 984. [CrossRef]
6. Jamil, M.; Siddiqui, N.A.; Ahmed, N.; Usman, M.; Umar, M.; Rahim, H.U.; Imran, Q.S. Facies Analysis and Sedimentary Architecture of Hybrid Event Beds in Submarine Lobes: Insights from the Crocker Fan, NW Borneo, Malaysia. *J. Mar. Sci. Eng.* **2021**, *9*, 1133. [CrossRef]
7. Shan, Y.; Wang, X.; Cui, J.; Mo, H.; Li, Y. Effects of Clay Mineral Composition on the Dynamic Properties and Fabric of Artificial Marine Clay. *J. Mar. Sci. Eng.* **2021**, *9*, 1216. [CrossRef]
8. Xiao, X.; Li, Y.; Tang, J.; Luo, F.; Shu, F.; Wang, L.; He, J.; Zou, X.; Chi, W.; Lin, Y.; et al. Coupling Relationship of Geomorphic Evolution and Marine Hydrodynamics in the Stage-Specific Development of Urban Bays: A Modelling Case Study in Quanzhou Bay (1954–2017), China. *J. Mar. Sci. Eng.* **2022**, *10*, 1677. [CrossRef]

9. He, C.; Sha, L.; Zhao, D.; Dai, L.; Li, Z.; Tang, J.; Li, X.; Li, D. Sedimentary Environmental Evolution of the Western Taiwan Shoal Area since the Late Pleistocene. *J. Mar. Sci. Eng.* **2021**, *9*, 1150. [CrossRef]

Disclaimer/Publisher's Note: The statements, opinions and data contained in all publications are solely those of the individual author(s) and contributor(s) and not of MDPI and/or the editor(s). MDPI and/or the editor(s) disclaim responsibility for any injury to people or property resulting from any ideas, methods, instructions or products referred to in the content.

Journal of
Marine Science and Engineering

MDPI

Article

The Sediments in the Beibu Gulf Reveal Dramatic Paleoenvironmental Changes and Climate Events over the Past ~20,000 Years

Yuchun Li [1], Tianlai Fan [1,*], Aihua Wang [2,3], Jun Zeng [4,5], Yubiao Lv [4,5], Mingwang Zhang [1] and Dajun Liu [1]

1 Guangxi Laboratory on the Study of Coral Reefs in the South China Sea, Coral Reef Research Center of China, School of Marine Sciences, Guangxi University, Nanning 530004, China; 15941454960@163.com (Y.L.); zmw0612@163.com (M.Z.); gxu_ldj@163.com (D.L.)
2 Nanjing Hongchuang Exploration Technology Service, Nanjing 210023, China; jsnjywa@163.com
3 Nanjing Center, China Geological Survey, Nanjing 210016, China
4 Guangxi Academy of Sciences, Nanning 530007, China; gdyelhbkj@163.com (J.Z.); lyuyubiao@163.com (Y.L.)
5 Institute of Beibu Gulf Marine Industry, Fangchenggang 538000, China
* Correspondence: fantl2004@hotmail.com

Abstract: The geochemical characteristics of a 2.1 m BBW25 core, collected from the Beibu Gulf, have been investigated in terms of the major and trace elements, organic matter, and $CaCO_3$ and AMS ^{14}C dating by XRF, ICP-OES, ICP-MS, and more. We have found through previous research that there are issues with unclear delineation of sedimentary evolution environments and inexact responses between chemical weathering intensity and major paleoclimate events in the Beibu Gulf. The AMS ^{14}C dating results indicate that the sedimentary age at the bottom was 19.24 ky b.p. $CaCO_3$, $\delta^{13}C$, C/N, and Sr/Ba indexes show a sedimentary environment change from terrestrial to marine environments and a "jump" of ~4000 years in continent–ocean changes. The evolution of the sedimentary environment of Beibu Gulf was divided into three environments and five sub-environments. The changes in chemical weathering intensity indicators recorded by the CIX and the Fe/Al ratio respond well to the East Asian monsoon cycle, the meltwater events, and the alternation of cold and warm events. This study explains the chemical weathering intensity and sedimentary environment in the BBW25 core by geochemical characteristics and further reveals the paleoenvironmental characteristics and possible driving mechanisms over the past ~20,000 years.

Keywords: Beibu Gulf; core sediment; geochemical characteristics; sedimentary environment; climate change

Citation: Li, Y.; Fan, T.; Wang, A.; Zeng, J.; Lv, Y.; Zhang, M.; Liu, D. The Sediments in the Beibu Gulf Reveal Dramatic Paleoenvironmental Changes and Climate Events over the Past ~20,000 Years. *J. Mar. Sci. Eng.* **2024**, *12*, 615. https://doi.org/10.3390/jmse12040615

Academic Editor: Gemma Aiello

Received: 27 February 2024
Revised: 26 March 2024
Accepted: 27 March 2024
Published: 2 April 2024

Copyright: © 2024 by the authors. Licensee MDPI, Basel, Switzerland. This article is an open access article distributed under the terms and conditions of the Creative Commons Attribution (CC BY) license (https://creativecommons.org/licenses/by/4.0/).

1. Introduction

As part of the South China Sea, the largest marginal sea in the western Pacific, the Beibu Gulf preserved information on changes in the marine environment with sediments since the late last glacial maximum (LGM), and its deposition processes are highly impacted by land–sea interactions, sea-level changes, East Asian monsoons (EAMs), and other processes. The Beibu Gulf, located on the northwestern edge of the South China Sea, is a semi-enclosed bay with an area of approximately 12.93×10^4 km^2, an average depth of 42 m, and a maximum depth of 100 m. It is located in the shelf area of the South China Sea and is connected to this sea through the Qiongzhou Strait. Due to its unique geographical location, diverse seabed topography, and complex ocean currents and hydrodynamic environment, its marine sediments preserve rich paleoclimatic information in East Asia [1–11].

Since the end of the last glacial maximum (LGM) (20,000 years before the present), the climate system of the Beibu Gulf has experienced the Heinrich1 (H1) cold, Bølling-Allerød (BA) warm, Meltwater Pulse 1B (MWP-1B), Younger Dryas (YD) cold, and Holocene rhythmic cooling (Bond) events. Prior studies have extensively investigated the geochemical characteristics of these sediments and provided important information on sea level, sea surface temperature (SST), and circulation changes; sediment source-to-sink processes; as well as chemical weathering intensity changes and coastal erosion processes in the surrounding continents. The acquisition of knowledge regarding the interactions between paleoclimate and continent–ocean changes is crucial for the comparative analysis and comprehensive understanding of present and future interconnections among these systems. Currently, there are disagreements regarding the delineation of sedimentary evolution environments and responses between the chemical weathering intensity and major paleoclimate events in the Beibu Gulf [12–19].

The C4 and B106 cores in the central and southern parts of the Beibu Gulf indicated that the Beibu Gulf was a terrestrial to marine–terrestrial alternating sedimentary environment prior to 13.4 ka BP. The STAT22 core indicated a marine–terrestrial alternating sedimentary environment at around 11.1 ka BP. The inconsistent paleoclimatic events in different latitude regions may reflect the lagging response of low-latitude regions to the global climate or the existence of regional special climate change mechanisms. In fact, in the South China Sea, there are differences in the Paleoclimatic events revealed by data from different cores, and even different research methods used at the same core can yield different results, possibly due to limitations in certain methods or environmental constraints. The TWS-1, ZK20, ZK001, and PC338 cores in the South China Sea indicated that the chemical weathering intensity in the northwest of the South China Sea was significantly higher in the Holocene than in the glacial stage, while the KT2 core indicated an opposite conclusion [20–26].

Through a review of previous research, it was found that the classification of sedimentary environments in the Beibu Gulf mainly focused on the Holocene or the late last deglaciation, without considering the early last deglaciation or the LGM. Simultaneously, the indicators for chemical weathering intensity and sedimentary environments have remained largely unchanged, and the new indicators developed in recent years have not been introduced. Therefore, this study discusses changes in the marine environment of the Beibu Gulf and high-resolution information on possible climate responses through investigating geochemical characteristics in the core sediments. Thus, our work aims at improving our understanding of the changes in the marine environment of marginal seas over the past 20,000 years and their correlation with global climate change.

2. Material and Methods

2.1. Materials

As shown in Figure 1, sampling point BBW25 is located in the offshore area southwest of Hainan Island (108°08.9166′ E, 18°07.2945′ N). A gravity piston sampler was used to sample the columnar core sediment, with an overlying water depth of 84 m and an overall length of approximately 2.1 m. The samples had high viscosity and exhibited a homogeneous fluid–soft plastic mud state. Occasionally, bivalve and gastropod debris could be seen in the upper part. The lithology was mainly clay with a small amount of fine sand. The lower sediment contained almost no shell debris and was composed of clay (Figure 1). The core samples were stored in a freezer (below −20 °C) in the cabin laboratory of the ship and then transferred to Guangdong Ocean University for cryopreservation. To avoid the effects of benthic disturbance, resuspension, and seafloor re-sedimentation, the core was cut into 210 samples (1 cm per sample). All samples were wrapped with pre-sanitized aluminum foil and sealed in sterile polyethylene bags. One quarter of each sample was taken for subsequent experiments in this study.

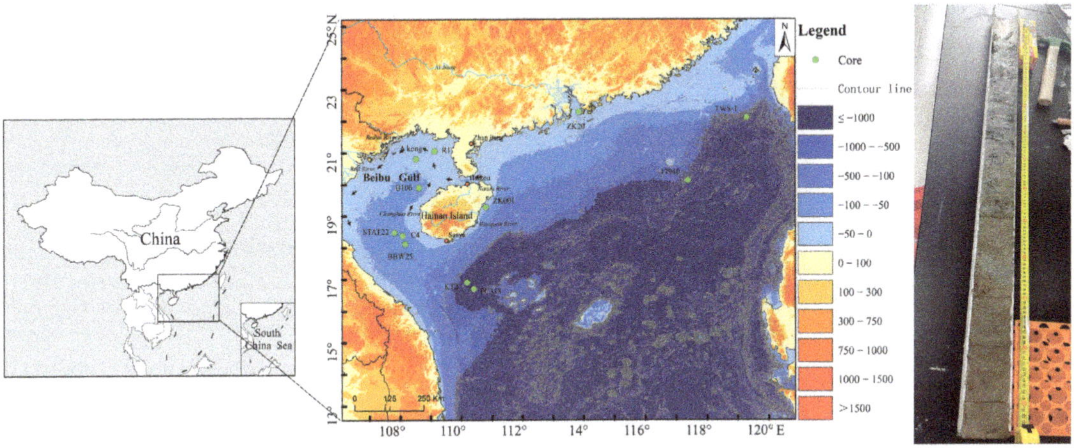

(a) (b)

Figure 1. BBW25 sediment core: (**a**) Topographic map of core BBW25 sampling location, surrounding areas, and the cores mentioned in the article; (**b**) core profile.

2.2. Methods

The samples were first placed in an oven at 60 °C for drying and then ground to below 200 mesh. Rock powder samples were sent to BETA Laboratory for AMS ^{14}C dating. The sample type was organic sediment, and the pre-treatment method was acid washing. The half-life of ^{14}C used for dating was 5568 years. The calibration program was BetaCal4.20, the calibration method was the HPD method, and the database used was MARINE20. Measured ages were corrected by ^{13}C fractionation to obtain conventional ages, and the conventional ages were corrected by the tree ring to obtain calendar ages. To facilitate the calculation, interval averages with high confidence (2σ, 95.4%) of the corrected ages were mainly used in the comparison of AMS ^{14}C ages. The corrected calendar ages were calculated on this basis. Chemical element contents of rock powder samples were measured using a (NITON XL3T) handheld X-ray fluorescence spectrometer (XRF—School of Marine Sciences, Guangxi University, Nanning, China). Major element contents of the processed samples were detected using (HORIBA) inductively coupled plasma optical emission spectroscopy (ICP-OES—Nanjing Hongchuang Exploration Technology Service Co. Ltd., Nanjing, China) and the processing and experimental processes were referred to Chemical analysis methods of silicate rocks Part 31: Determination of 12 components such as silica [27]. The rock powder samples were washed with 10% hydrochloric acid (Qingtian Biotechnology Co. Ltd., Nanning, China) for 24 h. The solid residues were rinsed with deionized water, centrifuged eight times for a neutral pH, and dried at 60 °C. After acid washing, TOC and TN contents of the rock powder samples were detected using the (vario MACRO cube) element analyzer (School of Marine Sciences, Guangxi University, Nanning, China). After acid washing, the samples were mixed with rod-shaped copper oxide and platinum wire as catalysts, then burnt at 800–850 °C in the experimental furnace with O_2 introduced. The generated CO_2 was released after condensation and purification in a liquid nitrogen cold trap for detection of organic carbon isotope ($\delta^{13}C_{V\text{-}PDB}$). $\delta^{13}C_{V\text{-}PDB}$ contents of the rock powder samples were detected using the (Vario EL III/Isoprime) elemental analyzer–isotope ratio mass spectrometer (EA-IRMS—Instrumental Analysis Center of Shanghai Jiao Tong University, Shanghai, China). The content of Sr and Ba selected extraction from the core was detected using (Agilent 7900) Inductively Coupled Plasma Mass Spectrometry (ICP-MS—Nanjing Hongchuang Exploration Technology Service Co. Ltd., Nanjing, China), while the processing and experimental procedures were referred to Chemical Analysis Methods of Silicate Rocks Part 30: Determination of 44 [28,29].

3. Results

3.1. Construction of the Stratigraphic Age and Sedimentation Rate Frameworks Based on AMS ^{14}C

Samples were taken for AMS ^{14}C testing at depths of 40, 75, 120, 160, and 210 cm in the core. These results are represented by "cal. ka BP", which means "calibrated thousand years before present", abbreviated as "ka BP". The age framework should be established based on these results, and the Bayesian age-depth model should be utilized to acquire more comprehensive and precise age information of the BBW25 sediment core [30]. The sedimentation rate of the upper profile is 0.008 cm/yr, and the lower profile is 0.102 cm/yr.

According to Figure 2 and Table 1, the age at the bottom of the sedimentary column in borehole BBW25 is 19.24 ka BP. In the sections above and below the burial depth of 119–120 cm, the age of the sediments changes significantly. The sedimentation rates of the upper and lower sections also have obvious changes, which can be attributed to changes in climate, sea level, or the source of the sediments [11].

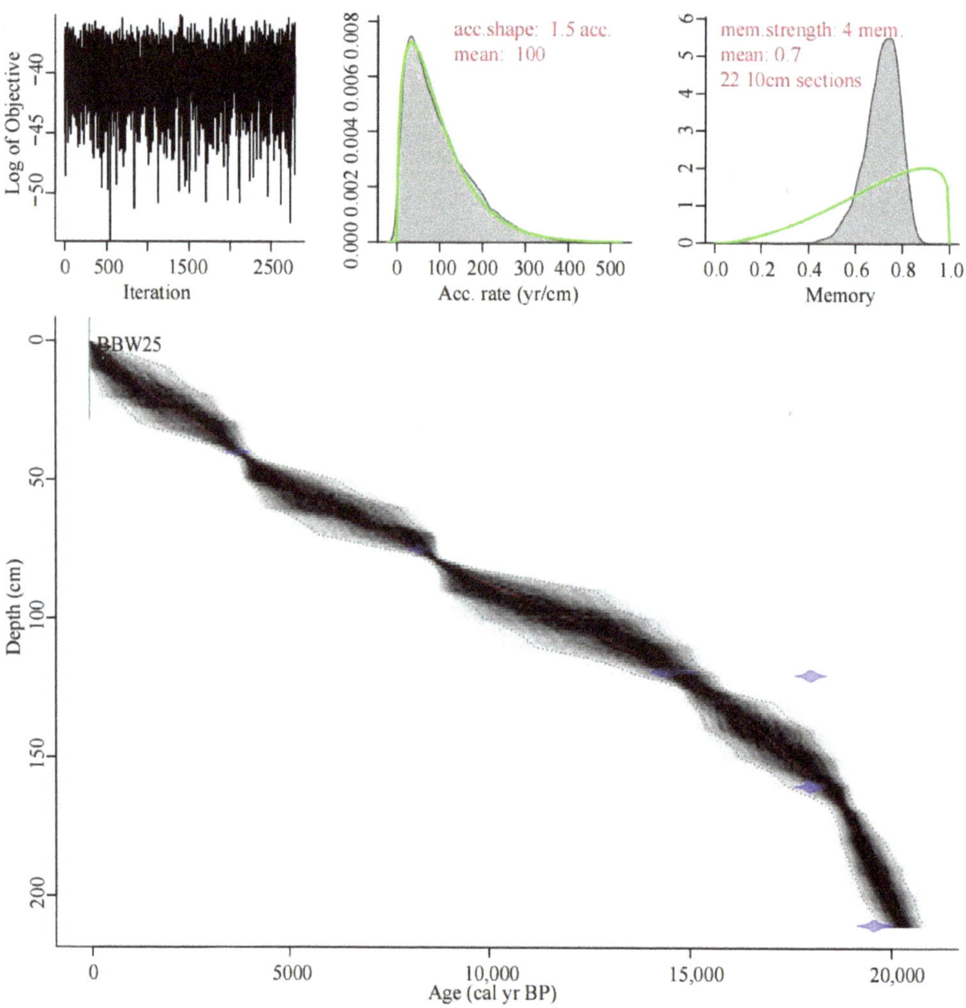

Figure 2. Bayesian age-depth model.

Table 1. BBW25 core AMS ^{14}C age results.

Depth (cm)	Conventional Radiocarbon Age	Age Correction (95.4% Probability)	Calibrated Average Calendar Age
40	3850 ± 30 BP	(72.5%) 2456–2271 cal BC 4405–4220 cal BP (22.9%) 2260–2204 cal BC 4209–4153 cal BP	4281 BP
75	7880 ± 30 BP	(95.4%) 6368–6068 cal BC 8317–8017 cal BP	8167 BP
119	12,890 ± 40 BP	(95.4%) 12,891–12,299 cal BC 14,840–14,248 cal BP	14,544 BP
120	15,850 ± 50 BP	(95.4%) 16,648–16,180 cal BC 18,597–18,129 cal BP	18,363 BP
160	15,290 ± 50 BP	(53%) 16,545–16,354 cal BC 18,494–18,303 cal BP (42.4%) 16,803–16,649 cal BC 18,752–18,598 cal BP	18,521 BP
210	16,700 ± 50 BP	(95.4%) 17,542–17,047 cal BC 19,491–18,996 cal BP	19,244 BP

3.2. Changes in Major Elements

Contents of major elements in core sediments also show a similar age-depth model pattern to that in the upper and lower profiles. Contents of major elements vary with the burial depth as shown in Figure 3. Among them, CaO and MgO contents exhibited patterns similar to those in the upper and lower sections in the age framework. The average content is ranked as $SiO_2 > Al_2O_3 > Fe_2O_3 > CaO > K_2O > Na_2O > TiO_2 > P_2O_5$. Among them, the SiO_2 content range is 50–70%; Al_2O_3 content is 8–15%; Fe_2O_3 content is 3–7%; CaO content is 0.2–8.5%; K_2O and MgO content is 1–4%; Na_2O content is 0.9–1.3%; TiO_2 content is 0.6–1.1%; P_2O_5 content is 0.09–0.2%; and MnO content is 0.02–0.07%. Figure 3a shows the results of the ICP-MS tests, which were performed at 10 cm intervals, for a total of 20 samples tested. Figure 3b shows the results of the XRF tests, which were performed at 1 cm intervals, for a total of 210 samples tested.

3.3. Changes in Organic Matter

Organic matter proxies and the content of Sr and Ba in the core sediments also show a similar age-depth model pattern to that in upper and lower sections. The variation in organic matter content with burial depth is shown in Figure 4, with a similar pattern of upper and lower sections in the age framework. The TOC content range of core sediment was 0.28–0.71%; TN content was 0.03–0.12%; and $\delta^{13}C_{V\text{-}PDB}$ (hereafter abbreviated as $\delta^{13}C$) content ranged between −25.5 and −18.54‰. Organic matter was tested at 2 cm intervals and a total of 105 samples were tested. The Sr content range of core sediment was 7–238.8%; and Ba content was 1.5–18.63%. Sr and Ba were tested at 5 cm intervals and a total of 42 samples were tested.

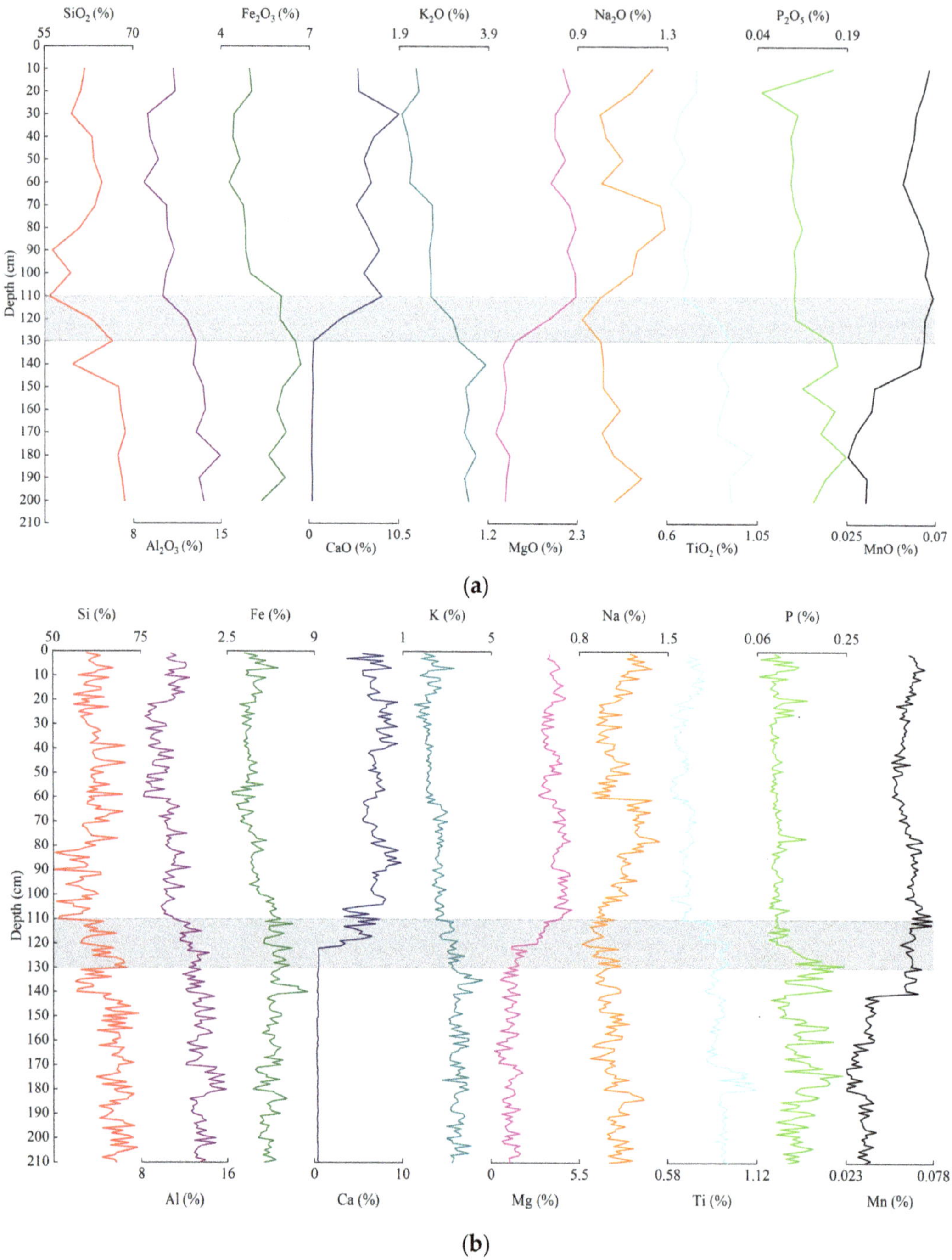

Figure 3. Measurement of major element contents in the BBW25 core: (**a**) ICP-OES; (**b**) XRF.

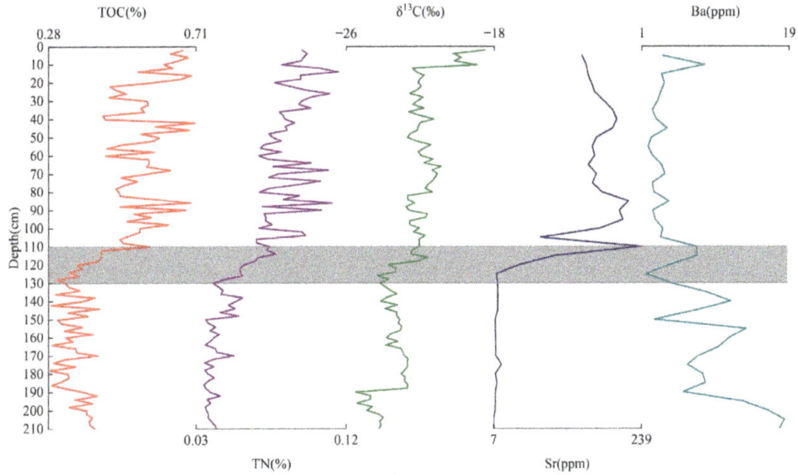

Figure 4. Organic matter proxies and the content of Sr and Ba in the BBW25 core.

3.4. Changes in Carbonate Contents

The carbonate content in the core sediment also shows a similar age-depth model pattern to that of the upper and lower sections. The variation in carbonate content with depth is shown in Figure 5. During the hydrochloric acid dissolution process of the organic matter pretreatment in core sediment, the mass of escaped CO_2 can be derived by calculating the change in sample mass before and after the hydrochloric acid dissolution. Through calculating the mass of emitted CO_2, we can calculate the percentage of CO_3^{2} and $CaCO_3$ in the sample. The variation trends of the three groups of data are relatively similar when comparing the CO_3^{2} content with the CaO and Ca contents measured by ICP-MS and XRF, respectively. Thus, it can be concluded that carbonate mostly existed in the form of calcium carbonate in the sediment. Due to significant errors, the XRF test results had to be corrected. Figure 5 shows that the Ca contents in sediments were relatively high with significant changes. Thus, the relative error can be ignored. Also, Figure 5 indicates that the XRF and ICP-MS test results have consistent changes, proving a good correspondence between the results. Therefore, in this study, the ICP-MS test result data are used to calibrate XRF test result data.

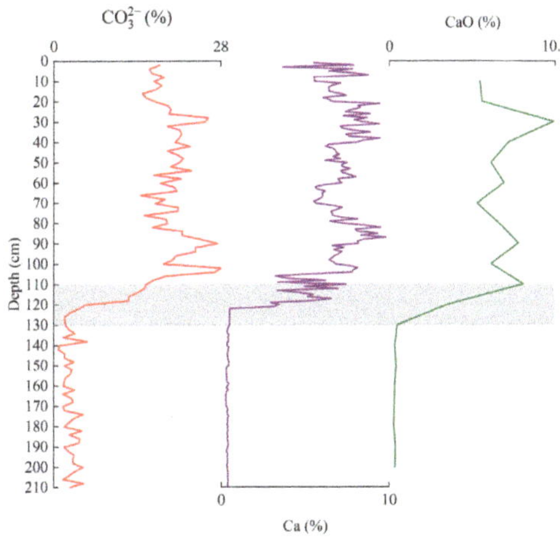

Figure 5. Comparison of carbonate and Ca contents in the BBW25 core.

4. Discussion

4.1. Response of Sedimentary Environments in the Beibu Gulf to Sea Level and Sea Surface Temperature (SST) in the South China Sea over the Past 20,000 years

4.1.1. Changes in Sea Level and SST and the Sedimentary Environment Indicators

As shown in Figure 6, previous studies have obtained information about the relative change in sea level in the South China Sea based on sea level changes in the Sunda shelf, Malacca Strait, and Red River Delta, as well as SST changes in the South China Sea based on changes in long-chain alkenones and $U^{k}{}_{37}$ in the 17940 core profile [9,31]. Sea level in the South China Sea was positively correlated with the mean SST. As temperatures increased, the ice age ended and glacial melting caused the sea level to rise. As shown in Figure 6, the sea level during the LGM in the South China Sea was estimated to be at least 110 m below its current position, resulting in complete exposure of the Beibu Gulf as terrestrial land. Subsequently, throughout the last deglaciation period, there was a gradual rise in sea level that led to the submergence of previously exposed land, transforming the sedimentary environment of the Beibu Gulf from terrestrial to marine. The BBW25 core sedimentary column was buried at a depth of approximately 84 m in modern times. At approximately 14.6–14.3 ka BP, the sea level was approximately between −87.5 and −82 m when the depth of the BBW25 core sedimentary column was 119–120 cm (depositional age about 14.54 ka BP). This was the location of significant changes in most geochemical characteristic indicators.

The Sr–Ba ratio is a geochemical indicator of paleosalinity in sediments and has been used to indicate the changes and evolution environments of transgression in the Beibu Gulf [22,32,33]. Previous studies in deltaic environments have found that Sr/Ba is <1 in freshwater (rivers), <1–3 in brackish water (delta front), <3–8 in saline water (prodelta), and >8 in normal seawater (shallow sea environment and residual sand) [34,35]. The C/N ratio and $\delta^{13}C$ are important indicators for determining the source of organic matter in sediments. Based on the ratio of total organic carbon and nitrogen contents in sediments, distinguishing between terrestrial and marine organic matter is possible. Generally, aquatic algae are rich in proteins and lipids, and their organic matter has a lower C/N ratio. Terrestrial higher plants are rich in their carbon content such as cellulose and lignin, while proteins comprise less than 20%. Their organic matter has a higher C/N ratio [36]. Previous studies on the TOC and TN of sediments in various sedimentary environments have found that the C/N ratio of marine sediments ranges between 4–7, while in terrestrial sediments it is ≥12. The C/N ratios for high-salinity and subtidal sediments are about 11.6 and 9.3, respectively. The C/N ratio of mixed marine and terrestrial sediments ranges between 8–12 [37–39]. Based on the C/N changes observed in the borehole in this study, the C/N ratios of 4–8 in the BBW25 core indicates marine sedimentation, while values >8 indicate marine–terrestrial alternating sedimentation. Previous studies on the $\delta^{13}C$ of sediments in a large number of different sedimentary environments revealed that the $\delta^{13}C$ of freshwater algae and terrestrial end-members was in the range of −25–−30‰; the $\delta^{13}C$ of marine algae and seawater end-members was in the range of −16–−23‰; and the $\delta^{13}C$ of sediments at the bottom of the ocean was in the range of −18–−24‰ [36,37,40–42]. Based on the $\delta^{13}C$ of BBW25 core sediments in this study, $\delta^{13}C$ of −18–−23‰ indicated a marine sedimentary environment, $\delta^{13}C$ of −23–−24‰ indicated a marine–terrestrial alternating sedimentary environment, and $\delta^{13}C$ of −24–−26‰ indicated a terrestrial sedimentary environment. The BBW25 core is mainly composed of clay particles, with no drastic changes in particle size. Previous studies on minerals in the Beibu Gulf have shown that when hydrodynamic forces are strong, kaolin tends to be present [43]. However, the kaolin content in the upper section of the core is higher than that in the lower section, and there is no kaolin in the lower section, indicating that the upper part of the core is in a marine sedimentary environment. The tectonic subsidence over the past 20,000 years has mainly been concentrated in the Qiongzhou Strait because the Beibu Gulf was a mainly terrestrial environment during the last glacial period, and the Qiongzhou Strait was mainly developed from 11 to 7 ka BP [11,44,45], while the drastic changes in geochemical characteristics of the BBW25 core occurred at about 14 ka BP. Thus, the changes in the sedimentary environment reflected by the geochemical characteristics of core sediments in the Beibu Gulf can be attributed to the rise in sea level observed in the South China Sea. Figure 5's research indicates that calcium existed in sediments in the form of calcium carbonate. Calcium carbonate can reflect the proportion of biogenic sources in sediments and can be used as a representative for marine inputs. Low contents were associated with an increased transfer of debris from the surrounding continent, while high contents were associated with increased inputs of marine biogenic sources [46].

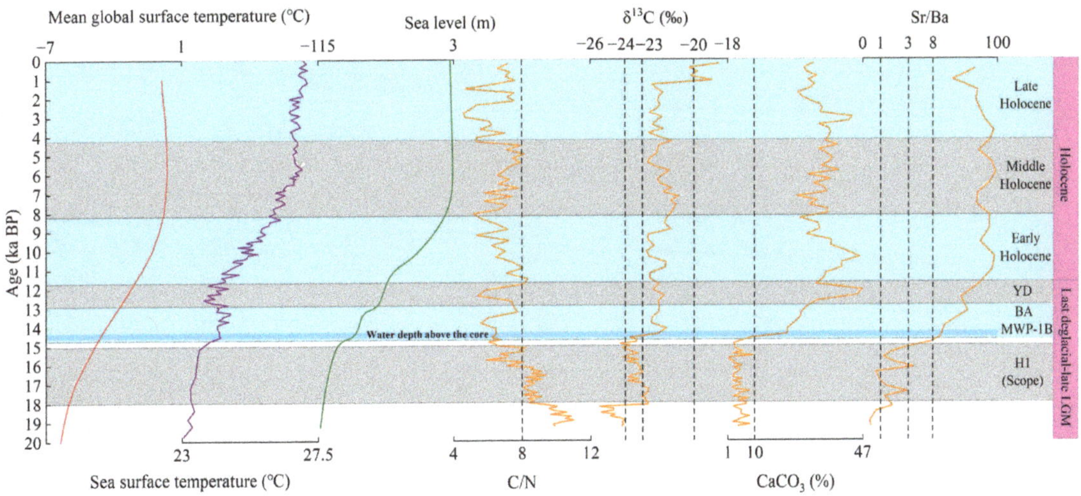

Figure 6. Comparison of sedimentary environment indicators in core over time with mean global surface temperature [47], sea surface temperature [9], and sea level change [31] in the South China Sea.

4.1.2. Establishment of a "Simplified Age-Depth Framework" and Phasing of Sedimentary Evolution

Table 1 shows that there is a "jump" in the age of the sediments at the depth of 119–120 cm, ~4000 years, making classifying sedimentary environments and analyzing chemical weathering intensity in the age-depth framework difficult. This phenomenon also occurred in the STAT22 core (both the STAT22 and BBW25 cores were located in the southern part of the Beibu Gulf). The ages at the burial depths of 290–295 cm and 305–310 cm in the STAT22 core were 11.11 and 17.99 ka BP, respectively, with an age "jump" of ~7000 years for two samples at consecutive burial depths. However, this phenomenon has not been explained. The classification of sedimentary environments in the Beibu Gulf based on the STAT22 core study also began from the Holocene [22]. Our study analyzed the sedimentation rates of seafloor sediments at the depths of 119–120, 160, and 210 cm, as well as the range of sedimentary environment indicators among the depths (sedimentary environment indicators) (Table 1 and Figure 7). It was found that the sedimentation rate of the 150–160 cm core section was 0.253 cm/yr, while the that of the 160–190 cm core section in the same sedimentary environment was 0.069 cm/yr. The sedimentation rates of the two depths differed significantly, which could be due to the mixing of new and old sediments during the last deglaciation, resulting in inaccurate age test results. Previous researchers who studied cores in the South China Sea proposed the theory of "tropical shelf weathering during the ice age" [26]. It was found that during the last ice age, the temperature drop in low-latitude tropical regions was relatively small, resulting in a decrease in sea level and widespread exposure of the continental shelf. Under the combined actions of suitable temperature and humidity, strong groundwater circulation, sufficient reaction time, and a large mineral surface area, the overlying loose deposits underwent further chemical weathering. At the same time, new land emerged during the LGM. The shelf sediments, which were previously deposited at high sea levels during the interglacial period, may have been re-weathered and transported by river erosion to the river basins near the northwest slope of the South China Sea. Starting from the deglaciation, the sea level rise caused a reduction in the exposed area of the continental shelf. The northwest of the South China Sea received more initial weathering products from river basins through ocean current suspension transport [24,25,48]. Studies on the R1 core near the Weizhou Island in the Beibu Gulf of the South China Sea found that fossils in the late Pleistocene strata (buried at a depth of about 1.5 m) contained Holocene marine sediments and transgression records [49].

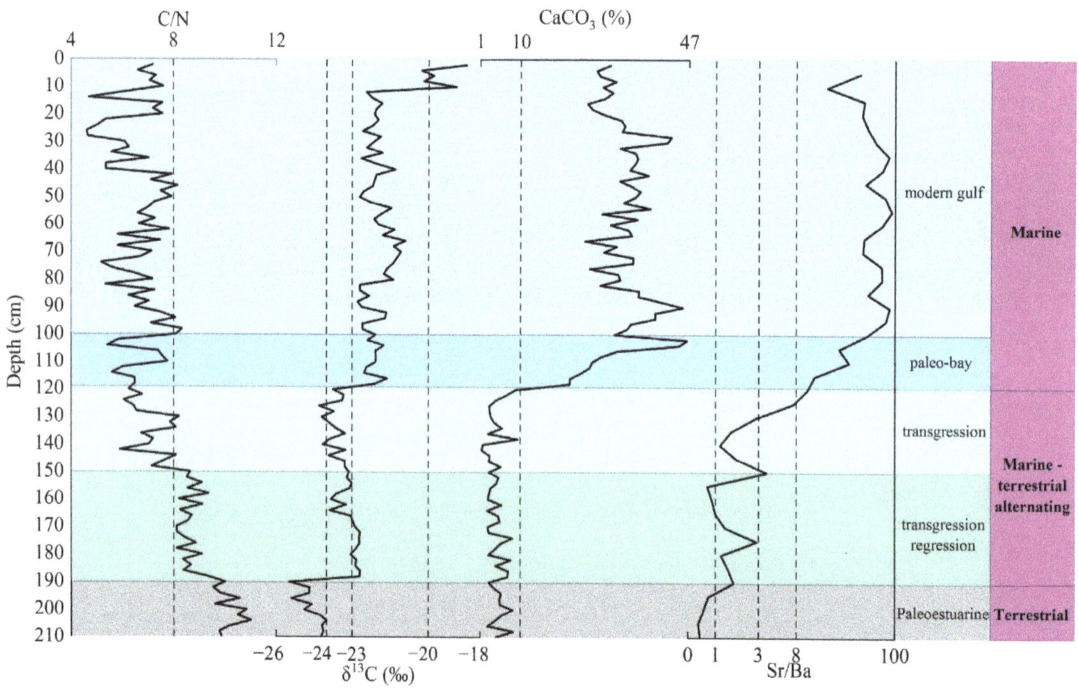

Figure 7. Trend of selected proxies with depth and paleoenvironmental restoration.

In summary, the sediment source timelines in the Beibu Gulf during the last deglaciation were complex. To facilitate the analysis of core sedimentary records, only the carbon-14 age data at burial depths of 119–210 cm was used to establish the "simplified age-depth framework" for the lower section of the core.

Based on the "jump" in calcium carbonate content, combined with the phased variation characteristics of C/N, Sr/Ba, and $\delta^{13}C$ (Figures 6 and 7), the evolution of the sedimentary environment in the Beibu Gulf was divided into the following three sedimentary environments and five sedimentary sub-environments:

(1) Terrestrial sedimentary environment—paleo-estuarine sub-environment (from 210 to 190 cm) (19.24–18.21 ka BP): This period corresponds to the end of the LGM and the beginning of the last deglaciation. The SST and sea level of the South China Sea did not exhibit a significant increase. During this period, the sea level in the South China Sea was observed to be lower than the current maximum water depth in the Beibu Gulf [50]. In the BBW25 core, from the burial depth of 190 cm to the bottom, the C/N ratio was relatively high; $\delta^{13}C$ was <-24‰ and $\delta^{13}C$ at the bottom of the core was <-25‰. This indicated that the organic matter input was mainly represented by freshwater plants or terrestrial

particulate matter. The Sr/Ba ratios were all <1, indicating that sedimentation originated from terrestrial environments. The ratios jointly indicated terrestrial deposits below 190 cm in the BBW25 core.

(2) Marine–terrestrial alternating sedimentary environment—transgression and regression sub-environment (from 190 to 150 cm) (18.21–16.15 ka BP): During this period, the SST and sea level in the South China Sea slowly rose, with seawater possibly entering the Beibu Gulf. At the burial depths of 190–150 cm in the core, the C/N ratio decreased compared to that in the terrestrial environment. At a ratio of 8–9, there was a decreasing and then increasing trend, which may record a transgression and regression. Sr/Ba and $\delta^{13}C$ showed that at burial depths of 190–166 cm, Sr/Ba was 1–3, which indicates brackish conditions. $\delta^{13}C$ was between −22 and −23‰, which corresponds to the seawater end-member organic matter input, indicating a transgression process at that time. Subsequently, at the burial depth of 150 cm, $\delta^{13}C$ decreased to between −23 and −24‰. There were both marine and terrestrial organic matter inputs. The decrease of Sr/Ba < 1 indicates a transition to terrestrial sedimentation, and a regression event was recorded. The age of this regression event was relatively consistent with the recorded age of the H1 cold event, which may be due to an impact of the H1 cold event on sea level.

(3) Marine–terrestrial alternating sedimentary environment—transgression sub-environment (from 150 to 119 cm) (16.15–14.54 ka BP): During this period, the SST and sea level in the South China Sea rose rapidly. The sea level rose from −103 m to approximately −86 m. The seawater entered from the mouth of the Nanbu Gulf, arriving at the current BBW25 at a depth of 84 m. The sedimentary record coincided with the sea level. In general, the trends of C/N and calcium carbonate in sediments were opposite [50]. As the sea level rose and seawater invaded the land, the carbonate content increased, and the C/N ratio decreased to below eight. This indicated that the growth of marine calcareous organisms was beginning to flourish, and the dilution effect of terrestrial debris may be weakened [51], leading to an increase in the carbonate content, while the terrestrial contribution of organic matter continued to decrease. $\Delta^{13}C$ was in the range of −23–−24‰, and Sr/Ba was in the range of 1–8, suggesting a continuous environment of transgression.

(4) Marine sedimentary environment—paleo-bay sub-environment (from 119 to 100 cm) (14.54–11.7 ka BP): This period was the beginning of the BA warm event and the end of the YD cold event as well as the occurrence of the MWP-1B event. The South China Sea's SST remained in a fluctuating state without significant increase, influenced by the alternation of warm and cold events. However, under the influence of the MWP-1B event [52], the sea level of South China Sea rose rapidly, from −86 m to approximately −56 m. In this core section, $\delta^{13}C$ organic matter rose above −23‰. The C/N ratio decreased, while the Sr/Ba ratio increased rapidly. All these proxies together indicated a clear shift towards marine environmental conditions. Subsequently, seawater continued to invade northward through narrow waterways, affecting the nearshore sedimentation in the Beibu Gulf.

(5) Marine sedimentation environment—modern gulf sub-environment (from 100 to 0 cm) (11.7–1.07 ka BP–present-day): The Holocene epoch witnessed a stabilization of both sea surface temperature (SST) and sea level in the South China Sea, following their earlier rise around 5.5 ka and 8 ka, respectively. It was not until approximately 8.5 ka BP that Hainan Island became completely detached from the Leizhou Peninsula, thereby connecting the Beibu Gulf to the South China Sea through the Qiongzhou Strait. During the early-to-middle Holocene (10–8.5 ka BP and 7–5 ka BP), SST variations were primarily influenced by the East Asian monsoon [53–57]. This phenomenon can also account for the significant fluctuations observed in the sedimentary environmental record indicators within the BBW25 core during the early and middle Holocene. The intensification of monsoon activity resulted in enhanced circulation within the Beibu Gulf, exacerbating terrestrial erosion processes and, consequently, amplifying dilution effects. Therefore, more terrestrial and marine materials were mixed into the sediment. As shown in Figure 6, drastic fluctuations were observed in the sedimentary environment record indicators at burial depths of 100–10 cm (11.7–1.07 ka BP) in the core. But they were all within the range

of marine sedimentary fluctuations, indicating the formation of the modern ocean. $\delta^{13}C$ rose above −20‰ at burial depths of 10–0 cm (1.07 ka BP–present-day), suggesting the direct input of organic matter to surface sediments by seawater. This indicates that the modern marine environment in the Beibu Gulf was formed around 1000 years ago.

Previous studies on the thickness of A kong core in the Beibu Gulf found that the depth of marine sediments in the Holocene were located above 1 m; the depth of marine–terrestrial transitional sediments were from 1 to 1.4 m; and the depth of terrestrial stratum sediments were from 1.4 to 2.7 m during the late Pleistocene. The Holocene transgression layer was thinner than the late Pleistocene sedimentary layer [58,59], which was consistent with the thickness recorded in the BBW25 core. The division of sedimentary environments in the Beibu Gulf by previous researchers was different from that of this study. After analyzing the total Sr/Ba and sporopollen in the STAT22 core, it was believed that the paleo-estuarine sub-environment in the Beibu Gulf was between 11.1 and 9 ka BP, the period of expansion from the paleo-estuarine sub-environment to the paleo-bay sub-environment and maintaining stability was 9–6–4 ka BP, and the period of the modern gulf sub-environment and maintaining stability was from 4–1 ka BP–the present-day. The comprehensive analysis of pollen, spores, and algae in the C4 and B106 cores suggested that the Beibu Gulf was in a terrestrial to marine-terrestrial alternating sub-environment before 13.4 ka BP; a paleo-bay sub-environment from 13.4 to 11.7 ka BP; and a modern marine sub-environment from 11.7 ka BP to the present day [9,21].

4.2. *Response of the Intensity of Terrestrial Weathering in the Potential Source Area to Climate Events in East Asia over the Past 20,000 Years*

Previous studies have demonstrated that chemical weathering can alter the elemental composition of sediments, both in terms of major and trace elements. These changes are primarily influenced by climate variations. The process of element migration within sediment source areas has the potential to reverse climate change patterns and provide insights into paleoclimate reconstruction. Geochemical characteristics and ratios of sedimentary elements serve as commonly employed indicators for quantifying the intensity of chemical weathering. Therefore, it is crucial to carefully select effective element tracers for assessing chemical weathering intensity and reconstructing paleoclimate events [60–63].

4.2.1. Source Area Analysis of Core Sediments

By comparing the findings from previous studies on the sediment source-to-sink process in the northwest South China Sea (Figure 8) with the core location examined in this study, it is inferred that the BBW25 core sediment originates from the Red River, Pearl River (including Qinjiang river within its water system), and Hainan Island. Previous research has indicated that the chemical weathering of sediments in these areas is influenced by the East Asian monsoon, exhibiting a similar intensity trend to monsoon precipitation records of East Asia, oxygen isotopes found in Greenland ice cores, and the SST of the northern South China Sea [26,64,65]. The chemical weathering records of the BBW25 core, covering the northwest South China Sea, are predominantly influenced by the East Asian monsoon. Therefore, we anticipate conducting a comparative analysis between the chemical weathering indicators of the BBW25 core and the indicators reflecting changes in the East Asian monsoon, aiming to investigate the significant climatic events recorded by the core since the late LGM.

4.2.2. Effective Chemical Weathering Intensity

The various geochemical elements in the BBW25 core sediments were greatly affected by the marine and terrestrial environments, making it difficult to conduct a comprehensive comparison of chemical weathering intensity. Therefore, selecting climate indicators that were less affected by sedimentary environments was crucial. As shown in Figure 9, major elements such as Al, Fe, K, and Na did not show a marked "jump" at the depth of 119–120 cm. Thus, these elements were less affected by the marine and terrestrial environmental conditions. Therefore, using these elements to establish weathering indicators was more effective.

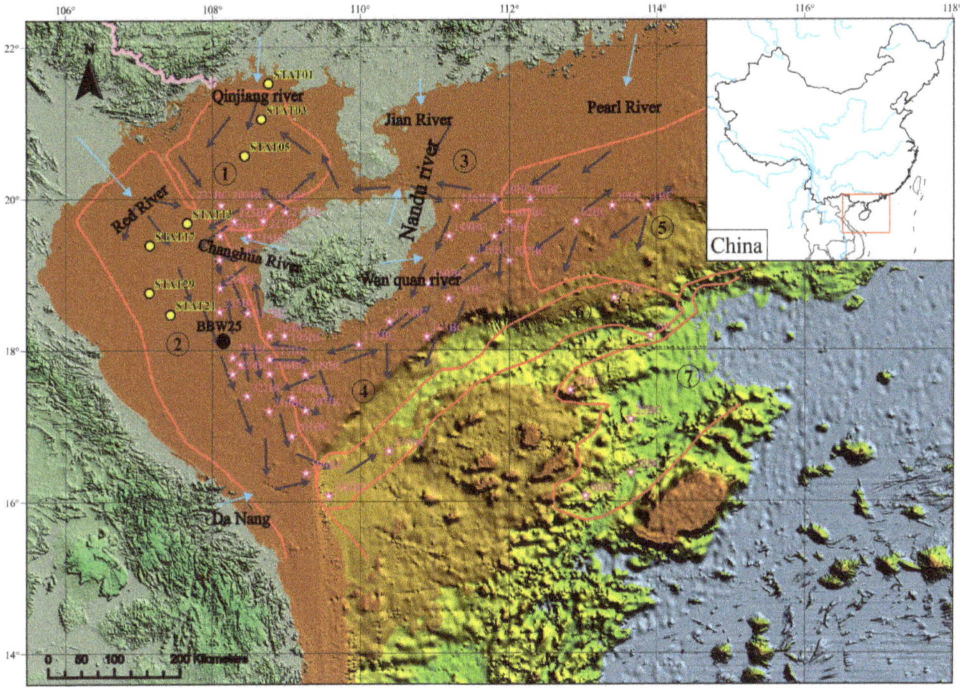

Figure 8. Source-to-sink processes in the northwestern part of South China Sea [66,67]. (1) Source: the starting point of the blue arrow is the source of terrigenous clastic sediments; (2) channel: the gray arrow direction is the main sediment transport direction and its pathway; (3) sinks: the circle number is the sedimentary block number: ① Beibu Gulf Basin; ② Yinggehai Basin; ③ shallow shelf around Hainan Island; ④ Qiongdongnan Basin; ⑤ Pearl River Estuary Basin; ⑥ Xisha Trough; ⑦ Sea Basin. (4) Black spot: BBW25 core; (5) yellow spots: STAT cores; pink stars: PC cores.

During chemical weathering, Al is a conservative and non-migratory element and remains unchanged compared to the mother rock and sediments. Na is highly mobile and readily departs from the mother rock and sediments. Therefore, Na is usually depleted in weathering products. K is easily leached from primary minerals but is resistant to leaching and is immobilized by secondary clay minerals in the weathering profiles. The modified chemical alteration index (CIX) can link the three elements (CIX = 100 × $Al_2O_3/(Al_2O_3 + Na_2O + K_2O)$). This index can well-eliminate the influence of carbonates and phosphates on silicates [68–72]. The Al/Fe ratio refers to the degree of aluminum removal and iron enrichment. The higher the value, the stronger the degree of weathering, indicating a humid and hot climate. On the contrary, a lower value indicates a relatively dry and cold climate [73–75]. This research chose the CIX and Al/Fe ratio as indicators of the intensity of chemical weathering in the sedimentary record in the core.

4.2.3. Response of Chemical Weathering Intensity Changes Recorded in Core to the Last Major Climate Event

Previous studies on the weathering records have indicated that the chemical weathering intensity of rocks under relatively warm and humid climates is higher than that during cold and dry periods [76–78]. When the sediment source remains stable, the rock type can serve as an indicator of weathering intensity. Kaolin is predominantly formed in a warm and humid acidic environment through extensive leaching of feldspar, mica, and pyroxene, thus representing a highly weathered mineral. In contrast, illite forms under dry and cold climatic conditions by undergoing weathering processes that involve potassium removal. If the climate becomes humid and hot and chemical weathering is thorough, potassium is taken away, and illite will further decompose into kaolinite [26]. The temperature during the late LGM–deglaciation stage was found to be lower than that of the Holocene, as indicated by the global average surface temperature curve in Figure 6. Additionally, it was observed that the upper section of the BBW25 core exhibited lower levels of illite and potassium content compared to those of the lower section, suggesting a correlation between chemical weathering intensity and variations in cold and warm climates. The chemical weathering recorded in the BBW25 core sediments can be attributed to the northwest region of the South China Sea, which was influenced by the East Asian monsoon [26]. Simultaneously, a comparison was made between the chemical weathering indicators under the "simplified framework" and the indicators for recording East Asian monsoon changes (Figure 9) to demonstrate their reliability.

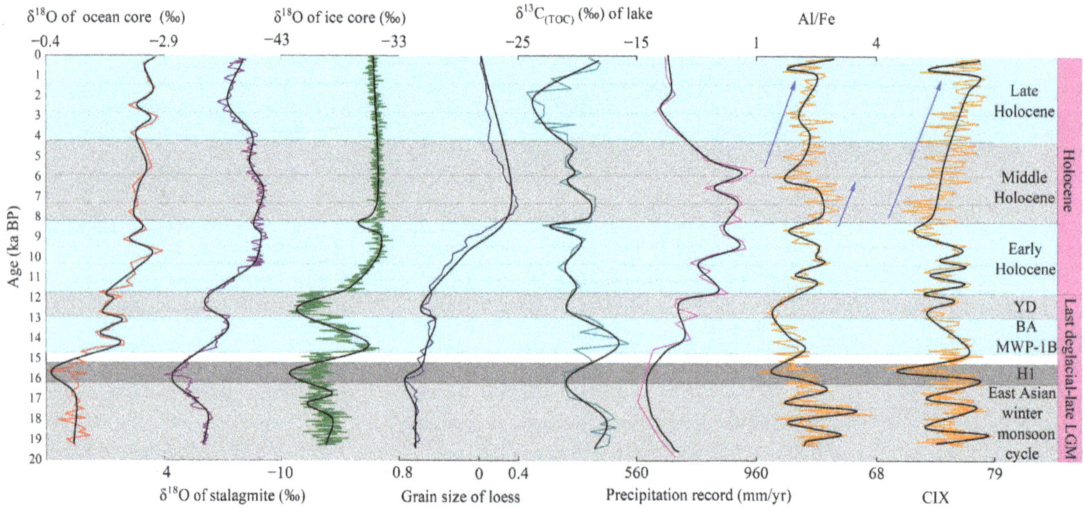

Figure 9. Comparison of age versus chemical weathering intensity curve in the core with $\delta^{18}O$ of 1144 core in South China Sea [79], $\delta^{18}O$ of stalagmites in Sanbao Cave [80], $\delta^{18}O$ of ice core in Greenland [81], grain size of loess in China [82], $\delta^{13}C$(TOC) of Lake Huguangmaer [83], and monsoon precipitation records of East Asia [84].

Based on the recorded changes in chemical weathering intensity in the core under the "simplified framework" and the analysis of major climate events since the end of the LGM (Figure 9), the BBW25 core records can be divided into the following three periods and eight climate events:

(1) Records of strong East Asian winter monsoon and weak cycles from the late LGM to the early deglaciation.

Previous studies have indicated that the South China Sea experienced a prevailing winter monsoon during the late last glacial maximum (LGM). The variations in chemical weathering intensity may serve as indicators of both intensified winter monsoon and reduced cyclical changes during this period. Consequently, the elevated terrestrial input characteristics likely reflect the substantial sea level drop and oceanic current dynamics driven by the robust winter monsoon during the late LGM [85–87].

(2) Records of alternating cold and warm climate events and meltwater events during the last deglaciation.

① The H1 cold event with cold and dry happened at high latitudes in the Northern Hemisphere, but with a slight increase in humidity in comparison to that in the late LGM [88,89]. The H1 cold event was clearly documented in the BBW25 core, with a distinct decline observed in the chemical weathering intensity indicated by the Fe/Al ratio and CIX. Simultaneously, $\delta^{18}O$ levels increased in South China Sea cores during the H1 cold event, suggesting that both the northern part of the South China Sea and southern China experienced a cold and arid climate during this period. The coarsening of particles in loess sediments, reduction in carbon isotopes within lakes, and decreased precipitation all indicate a dry and frigid climate across East Asia.

② The BA warm event happened because of the involvement of thermohaline circulation during the last deglaciation, and geological reconstruction data and numerical simulations confirmed this point. Greenland's temperature increased by about 10 °C during this event [90–95]. The $\delta^{18}O$ records of stalagmites and the South China Sea cores as well as the carbon isotope records of lacustrine sediments showed that the BA warm regions in the Northern Hemisphere. The rising trend in precipitation in East Asia at this time suggested a warm and humid climate. The enhanced chemical weathering intensity recorded in the BBW25 core during this stage was the response to the BA warm event. Meanwhile, the sedimentary environment indicator records responded well to the MWP-1A event. At this time, the significant rise in sea level led to a transition from terrestrial inputs to marine inputs. Therefore, it was believed that the low terrestrial input and high weathering intensity at this time can serve as indirect evidence for the synchronous start of the MWP-1A and BA warm events.

③ During the YD period, $\delta^{18}O$ of the Greenland ice core decreased, which indicated the cooling of high-latitude regions in the Northern Hemisphere. The SST in the South China Sea showed a decrease in temperature during the YD cold event compared to that during the BA warm event. The grain size of loess on the Loess Plateau became coarser, and $\delta^{18}O$ of stalagmites rose. The pollen records of lacustrine sediments indicated an increase in psychrophilic plants during the YD cold event [96–98]. The decrease in chemical weathering intensity recorded by BBW25 was a response to the YD cold event.

(3) Early Holocene climate oscillations and records of abrupt climate changes from the Middle to the Late Holocene.

① Previous studies have found that during the climate warming process towards the Holocene at the end of the last deglaciation, there were multiple climate oscillations in the early Holocene, manifested as transient dry and cold events; this was caused by the strongest summer monsoon event and winter monsoon strengthening event during the early Holocene ice age [15,87]. According to [14], three cooling events were found to occur at 9.4, 10.3, and 11.1 ka BP, respectively. The changes characteristics of the three cycles recorded by chemical weathering in the BBW25 core sediments from the late Younger Dryas event to early Holocene respond to the circle climate change events of the temperature decreasing and the winter monsoon strengthening during the ice age, while the temperature increasing and the summer monsoon strengthening during the interglacial period within a century scale of the early Holocene [99].

② There were several climate events from the Middle to the Late Holocene, among which the 8.2 ka BP event, as the beginning of the Middle Holocene, was possibly the strongest cold event since the Holocene; changes in the monsoon precipitation in the Northern Hemisphere were mainly caused by freshwater injection [100,101]. Figure 9 shows that the intensity of weathering recorded in the core had a significant decrease from 8.8 to 8 ka BP. After 8 ka BP, chemical weathering intensified, indicating the onset of a warm and humid climate.

③ A decline in chemical weathering recorded by the Al/Fe ratio from 7 ka BP to 5 ka BP may be related to the 5.9 ka BP event, which was a sudden cold climate change that occurred in the Middle Holocene; during this period, the dormant ENSO climate became active again [102,103]. This cold event has been recorded in lake sediments, loess sediments, ice core, and the change in sea level in China [104–106].

④ The most recent decline in chemical weathering in the core from modern times may have been caused by the Little Ice Age, with typical time of 0.4 ka BP and record age range of 0.6–0.15 ka BP in China [15,107]. The sedimentary data suggest that the Little Ice Age may have been the most severe cold event in the past three millennia. Glacier and lake records in the middle and high latitudes of the Northern Hemisphere indicate a cold climate, while stalagmite records in the low latitudes of the Northern Hemisphere suggest an arid climate [108–111].

The average values of CIX and Fe/Al ratio in the Holocene and the late LGM–deglaciation stage were 74.88 and 2.44, respectively, and 74.84 and 2.27, indicating that the chemical weathering intensity recorded in the study area during the Holocene was slightly higher than that during the late LGM—deglaciation stage.

5. Conclusions

According to the profile variation in the content of major and trace elements, TOC and TN, $CaCO_3$, and the value of $\delta^{13}C$ on the P36 core sediments combined with AMS 14C dating, the following points could be concluded:

(1) The BBW25 sediment core was deposited from the late Pleistocene. And taking the depth of 119 cm as the boundary, the BBW25 core could be divided into two sedimentary sections based on the change of geochemical characteristics: the low section and the upper section. The upper section is the depth of 0 to 119 cm which sedimentary ages are from present day to 14.54 ka BP. And the low section is the depth of 119 to 210 cm, whose sedimentary ages are from 14.54 to 19.24 ka BP. The reason for this phenomenon is due to a sedimentary environment change from terrestrial to marine environments.

(2) According to sedimentary environment indicators, the Beibu Gulf is divided into three sedimentary environments (terrestrial, marine–terrestrial alternating, and marine) and five sedimentary sub-environments (paleo-bay, transgression regression, transgression, paleo-bay, and modern gulf). By analyzing the sedimentary indicators in this study and previous research on other cores in the Beibu Gulf. The division of sedimentary environments in the Beibu Gulf obtained was more similar to the results obtained by previous researchers through a comprehensive analysis of C4 and B106 cores. This also indicated that the extracted Sr/Ba value was more accurate in dividing sedimentary environments than the total Sr/Ba value.

(3) The major climate events recorded by the CIX and the Al/Fe ratio in the BBW25 core since the late LGM include the East Asian winter monsoon cycle in the late LGM, the H1 cold event, the BA warm event, the MWP-1A event, the YD cold event, the 8.2 ka B.P. cold event, the 5.9 ka B.P. cold event, and the Little Ice Age event; the chemical weathering intensity recorded in the Holocene is slightly greater than that of the late LGM–deglaciation stage.

(4) In summary, the comparison of the sedimentary environments and chemical weathering intensity changes in BBW25 core sedimentary records with major climate events since the end of the LGM can help to prove the reliability of the simplified framework established in this paper. It can fully contribute to a better understanding of marine paleoenvironmental change in the Beibu Gulf and regional response of the marginal seas over to the sea level of South China Sea changes and the northern Hemisphere climate reconstruction since the late LGM, and which can be used as a reference for the future climate change research.

Author Contributions: Conceptualization, T.F.; methodology, Y.L. (Yuchun Li); software, Y.L. (Yuchun Li); validation, Y.L. (Yuchun Li); formal analysis, Y.L. (Yuchun Li); data curation, Y.L. (Yuchun Li) and T.F.; writing—original draft preparation, Y.L. (Yuchun Li); review and editing, Y.L. (Yuchun Li) and T.F.; providing partial experimental methods, A.W.; test funding support, J.Z. and Y.L. (Yubiao Lv); partial experimental work, M.Z. and D.L. Data and samples were collected from the open research cruise (number: NORC2021-11) supported by NSFC Shiptime Sharing Project 2020. All authors have read and agreed to the published version of the manuscript.

Funding: This research was funded by the Science Fund Project of Fangchenggang (2022BB141154, AB22013011).

Institutional Review Board Statement: Not applicable.

Informed Consent Statement: Not applicable.

Data Availability Statement: The data presented in this study are available on request from the first author or the corresponding author.

Acknowledgments: We thank Yue, Y.F., Wei L. and Zhang N.N. for their recommendations. We thank the anonymous reviewers for their comments and suggestions, which greatly improved the manuscript.

Conflicts of Interest: The authors declare no conflicts of interest.

References

1. Huang, J.; Jiang, F.; Wan, S.; Zhang, J.; Li, A.; Li, T. Terrigenous supplies variability over the past 22,000 yr in the southern South China Sea slope: Relation to sea level and monsoon rainfall changes. *J. Asian Earth Sci.* **2016**, *117*, 317–327. [CrossRef]
2. Li, M.; Ouyang, T.; Tian, C.; Zhu, Z.; Peng, S.; Tang, Z.; Qiu, Y.; Zhong, H.; Peng, X. Sedimentary responses to the East Asian monsoon and sea level variations recorded in the northern South China Sea over the past 36 kyr. *J. Asian Earth Sci.* **2019**, *171*, 213–224. [CrossRef]
3. Liu, Z.; Zhao, Y.; Colin, C.; Stattegger, K.; Wiesner, M.G.; Huh, C.A.; Zhang, Y.; Li, X.; Sompongchaiyakul, P.; You, C.F.; et al. Source-to-sink transport processes of fluvial sediments in the South China Sea. *Earth Sci. Rev.* **2016**, *153*, 238–273. [CrossRef]
4. Liu, F.; Yang, C.; Chang, X.; Liao, Z. Provenance discrimination of the last glacial sediments from the northeastern South China Sea and its paleoenvironmental indications. *Terr. Atmos. Ocean. Sci.* **2018**, *29*, 131–148. [CrossRef]
5. Hanebuth, T.; Stattegger, K.; Grootes, P.M. Rapid flooding of the Sunda Shelf: A late-glacial sea-level record. *Science* **2000**, *288*, 1033–1035. [CrossRef] [PubMed]
6. Yokoyama, Y.; Lambeck, K.; De Deckker, P.; Johnston, P.; Fifield, K.L. Timing of the Last Glacial Maximum from observed sea-level minima. *Nature* **2000**, *406*, 713–716. [CrossRef]
7. Rohling, E.J.; Grant, K.; Bolshaw, M.; Roberts, A.P.; Siddall, M.; Hemleben, C.; Kucera, M. Antarctic temperature and global sea level closely coupled over the past five glacial cycles. *Nat. Geosci.* **2009**, *2*, 500–504. [CrossRef]
8. Pelejero, C.; Grimalt, J.O. The correlation between the $_{37}$ k index and sea surface temperatures in the warm boundary: The South China Sea. *Geochim. Cosmochim. Acta* **1997**, *61*, 4789–4797. [CrossRef]
9. Pelejero, C.; Grimalt, J.O.; Heilig, S.; Kienast, M.; Wang, L. High-resolution $U^K{}_{37}$ temperature reconstructions in the South China Sea over the past 220 kyr. *Paleoceanography* **1999**, *14*, 224–231. [CrossRef]
10. Xia, P.; Meng, X.; Li, Z.; Zhi, P.; Zhao, M.; Wang, E. Late Holocene mangrove development and response to sea level change in the northwestern South China Sea. *Acta Oceanol. Sin.* **2019**, *38*, 111–120. [CrossRef]
11. Zhou, G.; Cao, X.; Xia, J.; Wang, S.; Song, Z. A Dramatic Marine Environment Change in the Beibu Gulf of the South China Sea around 3.2 kyr BP. *Lithosphere* **2022**, *2022*, 2632112. [CrossRef]
12. Wang, J.Y.; Bai, W.M.; Wang, Z.B.; Wang, M.H.; Li, B.J. Holocene climate evolution in eastern China and its correspondence with climate events. *Mar. Geol. Quat. Geol.* **2022**, *2*, 167–177.
13. Bond, G.; Kromer, B.; Beer, J.; Muscheler, R.; Evans, M.N.; Showers, W.; Hoffmann, S.; Lotti-Bond, R.; Hajdas, I.; Bonani, G. Persistent solar influence on North Atlantic climate during the Holocene. *Science* **2001**, *294*, 2130–2136. [CrossRef] [PubMed]
14. Bond, G.; Showers, W.; Cheseby, M.; Lotti, R.; Almasi, P.; DeMenocal, P.; Priore, P.; Cullen, H.; Hajdas, I.; Bonani, G. A pervasive millennial-scale cycle in North Atlantic Holocene and glacial climates. *Science* **1997**, *278*, 1257–1266. [CrossRef]
15. Wang, S.W. Holocene cold events in the North Atlantic: Chronology and climate impact. *Quat. Res.* **2009**, *29*, 1146–1153.
16. Li, X.; Liu, X.; Pan, Z.; Shi, Z.; Xie, X.; Ma, H.; Zhai, J.; Liu, H.; Xie, X.; Dai, A. Transient vegetation degradation reinforced rapid climate change (RCC) events during the Holocene. *npj Clim. Atmos. Sci.* **2023**, *6*, 125. [CrossRef]
17. Li, X.; Liu, X.; Pan, Z.; Xie, X.; Shi, Z.; Wang, Z.; Bai, A. Orbital-scale dynamic vegetation feedback caused the Holocene precipitation decline in northern China. *Commun. Earth Environ.* **2022**, *3*, 257. [CrossRef]
18. Zhou, S.Z.; Zhao, J.D.; Wang, J.; Xu, L.B.; Cui, J.X.; Ou, X.J.; Xie, J.M. Quaternary cryosphere-Long-scale study of global change. *J. China Acad. Sci.* **2020**, *4*, 475–483. [CrossRef]
19. Wu, K.; Shi, X.; Lou, Z.; Wu, B.; Li, J.; Zhang, H.; Cao, P.; Rahim Mohamed, C.A. Sedimentary responses to climate changes and human activities over the past 7400 Years in the western sunda shelf. *Front. Earth Sci.* **2021**, *9*, 631815. [CrossRef]
20. Sun, X.; Li, X.; Beug, H.J. Pollen distribution in hemipelagic surface sediments of the South China Sea and its relation to modern vegetation distribution. *Mar. Geol.* **1999**, *156*, 211–226. [CrossRef]
21. Li, Z.; Zhang, Y.; Li, Y.; Zhao, J. Palynological records of Holocene monsoon change from the Gulf of Tonkin (Beibuwan), northwestern South China Sea. *Quat. Res.* **2010**, *74*, 8–14. [CrossRef]
22. Huang, X.Q.; Liang, K.; Xi, L.; Xia, Z.; Zhang, S.Z. Holocene transgression evolution in Beibu Gulf and its relationship with climate and glaciers. *Acta Geologica Sinica* **2022**, *1*, 49–64.
23. Cui, Z.; Schulz-Bull, D.E.; Hou, Y.; Xia, Z.; Waniek, J.J. Geochemical characteristics and provenance of Holocene sediments (core STAT22) in the Beibu Gulf, South China Sea. *J. Coast. Res.* **2016**, *32*, 1105–1115. [CrossRef]

24. Li, M.; Ouyang, T.; Roberts, A.P.; Heslop, D.; Zhu, Z.; Zhao, X.; Tian, C.; Peng, S.; Zhong, H.; Peng, X.; et al. Influence of sea level change and centennial East Asian monsoon variations on northern South China Sea sediments over the past 36 kyr. *Geochem. Geophys. Geosyst.* **2018**, *19*, 1674–1689. [CrossRef]
25. Zhao, Z.X.; Wan, S.M.; Ju, M.S.; Pei, W.Q.; Jin, H.L.; Zhang, J.; Li, A.C. Sr-Nd isotope and rare earth element evidence of sediment sources and weathering evolution in the northwest of the South China Sea since the last glacial period. *Bull. Miner. Rock Geochem.* **2023**, *4*, 702–716+682–683. [CrossRef]
26. Wan, S.M.; Qin, L.; Yang, S.Y.; Zhao, D.B.; Zhang, J.; Jiao, D.F.; Li, T.G. Glacial shelf weathering and carbon cycle in the South China Sea. *Quat. Study* **2020**, *6*, 1532–1549.
27. *GB/T 14506.31-2019*; Chemical Analysis Methods of Silicate Rocks Part 31: Determination of 12 Components Such as Silica—Lithium Metaborate Melting-Inductively Coupled Plasma Atomic Emission Spectrometry. National Technical Committee for Standardization of Land and Resources: Geneva, Switzerland, 2019.
28. *GB/T 14506.30-2010*; Chemical Analysis Methods of Silicate Rocks Part 30: Determination of 44 Elements. National Technical Committee for Standardization of Land and Resources: Geneva, Switzerland, 2010.
29. Wang, A.H.; Ye, S.Y.; Liu, J.K.; Ding, X.G.; Li, H.L.; Xu, N.C. Discussion on the discrimination of marine sedimentary environment with different selective extraction methods of strontium and barium ratio—Taking the modern Yellow River Delta as an example. *Acta Sedimentol.* **2020**, *6*, 1226–1238. [CrossRef]
30. Blaauw, M.; Christen, J.A. Flexible paleoclimate age-depth models using an autoregressive gamma process. *Bayesian Anal.* **2011**, *6*, 457–474. [CrossRef]
31. Tanabe, S.; Hori, K.; Saito, Y.; Haruyama, S.; Kitamura, A. Song Hong (Red River) delta evolution related to millennium-scale Holocene sea-level changes. *Quat. Sci. Rev.* **2003**, *22*, 2345–2361. [CrossRef]
32. Armstrong-Altrin, J.S.; Machain-Castillo, M.L.; Rosales-Hoz, L.; Carranza-Edwards, A.; Sanchez-Cabeza, J.A.; Ruíz-Fernández, A.C. Provenance and depositional history of continental slope sediments in the Southwestern Gulf of Mexico unraveled by geochemical analysis. *Cont. Shelf Res.* **2015**, *95*, 15–26. [CrossRef]
33. Liu, Y.J. *Element Geochemistry*; Science Press: Beijing, China, 1984.
34. Wang, A.; Wang, Z.; Liu, J.; Xu, N.; Li, H. The Sr/Ba ratio response to salinity in clastic sediments of the Yangtze River Delta. *Chem. Geol.* **2021**, *559*, 119923. [CrossRef]
35. Dashtgard, S.E.; Wang, A.; Pospelova, V.; Wang, P.L.; La Croix, A.; Ayranci, K. Salinity indicators in sediment through the fluvial-to-marine transition (Fraser River, Canada). *Sci. Rep.* **2022**, *12*, 14303. [CrossRef] [PubMed]
36. Meyers, P.A. Preservation of elemental and isotopic source identification of sedimentary organic matter. *Chem. Geol.* **1994**, *114*, 289–302. [CrossRef]
37. Lamb, A.L.; Wilson, G.P.; Leng, M.J. A review of coastal palaeoclimate and relative sea-level reconstructions using $\delta^{13}C$ and C/N ratios in organic material. *Earth Sci. Rev.* **2006**, *75*, 29–57. [CrossRef]
38. Thornton, S.F.; McManus, J. Application of organic carbon and nitrogen stable isotope and C/N ratios as source indicators of organic matter provenance in estuarine systems: Evidence from the Tay Estuary, Scotland. *Estuar. Coast. Shelf Sci.* **1994**, *38*, 219–233. [CrossRef]
39. Milliman, J.D.; Qinchun, X.; Zuosheng, Y. Transfer of particulate organic carbon and nitrogen from the Yangtze River to the ocean. *Am. J. Sci.* **1984**, *284*, 824–834. [CrossRef]
40. Middelburg, J.J.; Nieuwenhuize, J. Carbon and nitrogen stable isotopes in suspended matter and sediments from the Schelde Estuary. *Mar. Chem.* **1998**, *60*, 217–225. [CrossRef]
41. Chmura, G.L.; Aharon, P. Stable carbon isotope signatures of sedimentary carbon in coastal wetlands as indicators of salinity regime. *J. Coast. Res.* **1995**, *11*, 124–135. Available online: http://www.jstor.org/stable/4298316 (accessed on 1 October 2023).
42. Khan, N.S.; Vane, C.H.; Horton, B.P. Stable carbon isotope and C/N geochemistry of coastal wetland sediments as a sea-level indicator. *Handb. Sea-Level Res.* **2015**, 295–311. [CrossRef]
43. Zhang, N. *The Mineralogical Characteristics of the Sediments and Its Environmental Significance in the Beibuwan Gulf*; China University of Geosciences: Beijing, China, 2015.
44. Ni, Y.G.Y.; Xia, Z.; Ma, S.Z. The opening of Qiongzhou Strait: Evidence from sub-bottom profiles. *Mar. Geol. Quat. Geol.* **2014**, *34*, 79–82.
45. Li, C.R.; Yang, X.; Fan, C.; Hu, L.; Dai, L.; Zhao, S. On the Evolution Process of the Beibu Gulf Basin and Forming Mechanism of Local Structures. *Acta Geol. Sin.* **2018**, *92*, 2028–2039. [CrossRef]
46. Kaboth-Bahr, S.; Bahr, A.; Yamoah, K.A.; Chuang, C.K.; Li, H.C.; Su, C.C.; Wei, K.Y. Rapid humidity changes across the Northern South China Sea during the last ~40 kyrs. *Mar. Geol.* **2021**, *440*, 106579. [CrossRef]
47. Snyder, C.W. Evolution of global temperature over the past two million years. *Nature* **2016**, *538*, 226–228. [CrossRef] [PubMed]
48. Wan, S.; Clift, P.D.; Zhao, D.; Hovius, N.; Munhoven, G.; France-Lanord, C.; Wang, Y.; Xiong, Z.; Huang, J.; Yu, Z.; et al. Enhanced silicate weathering of tropical shelf sediments exposed during glacial lowstands: A sink for atmospheric CO_2. *Geochim. Cosmochim. Acta* **2017**, *200*, 123–144. [CrossRef]
49. Cang, S.C.; Chen, L.R.; Dong, T.L. Study on evolution history of sedimentary environment since Pliocene in R_1 core of Beibu Gulf. *Mar. Geol. Quat. Geol.* **1992**, *4*, 53–58. [CrossRef]
50. Xu, D. *Sedimentary Records Since the Last Deglaciation Period in the Eastern Beibu Gulf and the Formation of Modern Sedimentary Pattern*; Graduate School of China Academy of Sciences (Institute of Oceanography): Beijing, China, 2014.

51. Zhang, L.L.; Chen, M.H.; Chen, Z.; Xiang, R.; Liu, J.G. Distribution of calcium carbonate content in surface sediments of the South China Sea and its influencing factors. *Geosci. J. China Geo Univ.* **2010**, *6*, 891–898.
52. Huang, E.Q.; Tian, J. Melting water events and abrupt climate change in the last deglaciation period. *Sci. Bull.* **2008**, *12*, 1437–1447.
53. Zhao, H.T. Origin and time of Qiongzhou Strait. *Mar. Geol. Quat. Geol.* **2007**, *27*, 33–40.
54. Yao, Y.; Harff, J.; Meyer, M.; Zhan, W. Reconstruction of paleocoastlines for the northwestern South China Sea since the Last Glacial Maximum. *Sci. China Ser. D Earth Sci.* **2009**, *52*, 1127–1136. [CrossRef]
55. Dykoski, C.A.; Edwards, R.L.; Cheng, H.; Yuan, D.; Cai, Y.; Zhang, M.; Lin, Y.; Qing, J.; An, Z.; Revenaugh, J. A high-resolution, absolute-dated Holocene and deglacial Asian monsoon record from Dongge Cave, China. *Earth Planet. Sci. Lett.* **2005**, *233*, 71–86. [CrossRef]
56. Huang, J.; Li, A.; Wan, S. Sensitive grain-size records of Holocene East Asian summer monsoon in sediments of northern South China Sea slope. *Quat. Res.* **2011**, *75*, 734–744. [CrossRef]
57. Wang, L.; Li, J.; Lu, H.; Gu, Z.; Rioual, P.; Hao, Q.; Mackay, A.W.; Jiang, W.; Cai, B.; Xu, B.; et al. The East Asian winter monsoon over the last 15,000 years: Its links to high-latitudes and tropical climate systems and complex correlation to the summer monsoon. *Quat. Sci. Rev.* **2012**, *32*, 131–142. [CrossRef]
58. Wang, P.X. *Collected Papers on Marine Microorganisms and Paleontology*; Ocean Press: Beijing, China, 1980.
59. Li, G.Z.; Bian, Y.H.; Wang, P.X. Holocene transgressive strata and their microfossil characteristics in the northeast of Beibu Gulf. *Trop. Ocean* **1988**, *2*, 63–70.
60. Yang, S.; Li, C.; Cai, J. Geochemical compositions of core sediments in eastern China: Implication for Late Cenozoic palaeoenvironmental changes. *Palaeogeogr. Palaeoclimatol. Palaeoecol.* **2006**, *229*, 287–302. [CrossRef]
61. Wei, G.; Liu, Y.; Li, X.; Shao, L.; Liang, X. Climatic impact on Al, K, Sc and Ti in marine sediments: Evidence from ODP Site 1144, South China Sea. *Geochem. J.* **2003**, *37*, 593–602. [CrossRef]
62. Wei, G.; Liu, Y.; Li, X.; Chen, M.; Wei, W. High-resolution elemental records from the South China Sea and their paleoproductivity implications. *Paleoceanography* **2003**, *18*, 1054. [CrossRef]
63. Murray, R.W.; Knowlton, C.; Leinen, M.; Mix, A.C.; Polsky, C.H. Export production and terrigenous matter in the Central Equatorial Pacific Ocean during interglacial oxygen isotope Stage 11. *Glob. Planet. Chang.* **2000**, *24*, 59–78. [CrossRef]
64. Wan, S.; Toucanne, S.; Clift, P.D.; Zhao, D.; Bayon, G.; Yu, Z.; Cai, G.; Yin, X.; Révillon, S.; Wang, D.; et al. Human impact overwhelms long-term climate control of weathering and erosion in southwest China. *Geology* **2015**, *43*, 439–442. [CrossRef]
65. Hu, D.; Clift, P.D.; Böning, P.; Hannigan, R.; Hillier, S.; Blusztajn, J.; Wan, S.; Fuller, D.Q. Holocene evolution in weathering and erosion patterns in the Pearl River delta. *Geochem. Geophys. Geosyst.* **2013**, *14*, 2349–2368. [CrossRef]
66. Qiu, Y.; Peng, X.C.; Wang, Y.M.; Huang, W.K.; Du, W.B. *Quaternary Erosion Process and Sedimentary Response in the Northern South China Sea*; Geology Press: Beijing, China, 2017.
67. Li, M.K. *Evolution and Driving Mechanism of Paleoclimate and Environment since 36 Kyr BP in the Northwest South China Sea*; University of Chinese Academy of Sciences (Guangzhou Institute of Geochemistry, China Academy of Sciences): Beijing, China, 2018.
68. Nesbitt, H.W.; Markovics, G. Weathering of granodioritic crust, long-term storage of elements in weathering profiles, and petrogenesis of siliciclastic sediments. *Geochim. Cosmochim. Acta* **1997**, *61*, 1653–1670. [CrossRef]
69. Li, X.H.; Wei, G.; Shao, L.; Liu, Y.; Liang, X.; Jian, Z.; Sun, M.; Wang, P. Geochemical and Nd isotopic variations in sediments of the South China Sea: A response to Cenozoic tectonism in SE Asia. *Earth Planet. Sci. Lett.* **2003**, *211*, 207–220. [CrossRef]
70. Yang, S.; Jung, H.S.; Li, C. Two unique weathering regimes in the Changjiang and Huanghe drainage basins: Geochemical evidence from river sediments. *Sediment. Geol.* **2004**, *164*, 19–34. [CrossRef]
71. Wei, G.; Li, X.H.; Liu, Y.; Shao, L.; Liang, X. Geochemical record of chemical weathering and monsoon climate change since the early Miocene in the South China Sea. *Paleoceanography* **2006**, *21*, PA4214. [CrossRef]
72. Garzanti, E.; Padoan, M.; Setti, M.; López-Galindo, A.; Villa, I.M. Provenance versus weathering control on the composition of tropical river mud (southern Africa). *Chem. Geol.* **2014**, *366*, 61–74. [CrossRef]
73. Wei, G.; Liu, Y.; Li, X.H.; Shao, L.; Fang, D. Major and trace element variations of the sediments at ODP Site 1144, South China Sea, during the last 230 ka and their paleoclimate implications. *Palaeogeogr. Palaeoclimatol. Palaeoecol.* **2004**, *212*, 331–342. [CrossRef]
74. Wang, X. *Particle Size Characteristics and Geochemical Analysis of Surface Sediment on the Seabed of Panjin Port*; Liaoning Normal University: Dalian, China, 2018.
75. Kronberg, B.I.; Nesbitt, H.W.; Lam, W.W. Upper Pleistocene Amazon deep-sea fan muds reflect intense chemical weathering of their mountainous source lands. *Chem. Geol.* **1986**, *54*, 283–294. [CrossRef]
76. West, A.J. Thickness of the chemical weathering zone and implications for erosional and climatic drivers of weathering and for carbon-cycle feedbacks. *Geology* **2012**, *40*, 811–814. [CrossRef]
77. Catalan, J.; Pla-Rabés, S.; García, J.; Camarero, L. Air temperature-driven CO2 consumption by rock weathering at short timescales: Evidence from a Holocene lake sediment record. *Geochim. Cosmochim. Acta* **2014**, *136*, 67–79. [CrossRef]
78. West, A.J.; Galy, A.; Bickle, M. Tectonic and climatic controls on silicate weathering. *Earth Planet. Sci. Lett.* **2005**, *235*, 211–228. [CrossRef]
79. Bühring, C.; Sarnthein, M.; Erlenkeuser, H. Toward a high-resolution stable isotope stratigraphy of the last 1.1 m.y.: Site 1144, South China Sea. *Proc. ODP Sci. Results* **2014**, *184*, 1–29. [CrossRef]
80. Wang, Y.; Cheng, H.; Edwards, R.L.; Kong, X.; Shao, X.; Chen, S.; An, Z. Millennial-and orbital-scale changes in the East Asian monsoon over the past 224,000 years. *Nature* **2008**, *451*, 1090–1093. [CrossRef]

81. North Greenland Ice Core Project Members. High-resolution record of Northern Hemisphere climate extending into the last interglacial period. *Nature* **2004**, *431*, 147–151. [CrossRef] [PubMed]
82. Yang, S.; Ding, Z. A 249 kyr stack of eight loess grain size records from northern China documenting millennial-scale climate variability. *Geochem. Geophys. Geosyst.* **2014**, *15*, 798–814. [CrossRef]
83. National Earth System Science Data Sharing Service Platform. *China Sci. Technol. Resour. Guide* **2017**, *49*, 112. Available online: https://www.geodata.cn/ (accessed on 1 October 2023).
84. Beck, J.W.; Zhou, W.; Li, C.; Wu, Z.; White, L.; Xian, F.; Kong, X.; An, Z. A 550,000-year record of East Asian monsoon rainfall from 10Be in loess. *Science* **2018**, *360*, 877–881. [CrossRef] [PubMed]
85. Huang, C.Y.; Liew, P.M.; Zhao, M.; Chang, T.C.; Kuo, C.M.; Chen, M.T.; Wang, C.H.; Zheng, L.F. Deep sea and lake records of the Southeast Asian paleomonsoons for the last 25 thousand years. *Earth Planet. Sci. Lett.* **1997**, *146*, 59–72. [CrossRef]
86. Wang, L.; Sarnthein, M.; Erlenkeuser, H.; Grimalt, J.; Grootes, P.; Heilig, S.; Ivanova, E.; Kienast, M.; Pelejero, C.; Pflaumann, U. East Asian monsoon climate during the Late Pleistocene: High-resolution sediment records from the South China Sea. *Mar. Geol.* **1999**, *156*, 245–284. [CrossRef]
87. Mao, S.Y.; Zhu, X.W.; Wu, N.Y.; Sun, Y.G.; Guan, H.X. Records of Ice Melting Water and Climate Events Since the Last Glacial Maximum in the Northern South China Sea: Indications of Long Chain Fatty Alcohol Terrestrial Input. *J. Trop. Oceanogr.* **2015**, *2*, 52–65.
88. Li, M.K.; Ouyang, T.P.; Zhu, Z.Y.; Tian, C.J.; Peng, S.S.; Qiu, Y.; Peng, X.C.; Zhong, H.X.; Chen, H.J. Magnetic records of the H1 event on the northwest slope of the South China Sea. *Quat. Stud.* **2019**, *39*, 927–937. [CrossRef]
89. Yang, S.L.; Dong, X.X.; Xiao, J.L. History of East Asian Monsoon Changes Since the Last Glacial Maximum—Geological Records of Northern China. *Chin. Sci. Earth Sci.* **2019**, *49*, 1169–1181.
90. Cheng, J. *Research on the Phenomenon of "Overshoot" of Temperature and Salinity Circulation during Bølling-Allerød Warming Events*; Nanjing University of Information Science and Technology: Nanjing, China, 2013.
91. Petit, J.R.; Jouzel, J.; Raynaud, D.; Barkov, N.I.; Barnola, J.M.; Basile, I.; Bender, M.; Chappellaz, J.; Davis, M.; Delaygue, G.; et al. Climate and atmospheric history of the past 420,000 years from the Vostok ice core, Antarctica. *Nature* **1999**, *399*, 429–436. [CrossRef]
92. Cuffey, K.M.; Clow, G.D. Temperature, accumulation, and ice sheet elevation in central Greenland through the last deglacial transition. *J. Geophys. Res. Ocean.* **1997**, *102*, 26383–26396. [CrossRef]
93. Alley, R.B. Ice-core evidence of abrupt climate changes. *Proc. Natl. Acad. Sci. USA* **2000**, *97*, 1331–1334. [CrossRef] [PubMed]
94. McManus, J.F.; Francois, R.; Gherardi, J.M.; Keigwin, L.D.; Brown-Leger, S. Collapse and rapid resumption of Atlantic meridional circulation linked to deglacial climate changes. *Nature* **2004**, *428*, 834–837. [CrossRef] [PubMed]
95. Liu, Z.; Otto-Bliesner, B.L.; He, F.; Brady, E.C.; Tomas, R.; Clark, P.U.; Carlson, A.E.; Lynch-Stieglitz, J.; Curry, W.; Brook, E.; et al. Transient Simulation of Last Deglaciation with a New Mechanism for Bølling-Allerød Warming. *Science* **2009**, *325*, 310–314. [CrossRef] [PubMed]
96. Chen, F.; Xu, Q.; Chen, J.; Birks, H.J.B.; Liu, J.; Zhang, S.; Jin, L.; An, C.; Telford, R.J.; Cao, X.; et al. East Asian summer monsoon precipitation variability since the last deglaciation. *Sci. Rep.* **2015**, *5*, 11186. [CrossRef] [PubMed]
97. Ji, M.; Shen, J.; Wu, J.; Wang, Y. Paleovegetation and paleoclimate evolution of past 27.7 cal ka BP recorded by pollen and charcoal of lake Xingkai, Northeastern China. In *Earth Surface Processes and Environmental Changes in East Asia*; Kashiwaya, K., Shen, J., Kim, J., Eds.; Springer: Tokyo, Japan, 2015; pp. 81–94. [CrossRef]
98. Wu, J.; Liu, Q.; Wang, L.; Chu, G.Q.; Liu, J.Q. Vegetation and climate change during the last deglaciation in the Great Khingan Mountain, Northeastern China. *PLoS ONE* **2016**, *11*, e0146261. [CrossRef] [PubMed]
99. Wang, B.Y. *Research on the Instability and Suitable Period of the Early Holocene Monsoon Climate Recorded by Chinese Stalagmites*; Southwest University: Chongqing, China, 2019.
100. O'Brien, S.R.; Mayewski, P.A.; Meeker, L.D.; Meese, D.A.; Twickler, M.S.; Whitlow, S.I. Complexity of Holocene climate as reconstructed from a Greenland ice core. *Science* **1995**, *270*, 1962–1964. [CrossRef]
101. Zhang, H.L.; Pu, X.Q. Stalagmite records of climate evolution and cold dry events in the Xundian area of Yunnan during the Middle Holocene. *J. Earth Sci.* **2011**, *32*, 95–100. [CrossRef]
102. Sandweiss, D.H.; Maasch, K.A.; Anderson, D.G. Transitions in the mid-Holocene. *Science* **1999**, *283*, 499–500. [CrossRef]
103. Spindler, K. *The Man in the Ice: The Discovery of a 5000-Year-Old Body Reveals the Secrets of the Stone Age*; Harmony Books: New York, NY, USA, 1994.
104. Zhang, Z.K.; Wang, S.M.; Wu, R.J. Environmental Evolution and Southwest Monsoon Changes of Sedimentary Records of Erhai Lake in the Middle Holocene. *Sci. Bull.* **1998**, *19*, 2127–2128.
105. Yang, H.R.; Xie, Z.R. Climate fluctuations and sea level fluctuations in eastern China over the past 20000 years. *Ocean Lakes* **1984**, *1*, 1–13.
106. Huang, C.C.; Zhou, J.; Pang, J.; Han, Y.; Hou, C. A regional aridity phase and its possible cultural impact during the Holocene Megathermal in the Guanzhong Basin, China. *Holocene* **2000**, *10*, 135–142. [CrossRef]
107. Zhu, K.Z. Preliminary Study on Climate Change in China over the Past 5000 Years. *Meteorolog. Sci. Technol. Data* **1973**, *S1*, 2–23. [CrossRef]
108. Moros, M.; Andrews, J.T.; Eberl, D.D.; Jansen, E. Holocene history of drift ice in the northern North Atlantic: Evidence for different spatial and temporal modes. *Paleoceanography* **2006**, *21*, PA2017. [CrossRef]

109. Bond, G.C.; Showers, W.; Elliot, M.; Evans, M.; Lotti, R.; Hajdas, I.; Johnson, S. The North Atlantic's 1–2 kyr climate rhythm: Relation to Heinrich events, Dansgaard/Oeschger cycles and the Little Ice Age. *Geophys. Monogr. Am. Geophys. Union* **1999**, *112*, 35–58. [CrossRef]
110. Holzhauser, H.; Magny, M.; Zumbuühl, H.J. Glacier and lake-level variations in west-central Europe over the last 3500 years. *Holocene* **2005**, *15*, 789–801. [CrossRef]
111. Wang, Y.; Cheng, H.; Edwards, R.L.; He, Y.; Kong, X.; An, Z.; Wu, J.; Kelly, M.J.; Dykoski, C.A.; Li, X. The Holocene Asian monsoon: Links to solar changes and North Atlantic climate. *Science* **2005**, *308*, 854–857. [CrossRef]

Disclaimer/Publisher's Note: The statements, opinions and data contained in all publications are solely those of the individual author(s) and contributor(s) and not of MDPI and/or the editor(s). MDPI and/or the editor(s) disclaim responsibility for any injury to people or property resulting from any ideas, methods, instructions or products referred to in the content.

Journal of
Marine Science and Engineering

MDPI

Article

Experimental Analysis of the Changes in Coral Sand Beach Profiles under Regular Wave Conditions

Zhen Yao [1], Jie Chen [1,2,3,*], Changbo Jiang [1,2,3], Hai Liang [1], Zhiyuan Wu [1,2,3], Bin Deng [1,2,3], Yuannan Long [1,2,3] and Chen Bian [1]

[1] School of Hydraulic and Environmental Engineering, Changsha University of Science & Technology, Changsha 410114, China; zyao@stu.csust.edu.cn (Z.Y.); jcb36@163.com (C.J.); 18969119559@163.com (H.L.); zwu@csust.edu.cn (Z.W.); dengbin07@csust.edu.cn (B.D.); lynzhb@csust.edu.cn (Y.L.); biancheng19@163.com (C.B.)

[2] Key Laboratory of Dongting Lake Aquatic Eco-Environmental Control and Restoration of Hunan Province, Changsha 410114, China

[3] Key Laboratory of Water-Sediment Sciences and Water Disaster Prevention of Hunan Province, Changsha 410114, China

* Correspondence: chenjie166@163.com

Abstract: This study utilized 50 laboratory experiments to document the evolution of coral beaches under varying regular wave conditions, including five distinct wave periods and ten wave heights. Both the type of equilibrium beach and the shape of sand bars were used to represent beach evolution. The evolution of coral sand beaches was then compared to quartz sand beaches. The experimental results show that the predicted (modeled) equilibrium profile of a quartz sand beach was not applicable to coral sand beaches. Compared to sand bars on quartz sand beaches, the distance from bar crests to the beach berm in coral sand beaches was greater, whereas the erosional depth of sand troughs was deeper. However, the grain size distribution of sand associated with the coral sand beach under wave action was consistent with Celikoglu's law. Both an equilibrium beach profile classification model and a sand bar shape prediction model for coral sand beaches were developed based on the experimental data.

Keywords: reef islands; coral sand beach; equilibrium beach profile; coral sand bar

Citation: Yao, Z.; Chen, J.; Jiang, C.; Liang, H.; Wu, Z.; Deng, B.; Long, Y.; Bian, C. Experimental Analysis of the Changes in Coral Sand Beach Profiles under Regular Wave Conditions. *J. Mar. Sci. Eng.* **2024**, *12*, 287. https://doi.org/10.3390/jmse12020287

Academic Editor: Gemma Aiello

Received: 16 January 2024
Revised: 31 January 2024
Accepted: 2 February 2024
Published: 5 February 2024

Copyright: © 2024 by the authors. Licensee MDPI, Basel, Switzerland. This article is an open access article distributed under the terms and conditions of the Creative Commons Attribution (CC BY) license (https://creativecommons.org/licenses/by/4.0/).

1. Introduction

Coral reef islands form through the accumulation of reef-derived carbonate sediment (i.e., debris from coral and other marine organism) of sand size (referred to herein as coral sand) [1] on coral reef platforms by wave and current processes [2,3] (Figure 1). Reefs and reef islands support important social and ecological systems. For example, coral reefs serve as natural breakwaters [4] and provide economic services by allowing for urban development [5] and tourism [6]. According to a recent global survey [7], coral reefs generate USD 11.5 billion per year in tourism revenues and USD 6.8 million per year in fisheries revenues and provide USD 10.7 billion in shoreline protection. However, exactly how islands and the communities that inhabit them will respond over the next century is still unclear, particularly given the potential sea-level rises (SLR) initiated by global warming [8].

According to projections by Slangen et al. [9] and Kopp et al. [10], the global sea level will increase by over 2 m by 2100. The rate of SLR is projected to exceed the rate of vertical accretion of coral reef platforms (Figure 1) [11]. As a result, the hydrodynamic conditions of reef flats will change, affecting the entire coral reef island [12]. Previous studies have concluded that without any human armoring of shorelines, these reefs will become uninhabitable, forcing people to migrate [13]. However, a survey of 709 atoll islands in the Pacific and Indian Oceans showed that 73.1% of the reef islands have been stable in size over the past few decades; 15.5% and 11.4% had increased or decreased in size, respectively [14]. Thus, these recent studies [15,16] indicate that coral reef islands are morphodynamically resilient landforms and that the risk of inundation as a result of SLR is (and will be) countered by adjustments in the morphology of ocean and lagoonal beaches (Figure 1: num 4 and 7) associated with the coral reef islands. This natural adaptation of the reef islands suggests that new predictions of the future trajectory of coral island beaches in response to the impacts of SLR are needed [16].

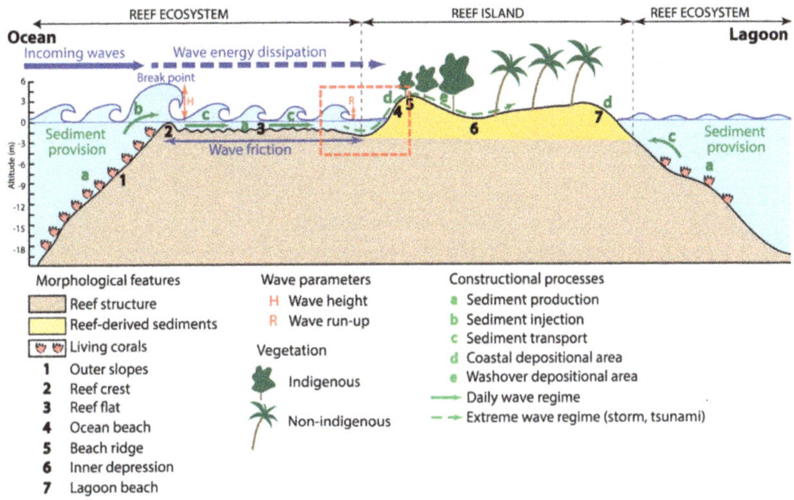

Figure 1. Conceptual diagram of the coral reef ecosystem. The red dashed area represents the research area in this paper. Reproduced from Duvat et al. [14].

Masselink et al. [16] argue that coral reef islands need to be considered as a separate type of island beach because of their different morphological characteristics. Therefore, a recent trend is to quantify the responses of sandy beaches associated with coral islands to varying hydrodynamic conditions. Unlike coral sand beaches, quartz sand beaches have been extensively studied, and a series of empirical formulas have been proposed through quantitative analyses to provide guidance for marine engineering projects [17–21].

Quartz sand beach evolution experiments, in which they are subjected to the continuous action of waves, show that beach profiles develop an equilibrium state, referred to as the equilibrium beach profile. A large number of studies have analyzed the morphology of balanced beaches and constructed functions to describe them [19,20,22]. Recently, Marini et al. [23,24] extended Dean's model [19] to beaches of any profile in the presence or absence of submerged breakwaters/bars. In addition, the transport patterns of quartz sand have been analyzed using laboratory experiments and field surveys [17,25,26], allowing for the classification of balanced shorelines on the basis of scouring and siltation [22,27,28]. Researchers have found from these studies that the presence of sand dams protects the coast [29]. This has led scholars to study the morphological parameters [30], formation theories [31], and movement trends of sandbars [32].

An interesting question arises as to whether the theoretical and empirical formulas developed for quartz sand beaches can be directly applied to coral sand beaches. The same problem arises with respect to the settling of coral sand particles [33], their threshold of motion [34,35], and their subsequent dispersion [36]. These findings suggest that coral sand particles, due to their special bio-skeletal structure and density, deviate significantly from the values calculated for these parameters using predictive models for quartz sand.

For this study, wave flume experiments were conducted in which a coral sand beach was subjected to varying regular wave conditions to assess differences in how quartz and coral sand beaches respond to wave conditions. It provides a new set of experimental data on the evolution of coral sand beaches. The evolution of the beach with respect to wave conditions was characterized on the basis of its classified equilibrium beach profile and sand bar morphology, both of which were compared to the evolution of quartz sand beaches. The distribution of coral sand grain sizes during beach evolution was also investigated. The experimental data were ultimately used to modify the equations for an equilibrium shoreline classification prediction model developed by Sunamura and Horikawa [28] and the prediction equations for bar parameters by Günaydın and Kabdaşlı [30].

2. Methodology

Since the 1940s, many researchers have investigated the morphology and evolution of shoreline profiles using experimental approaches and field measurements. The most widely used method of classifying equilibrium beach profiles was proposed by Sunamura and Horikawa [28]. As shown in Figure 2, they suggest that any beach profile can be classified into three types: (1) type A profiles, which are associated with storm or winter beaches that form erosional bars in the direction of the sea, (2) type B profiles, which are found on transitional beaches (the shoreline advances and sand piles up offshore), and (3) type C profiles, which are found on summer beaches (the shoreline progrades and no sand deposition takes place offshore). In this study, we refer to Type A as an erosional beach, Type B as a transient beach, and Type C as a depositional beach.

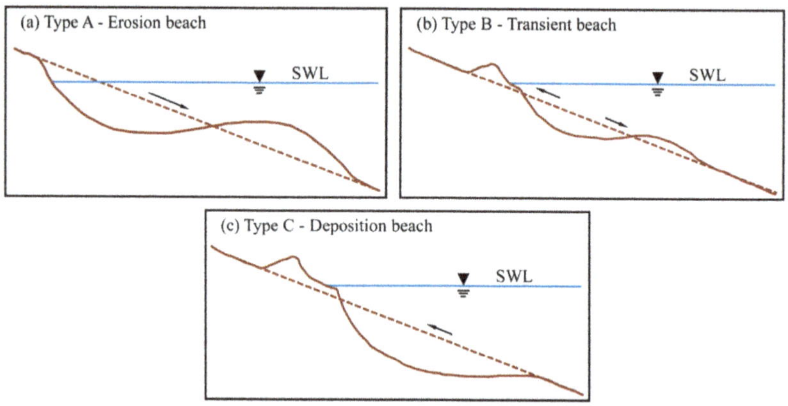

Figure 2. Classification of equilibrium beach profiles (after Sunamura and Horikawa [28]). The brown dashed line bit the initial beach profile, the brown solid line is the profile when the beach reaches equilibrium under wave action, the blue solid line is the still-water level, and the black arrow is the direction of sediment movement.

In the study of Sunamura and Horikawa [28], parameters that may control the motion of oblique particle clouds dictate the type of beach profile, and can be expressed by:

$$\text{Type of equilibrium beach profiles} = f(H, L, D_{50}, \beta) \quad (1)$$

where H is the wave height in the deep sea, L is the wavelength in the deep sea, D_{50} is the mean diameter of sediment particles, and β is the foreshore slope. Furthermore, Sunamura and Horikawa [28] argue that the dimensionless parameters can be defined as H/L, D_{50}/L, and tanβ; these parameters can be used to develop a predictive model for the classification of equilibrium beach profiles (Table 1). Based on the results of Sunamura and Horikawa [28], Xu [37] proposed a relationship (Equation (2)) between the dimensionless parameter containing the settling velocity and the classification of the shoreline profile. Parameters used to characterise sediment movement contain, in addition to the median particle size, D_{50}, the settling velocity of the particles, ω. This relationship can be expressed as:

$$\text{Type of equilibrium beach profiles} = f\left(\frac{H}{l}, \frac{\sqrt{gH}}{\omega}, \frac{gHT}{v}, \frac{\sqrt{gHT}}{d}, \tan\beta\right) \quad (2)$$

where g is the constant of gravitational acceleration, ν is the kinematic fluid viscosity, H/L is the wave steepness, $(gH)^{0.5}/\omega$ is the relative strength of wave action on the particle, gHT/ν is the same as the Reynolds number in dimension ($\frac{gHT}{\nu} \sim \frac{uL}{\nu}$), and $\frac{\sqrt{gHT}}{d}$ is the relative roughness at the sediment bed. Based on the findings of Jonsson [38], Xu [37] combined $\frac{gHT}{\nu}$ and $\frac{\sqrt{gHT}}{d}$, which represents the bed friction coefficient under wave conditions. Wu [39] believes that the impinging jet formed by the plunging waves is the main driving force for sediment transport initiation. Thus, as shown in Table 1, he added wave breaking parameters ($\xi = \frac{\tan\beta}{(H/L)^{0.5}}$) to the equilibrium beach profile classification model.

Table 1. Most used classification models for equilibrium quartz sand beach profiles.

Ref.	Erosional Beach	Transient Beach	Depositional Beach	Equation No.
Sunamura (1975) [28]	$(H/L)(D_{50}/L)^{-0.67}(\tan\beta)^{0.27} > 8$	$4 < (H/L)(D_{50}/L)^{-0.67}(\tan\beta)^{0.27} < 8$	$(H/L)(D_{50}/L)^{-0.67}(\tan\beta)^{0.27} < 4$	(3)
Xu (1988) [37]	$(\frac{H}{L})^{0.5}\left(\frac{\sqrt{gH}}{\omega}\right)(f_w + \tan\beta) > 0.29$	$(\frac{H}{L})^{0.5}\left(\frac{\sqrt{gH}}{\omega}\right)(f_w + \tan\beta) < 0.29$	$0.22 < (\frac{H}{L})^{0.5}\left(\frac{\sqrt{gH}}{\omega}\right)(f_w + \tan\beta) < 0.35$	(4)
Wu (2014) [39]	$\xi > 0.556\frac{(gH)^{0.5}}{\omega}(f_w + \tan\beta)$	$\xi < 0.556\frac{(gH)^{0.5}}{\omega}(f_w + \tan\beta)$	$0.394\frac{(gH)^{0.5}}{\omega}(f_w + \tan\beta) \le \xi \le 0.867\frac{(gH)^{0.5}}{\omega}(f_w + \tan\beta)$	(5)

For an erosional profile, the sand bar is subjected to coupled hydrodynamic and sediment transport processes which change its shape. In order to further investigate the effect of waves on the shape of sand bars, researchers began to define the shape of sand dams. The most common way to define a sand dam is shown in Figure 3. The geometric characteristics of a sand bar are defined by the distance from the equilibrium point to the original point (X_{be}); the water depth at the equilibrium point (h_{be}); the distance from the crest point to the original point (X_{bc}); the water depth at the crest point (h_{bc}); the distance from the closure point to the crest point (X_{bd}); and the water depth at the closure point (h_{bd}). In this paper, we do not consider the effect of coral sand grain size on coral sand bars; thus, we only compare the models of Silvester and Hsu [40] as well as Günaydın and Kabdaşlı [30] (Table 2).

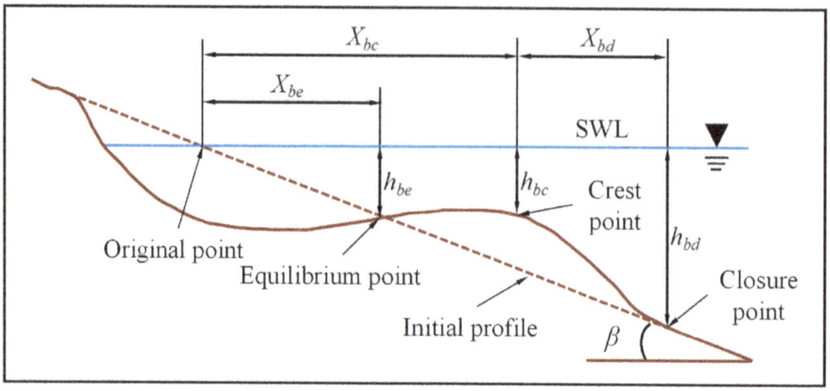

Figure 3. Schematic diagram showing the shape characteristics of a sand bar.

Table 2. Currently used equations for the determination of bar parameters.

Location	Geometric Parameters	Silvester and Hsu [40]	Equation No.
Equilibrium point	X_{be}	$X_{be}/L_0 = 0.96(H/L)/\tan\beta$	(6)
	h_{be}	\	
Bar crest	X_{bc}	$X_{bc}/L_0 = 0.022 + 1.508\left(\frac{H}{L}\right)/\tan\beta + 0.14\left[\left(\frac{H}{L}\right)/\tan\beta\right]^2$	(7)
	h_{bc}	$h_{bc}/(L\tan\beta) = 0.0269 + 0.391X_{bc}/L$	(8)
Closure point	X_{bd}	\	
	h_{bd}	\	
Location	**Geometric Parameters**	**Günaydın and Kabdaşlı [30]**	**EquationNo.**
Equilibrium point	X_{be}	$X_{be}/L = 113.98(\tan\beta\sqrt{H/L})^{1.9762}$	(9)
	h_{be}	$h_{be}/L = 11.87\left[\tan\beta(H/L)^{0.5}\right]^{1.7626}$	(10)
Bar crest	X_{bc}	$X_{bc}/L = 64.966(\tan\beta\sqrt{H/L})^{1.6754}$	(11)
	h_{bc}	$h_{bc}/L = 3.2041(\tan\beta\sqrt{H/L})^{1.413}$	(12)
Closure point	X_{bd}	$X_{bd}/L = 1.4843(h_{bd}/X_{bd})^{0.91}$	(13)
	h_{bd}	$h_{bd}/L = 102.33\left[\tan\beta(H/L)^{0.5}\right]^{1.1813}$	(14)

3. Experimental Setup

All experiments were conducted in a glass-walled wave flume in the hydraulics laboratory at the Changsha University of Science and Technology, China. The wave flume is 40.0 m long, 0.5 m wide, and 0.8 m high. A piston-type wave generator was installed at one end of the flume, and a coral sand beach was constructed in the wave flume with its toe 22.0 m away from the wave generator.

Coral sand particles from the South China Sea were used to create the coral sand beach profile (Figure 1). This carbonate sand consists mainly of coral (95%) and the remains of nearby organisms (e.g., gastropods and bivalves; 5%,), with a median diameter (D_{50}) of 0.585 mm, a coefficient of nonuniformity (D_{75}/D_{25}) of 1.7, and a coefficient of graduation ($D_{30}{}^2/(D_{60}D_{10})$) of 0.942.

The experimental setup was developed to reproduce the action of waves over a coral sand beach environment. Considering the size of the laboratory flume and the wave-making capacity of the wave maker, the experimental coral reef flat was generalized, but allows for the evolution of the coral sand beach (Figure 1). A coral sand profile was built over the flume platform. It was characterized by a slope of 5.5° (1:10), a length of 5.00 m, and a height of 0.50 m (Figure 4). The water level was maintained at 0.3 m above the platform during all experiments. The experiment did not simulate the specific environment of the coral sand beach on the reef flat but was developed such that its parameters could be adjusted to scale the physical model to a modern occurrence of a coral sand beach. As shown in Table 3, the hydrodynamic characteristics over the reef flat exhibited a range in wave heights from 0.0 to 1.0 m, a water depth between 0.3 and 3.0 m, and a wave period ranging from 2.0 to 20 s. Yu et al. [1] analyzed the grain sizes of coral reef sediments from 25 different areas located along the Chinese Nansha Islands. They found that the composition of sand mostly consisted of gravelly coarse sand; the gravel (>2 mm in diameter) makes up 30–50% to, in some areas, 50–80% of the sediment. Research conducted on Lady Elliot Island [41] found that the slope of the coral sand beach on a reef flat is approximately 1:10.

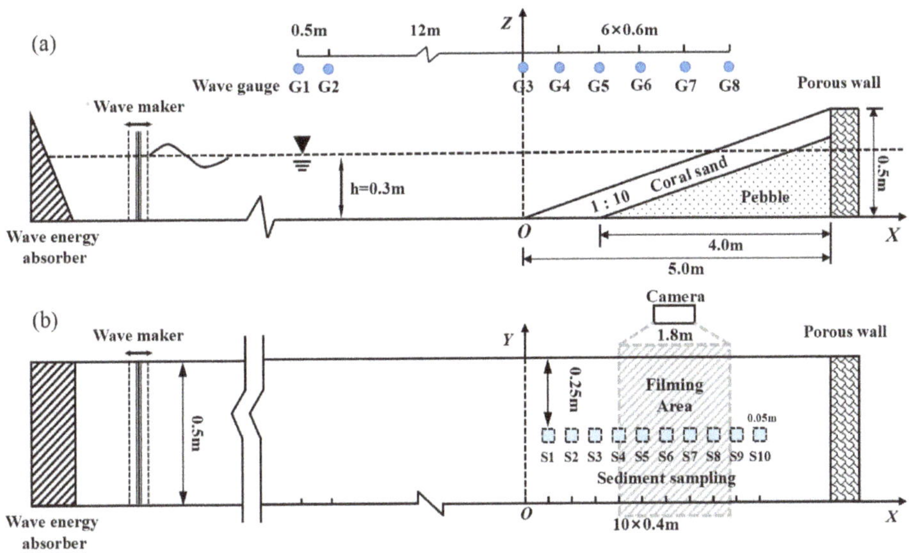

Figure 4. Schematic diagram of the experimental setup. (**a**) Plan view; (**b**) side view. Photographs were obtained from the grey stippled area.

Table 3. Summary of hydraulic characteristics measured over the coral reef flats.

Ref.	Research Location	Wave Height (m)	Water Depth (m)	Wave Period (s)
Hardy and Young [42]	The Great Barrier Reef	0.2~1.3	0.24~2.97	2.1~9.5
Nelson [43]	John Brewer Reef	0.1~0.7	0.8~2.7	3.0~5.2
Lowe et al. [44]	Kaneohe Bay	0.4~0.6	1.2~2.3	/
Vetter et al. [45]	Ipan, Guam	0.0~0.6	0.3~0.6	/
Taebi et al. [46]	Ningaloo Reef	0.0~0.5	1.0~2.0	
	The College of the Marshall Islands	0.0~0.3	0.4~0.8	6~23.7
Becker et al. [47]	Roi-Namur	0.0~0.2	0.4~0.8	4.3~20.9
	Ipan, Guam	0.0~0.1	0.5~0.7	4.9~19.6
Pomeroy et al. [48]	Ningaloo Reef	0.0~0.2	1.0~2.0	10~20
Lentz et al. [49]	Red Sea	0.0~0.2	0.3~1.2	4~8

In this paper the main dynamic factor of sediment movement is wave action. Therefore, scaling the experimental model mainly considered two aspects of wave motion, including similarity and sediment motion similarity under wave action. Thus, the geometric scale factor ($N_L = H_{Prototype}/H_{model} = 4$) and the time scale factor followed the Froude criterion ($N_T = T_{Prototype}/T_{model} = N_L{}^{0.5} = 2$) and were set following the Froude similitude criterion. Table 4 provides the wave conditions in the experiments, each of which was run for 3 h (which was sufficient time to develop an equilibrium profile). The sediment sample area is shown in Figure 4, and it was taken at the depth of 2 cm (to the seabed).

Table 4. Summary of scale-reduction factors for the experiments and utilized test conditions.

	Parameters	Natural Range	Scale Factor	Experimental Set Value
Sediment	Particle size, D (mm)	>2	4	0.585
Hydraulics	Wave height, H (m)	0.0–1.0	4	0.04, 0.05, 0.07, 0.08, 0.11, 0.12, 0.13, 0.14, 0.15, 0.16
	Water depth, h (m)	0.3–3.0	4	0.3
	Wave period, T (s)	2.0–20	2	1.4, 1.5, 1.6, 1.7, 1.8

Wave gauges (ULS 80D, General Acoustics, Germany) were used to record the wave data, whereas a topographic surveying system (RUI-III, Wuhan University, Wuhan, China) was used to survey the topo-bathymetry of the beach profile. A camera (Sony IXM 586, Sony, Japan) was used to record the morphological evolution at one-second intervals during the tests. The particle size distribution of the coral sand particles was measured using a laser particle sizer (Malvern Instruments MasterSize2000, Malvern Panalytical, United Kingdom: It uses the technique of laser diffraction to measure the size of particles. It does this by measuring the intensity of light scattered as a laser beam passes through a dispersed particulate sample).

4. Results

4.1. Evolution of the Coral Sand Beach Profile

For all experiments, our observations revealed that a large amount of coral sand as suspended sediment entered the water column from the sediment bed and started to move as the waves entered the breaking phase. Six snapshots extracted from the video recordings are shown in Figure 5 for waves characterized by h = 0.3 m, H = 0.07 m, and T = 1.8 s. Each of these images shows the same field of view that starts at X = 1.5 m and ends at X = 2.9 m. Figure 5a shows that the surface of the water remains still (calm) and the water is clear before the waves reach the coral sand beach. Figure 5b subsequently shows that when the wave peak reaches x = 2.0 m, the waveform begins to change. A turbid area appears in front of the wave peak (the starting point of the turbid area is located at X = 2.05 m, Figure 5b), indicating that the coral sand has become suspended in the water. The wave begins to break when the peak reaches x = 2.3 m, at which point air bubbles enter the rolled wave, and the starting position of the turbid region moves from X = 2.05 m to X = 2.15 m. As the seawater recedes from the beach, the starting position of the turbid region moves (Figure 5e,f) until it is stabilized at X = 1.6 m. In the wave breaking zone, the bed is scoured to form a sand trough. Meanwhile, a beach berm is formed at the shoreward end and a smaller sand bar is formed at the offshore end. As time increases, the area of the sand trough expands until the morphology of the coral sand beach profile reaches an equilibrium (constant) state.

The initial coral sand beach profile and those after running several hours are plotted in Figure 6 for two cases. In Figure 6b, local scour is observed between X = 1.3 and 2.8 m. The extent of the sand trough increased with the duration of the test, and the depth of the trough reached an equilibrium profile after 300 min. The beach berm appeared between X = 3.0 and 3.8 m. According to the classification of Sunamura and Horikawa [28] (Figure 2), the developed equilibrium beach profile represents a deposition beach. In Figure 6a, the initial beach reached equilibrium after 300 min. A beach berm was formed by siltation between X = 2.8 and 3.8 m, whereas a sand trough was formed by scouring between X = 1.3 and 2.8 m.

When the beach profile reaches equilibrium, bed sands are sampled (blue area in Figure 4) and changes in bed sand characteristics are recorded. The results are shown in Figure 7. The median grain size of the sediment near the top of the shoulder beach and the bottom of the sand channel becomes larger compared to the initial sediment. The median grain size of the sediment near the wave-facing side of the shoulder beach and the sand pattern in the offshore zone becomes smaller. Combined with a series of snapshots, Figure 5 shows that the turbulence increased after wave breaking and a large number of vortices appeared leading to sediment starting. Coarse-grained sediment settles after initiation due to the lack of current carrying capacity, meaning that the median size of the sediment near the sand trough is larger. Fine-grained sediment becomes suspended in the water column with the wave offshore transport movement and finally settles in the offshore area, meaning the median particle size of sediment near the offshore sand grain is smaller. At the same time, the upwelling current carries the bed sediment to the shore, and the coarse-grained sediment is carried to the top. As the upwelling velocity decreases, the coarse-grained sediment is silted up, so the median particle size of the sediment near the beach berm is larger.

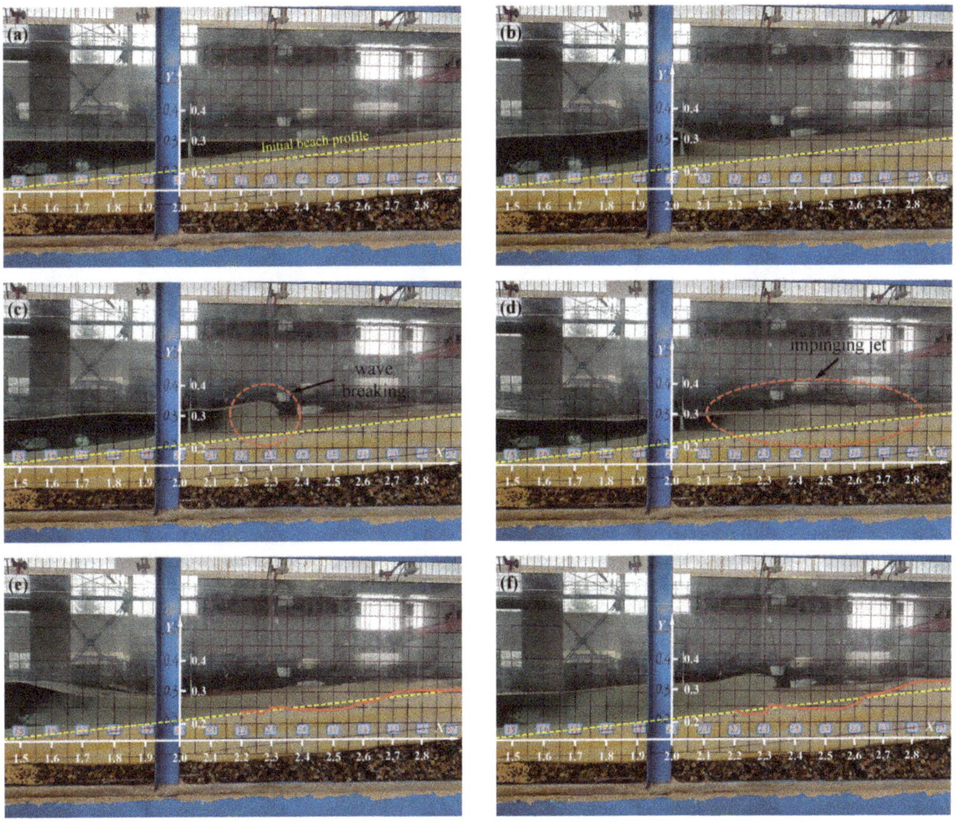

Figure 5. Snapshots (25 fps) of the wave breaking process and the accompanying change in the coral sand beach profile for a water depth of h = 0.3 m, a wave height H = 0.07 m, and wave period of T = 1.8 s. (**a**) t = 0 s, (**b**) t = 8.84 s, (**c**) t = 9.56 s, (**d**) t = 10.8 s, (**e**) t = 10 min, (**f**) t = 30 min. The yellow dashed line is the initial beach profile and the red solid line is the beach profile under wave action.

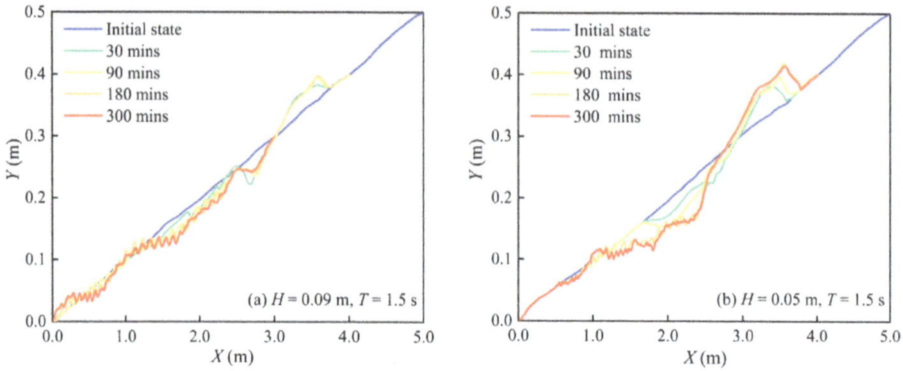

Figure 6. Changes in the coral sand beach profile when affected by regular waves between 30 and 300 min. (**a**) $H = 0.09$ m, $T = 1.5$ s; (**b**) $H = 0.05$ m, $T = 1.5$ s.

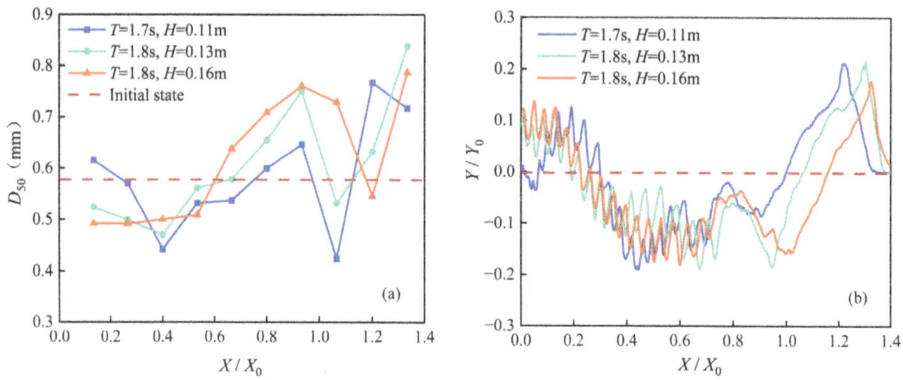

Figure 7. Changes in sediment grain size distributions and the bed elevation after being subjected to regular waves (**a**,**b**). X_0 and Y_0 represent the location of the original point. The value of Y/Y_0 indicates siltation and scour ($Y/Y_0 > 0$ means siltation, $Y/Y_0 < 0$ means scour). The value of X/X_0 indicates the position relative to the mean water level (i.e., when $X/X_0 > 1.0$ the position is above the mean water level).

4.2. Classification of Equilibrium Coral Sand Beach Profiles

Independent tests of beach classification (Equations (3)–(5)) were performed using our experimental data. Results of the observed experimental data (shown in Figure 8) were classified using the types of equilibrium beach profiles proposed by Sunamura and Horikawa (1975) (Figure 2). The distribution of the experimental classification data is similar to the results of Sunamura and Horikawa [28], Rector [27], and Watts [50]. The constant terms in Table 1 can be referred to as the coefficients of determination, α. Moreover, the coefficient of determination, α, 4, 8 (the dash line shown in Figure 8a), proposed by Sunamura and Horikawa [28] did not effectively classify the coral sand beach types. For beaches characterized by siltation (type C), the coefficient of determination, α, is smaller than the actual value. Moreover, the coefficient of determination, α, is greater than the actual value for erosional beaches. This indicates that the transitional area of coral beaches is smaller than that associated within quartz sand beaches.

When using the classification method of Xu [37] to predict the coral sand beach type, the erosional beach can be accurately classified, but there are errors in the prediction of the siltation and transitional beaches. As shown in Figure 8b, when $\alpha > 3.5$, this area includes data points for both siltation and transition beaches. When α is between 2.9 and 4.5, both transitional and erosional beaches are present. Compared with the classification model of Wu [39], $\alpha = 0.867$ can distinguish siltation beaches and transitional beaches. However, within the interval of $0.394 < \alpha < 0.867$, some of the siltation beaches were incorrectly predicted to be transition beaches.

Comparing the trends in the experimental data on the different regime maps, (H_0/L_0~$(D/L_0)^{0.67}i^{-0.27}$ and $(H_0/L_0)^{-0.5}$~$(gH)^{0.5}(f_w + i)/\omega$), the slope of the dividing line is most similar to the trends observed on a phase diagram for the same type of experimental data in Sunamura and Horikawa [28]. Therefore, herein we propose an equilibrium beach profile type prediction model that is applicable to coral sand beaches by parameter correcting the predictive classification model of Sunamura and Horikawa [28] (Figure 8a), where:

$$(H_0/L_0)(D/L_0)^{-0.67}(\tan\beta)^{0.27} > 7.35, \text{ Type A-Erosional Beach} \tag{15}$$

$$5.05 < (H_0/L_0)(D/L_0)^{-0.67}(\tan\beta)^{0.27} < 7.35, \text{ Type B-Transient Beach} \tag{16}$$

$$(H_0/L_0)(D/L_0)^{-0.67}(\tan\beta) < 5.05, \text{ Type C-Depositional Beach} \tag{17}$$

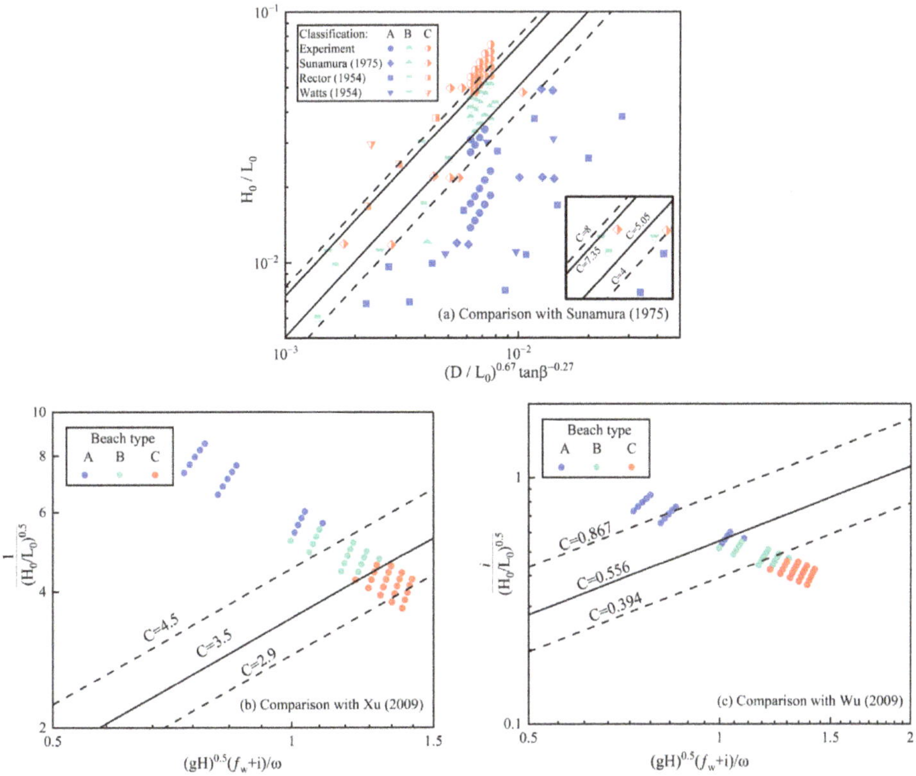

Figure 8. Comparison of the experimental results with the different limits of classification for equilibrium beach profiles. (**a**) Classification by means of Equation (3) of Sunamura and Horikawa (1975) [28]; (**b**) classification conducted using Equation (4) of Xu (1988) [37]; (**c**) classification using Equation (5) of Wu (2014) [39].

4.3. Geometric Characteristics of Offshore Coral Sand Bars

Figure 9 compares the distance from the wave crest point to the original point (X_{bc}) and the water depth at the bar crest (h_{bc}) with the value derived from the quartz sand prediction equation used by Günaydın and Kabdaşlı [30]. The water depth at the bar crest is greater for the coral sand bar compared to the quartz sand bar under the same regular wave action. Coral sand bars are located further from the berm. Thus, the depositional location of coral sand particles is further from the berm than for a quartz sand beach. Moreover, with an increase in the dimensionless parameter, tanβ $(H_0/L_0)0.5$, Günayd's model predictions follow the same trend as the measured values obtained during our experiments.

In order to compare the applicability of the predictive model for a quartz sand bar to a coral sand bar, the predictive models proposed by Günaydın and Kabdaşlı [30] and Silvester and Hsu [40] were parametrically corrected. The equations determined by non-linear regression and their correlation coefficients are shown in Table 5. Figures 10–12 show the experimental geometric characteristics of coral sand bars and the modified predictive models based on quartz sand bars in relation to the unfactored particle diameters. The figures illustrate that the functional relationship (polynomial structure) obtained through the experiments on the quartz sand bars also applies to the features of coral sand bars.

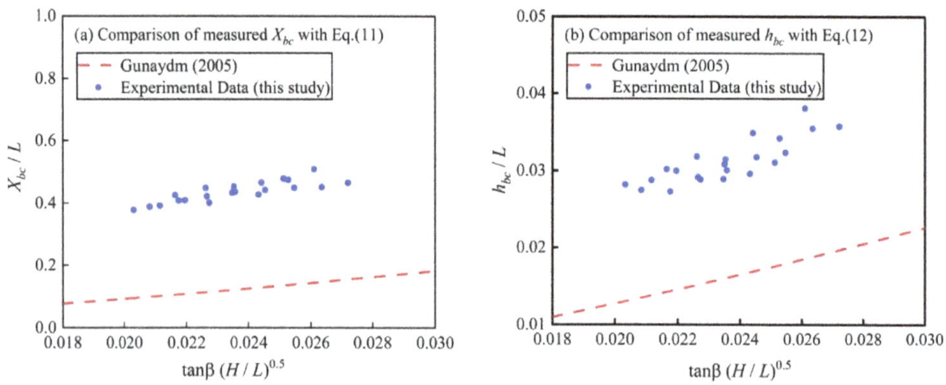

Figure 9. Comparison of our experimental data with the modeled data predicting the distance from the crest point to the original point (X_{bc}) and water depth at the bar crest (h_{bc}) proposed by Günaydın and Kabdaşlı [30].

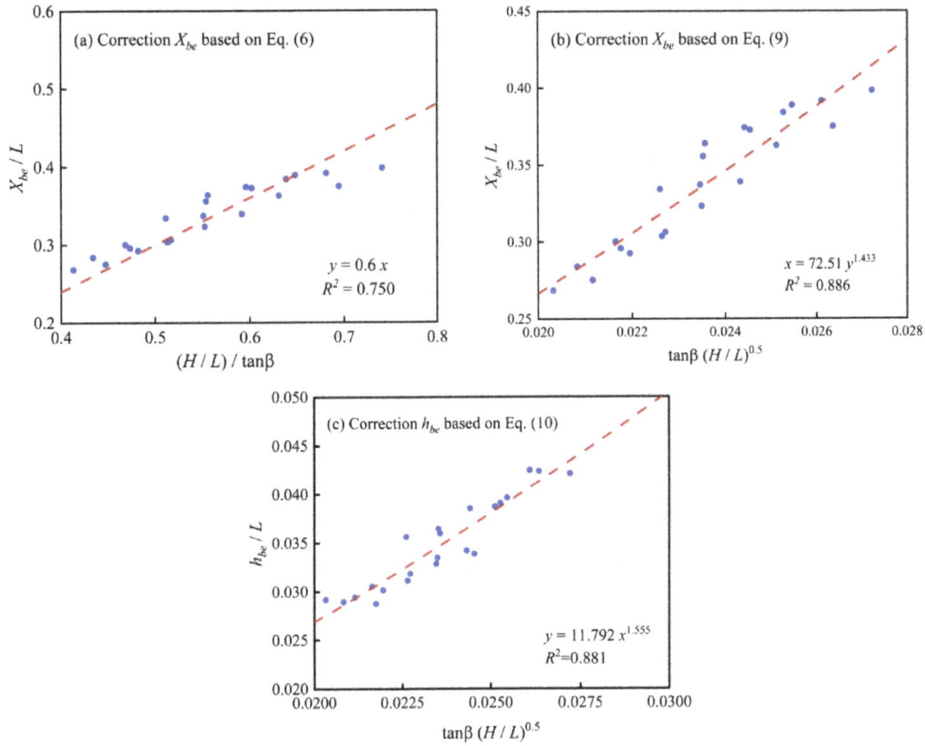

Figure 10. Comparison of the experimental results of the geometric characteristics of the equilibrium point generated herein with the modified prediction models. The red dashed line is the result of the corrected equation.

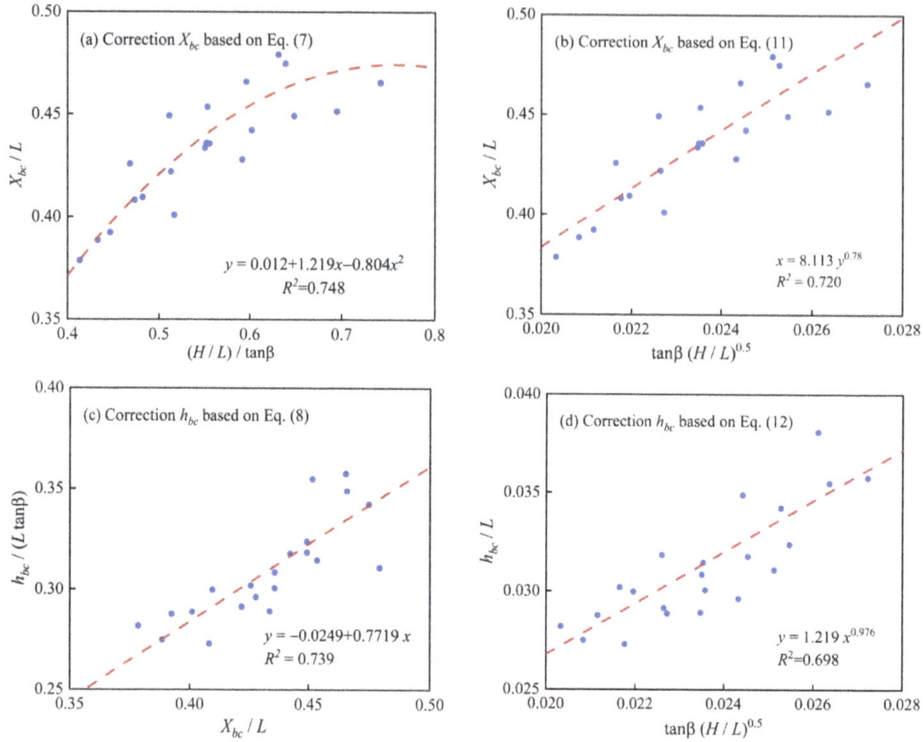

Figure 11. Comparison of geometric characteristics of bar crests from the experimental data with the modified prediction models. The red dashed line is the result of the corrected equation.

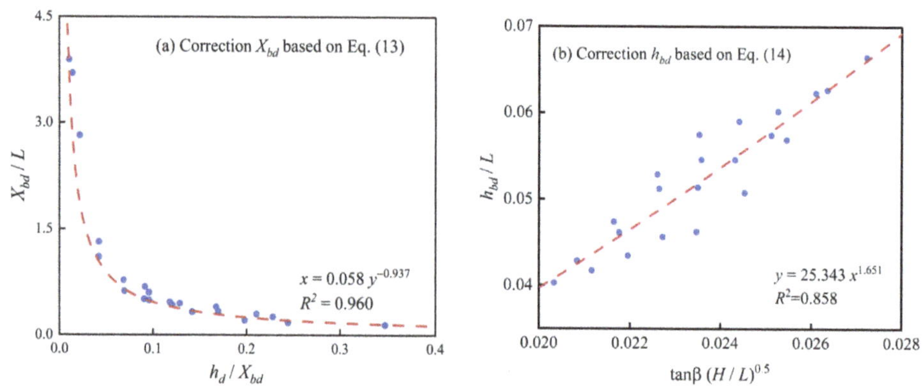

Figure 12. Comparison of the geometric characteristics of the closure point of the experimental results with the modified prediction models. The red dashed line is the result of the corrected equation.

The regression coefficients for each of the predictive equations are shown in Table 5 (columns 4 and 6). A comparison of the results of the corrected model developed by Günaydın and Kabdaşlı [30] with that of Silvester and Hsu [40] shows that the structure of the model of Silvester and Hsu [40] is applicable to the geometrical characteristics of the crest of the coral sand bar. The model structure of Günaydın and Kabdaşlı [30] was applied to the geometric characteristics of the equilibrium and closure points of coral sand bars.

Table 5. Modifications of equations for geometric parameters of coral sand bars based on models provided by Günaydın and Kabdaşlı [30] and Silvester and Hsu [40].

Location	Geometric Parameters	Silvester and Hsu [40]	R^2	Günaydın and Kabdaşlı [30]	R^2
Equilibrium point	X_{be}	$X_{be}/L = 0.6(H/L)/\tan\beta$	0.75	$X_{be}/L = 72.51(\tan\beta\sqrt{H/L})^{1.433}$	0.89
	h_{be}	\		$h_{be}/L = 11.792\left[\tan\beta(H/L)^{0.5}\right]^{1.555}$	0.88
Bar crest	X_{bc}	$X_{bc}/L = 0.012 + 1.219(\frac{H}{L})/\tan\beta - 0.804[(\frac{H}{L})/\tan\beta]^2$	0.75	$X_{bc}/L = 8.113(\tan\beta\sqrt{H/L})^{0.78}$	0.72
	h_{bc}	$h_{bc}/(L\tan\beta) = -0.0249 + 0.7719X_{bc}/L$	0.74	$h_{bc}/L = 0.219(\tan\beta\sqrt{H/L})^{0.976}$	0.70
Closure point	X_{bd}	\		$X_{bd}/L = 0.058(h_{bd}/X_{bd})^{-0.937}$	0.96
	h_{bd}	\		$h_{bd}/L = 25.343\left[\tan\beta(H/L)^{0.5}\right]^{1.651}$	0.86

5. Discussions

During the initial stages of beach evolution under regular wave action, the waves gradually deform by creeping up the slope, creating a wave-generated flow at the bed surface and driving coral sands on the beach into the water column (Table 4, Figure 5b). It is worth noting that the turbid zone occurs before the wave crest and the turbidity is low during this phase. When the wave reaches the region shown in Figure 5c, it begins to break and form an impinging jet [51]. Under the action of the impinging jet, irregular vertical vortices appear at the crest of the wave and descended obliquely [52]. These vortices drive a large amount of coral sand upward, creating the turbid zone, including that at the back of the wave crest. As the impinging jet enters the water column, tumbling occurrs on the free surface (Figure 1d), and a region of high turbidity occurrs between x = 2.45 and 2.75, which indicates that a large number of coral sand particles are lifted into the water column at this stage. This type of sediment movement has been observed in previous studies of quartz sand banks [17], indicating that coral sand beaches are similarly affected by wave movement mechanisms as quartz sand beaches. In addition, as suggested by Çelikoğlu et al. [53], the smaller particles of sediment are easily entrained by a current caused by infiltration into the pore space between the coarse particles. During the return phase of the current, a thin, high-speed layer of water removes coarse particles from the bed surface sediment, while the fine particles remain stationary because of the shading effects by coarse particles. As a result, the median particle size of the bed sediment near the shoreline becomes smaller. In the offshore area, due to the emergence of the breaking waves, strong turbulence leads to a decrease in the sand-carrying capacity of the water current, and the coarse particles of sediment settle rapidly. Thus, the median particle size of sediment in the siltation area becomes larger. This suggests that the pattern of bed sediment sorting of coral sandy beaches under regular wave action is similar to that of quartz sand beaches.

However, the results of equilibrating beach classification show that the existing predictive model for the classification of quartz sand beaches is not applicable to coral sand beaches. This is because the dimensionless parameters from Sunamura and Horikawa [28] consider only the effects of wave steepness, wave intensity, and bank slope while ignoring the characteristics of sediment movement (starting flow velocity, settling velocity, etc.). Previous studies [33–35] have shown that coral sands are irregular in shape and have a high porosity, which leads to different settling velocities, starting velocities, and diffusion patterns for coral sand particles in comparison to quartz sand. These particle properties in turn affected the rate of coral sand transport from the shore to offshore and cause variations in the coral sand shoreland delineation boundary coefficients. It is worth noting that while the classification prediction models of Wu [39] and Xu [37] include particle settling velocity, the classification of beach type is unclear (Figure 4b,c). This may be due to the fact that particle settling velocity formulas are not applicable to coral sand particles and that the existing settling formulas for coral sand particles are single particle-dominated. The applicability of these formulas for a large number of particles in motion needs to be investigated. Meanwhile, it is worth noting that the R^2 of the modified bar top prediction model is only around 0.7 to 0.8. This suggests that the two forms of formulations do not fully correspond to the physical evolution of the coral sand beach. This phenomenon also occurs when modeling the prediction of coral sand particle initiation velocities [34,35] and settling motions [33]. Moreover, beach profile changes are the result of the complex motions of multiple particles, so in this paper only parameter corrections are used to propose a prediction model applicable to coral sand beach profiles. The predictive models are only applicable to experimental data in regular wave conditions. In the future, it will be necessary to collect field data to test the applicability of the predictive models.

6. Conclusions

In this study, a set of laboratory experiments were conducted to investigate changes in coral sand beach profiles under specified regular wave conditions. A total of 50 runs were performed with regular waves consisting of ten different wave heights and four wave periods. The water depth and initial slope angle, $\tan\beta$, were assumed to be constant and equal to 0.3 m and 0.1, respectively, limiting our understanding of sediment deposition on coral sand beaches by regular waves. The following conclusions were drawn from the experiments:

(1) Observations of hydrodynamic processes and the movement of coral sand on the bed surface suggest that the evolution of a coral sand beach is similar to the evolution of a quartz sand beach. The morphological characteristics of the sand bar show that the erosion depth of a coral sand beach is deeper than that of a quartz sand beach. The location of sand bar formation is further from the horizontal plane than the quartz sand beach. These differences are related to sediment transport and depositional processes, such as the starting flow rate and settling velocity of coral sand particles, among other parameters.

(2) The results show that the predictive model for the classification of the type of quartz beach profile is not applicable to coral sand beaches. A classification prediction model for coral sand beaches was proposed through parameter modification.

(3) Geometrical features of coral sand bars on erosional beaches were obtained and compared with a prediction model for a quartz sand bar. The results show that the prediction model for a quartz sand bar is not applicable to a coral sand bar. A prediction model for a coral sand bar was proposed through parameter modification.

Although this paper investigates the effect of regular waves on the evolution of coral sand beaches and compares the prediction models of previous quartz sand beaches, only the evolution of stable coral sand islands and one water depth were considered in this paper. Two other commonly observed cases in which coral sand islands are located in tidal shoals and the sandbar stage were not considered. In order to further study the evolutionary mechanism of coral sand islands under SLR, the model should be generalized for these two stages in the future. That is, beach evolution should be studied when the sand island is becoming submerged and once it becomes completely submerged.

Author Contributions: Conceptualization, J.C. and Z.Y.; methodology, H.L. and Z.Y.; software, Z.Y.; validation, C.B.; formal analysis, H.L.; investigation, J.C. and H.L.; resources, J.C.; data curation, Z.Y. and H.L.; writing—original draft preparation, Z.Y.; writing—review and editing, Z.Y.; visualization, Z.Y.; supervision, J.C.; project administration, C.J., Z.W., B.D. and Y.L.; funding acquisition, J.C. and C.J. All authors have read and agreed to the published version of the manuscript.

Funding: This research was funded by the National Key Research and Development Program of China (Grant No. 2021YFB2601100), the National Natural Science Foundation of China (Grant No. 52271257) and the Natural Science Foundation of Hunan Province (Grant No. 2022JJ10047).

Institutional Review Board Statement: Not applicable.

Informed Consent Statement: Not applicable.

Data Availability Statement: The data (experimental results: the characteristics of coral sand bars) of this study that support its findings are openly available in Baidu Netdisk at [https://pan.baidu.com/s/14zRRZK4HtpTeoIrNgZiocw?pwd=nu3l, accessed on 4 February 2024].

Conflicts of Interest: The authors declare no conflicts of interest.

References

1. Yu, H.; Sun, Z.; Tang, C. Physical and Mechanical Properties of Coral Sand in the Nansha Islands. *Mar. Sci. Bull.* **2006**, *8*, 31–39.
2. Anna, F. Encyclopedia of Modern Coral Reefs: Structure, Form and Process. *Ref. Rev.* **2011**, *25*, 39–40.
3. Woodroffe, C.D. Reef-Island Topography and the Vulnerability of Atolls to Sea-Level Rise. *Glob. Planet. Chang.* **2008**, *62*, 77–96. [CrossRef]
4. Beetham, E.; Kench, P.S. Predicting Wave Overtopping Thresholds on Coral Reef-Island Shorelines with Future Sea-Level Rise. *Nat. Commun.* **2018**, *9*, 3997. [CrossRef]
5. Silva, I.R.; Rossi, J.C.; Nascimento, H.M.; Siqueira, T.G. Geoenvironmental Characterization and Urbanization of the Beaches on the Islands of Tinharé and Boipeba, South Coast of the State of Bahia, Brazil. *J. Coast. Res.* **2009**, *II*, 1297–1300.
6. Spalding, M.; Burke, L.; Wood, S.A.; Ashpole, J.; Hutchison, J.; zu Ermgassen, P. Mapping the Global Value and Distribution of Coral Reef Tourism. *Mar. Policy* **2017**, *82*, 104–113. [CrossRef]
7. Burke, L.; Reytar, K.; Spalding, M.; Perry, A. *Reefs at Risk Revisited*; World Resources Institute: Washington, DC, USA, 2011.
8. Albert, S.; Leon, J.X.; Grinham, A.R.; Church, J.A.; Gibbes, B.R.; Woodroffe, C.D. Interactions between Sea-Level Rise and Wave Exposure on Reef Island Dynamics in the Solomon Islands. *Environ. Res. Lett.* **2016**, *11*, 054011. [CrossRef]
9. Slangen, A.B.A.; Carson, M.; Katsman, C.A.; Van de Wal, R.S.; Köhl, A.; Vermeersen, L.L.; Stammer, D. Projecting Twenty-First Century Regional Sea-Level Changes. *Clim. Change* **2014**, *124*, 317–332. [CrossRef]
10. Kopp, R.E.; Horton, R.M.; Little, C.M.; Mitrovica, J.X.; Oppenheimer, M.; Rasmussen, D.J.; Strauss, B.H.; Tebaldi, C. Probabilistic 21st and 22nd century sea-level projections at a global network of tide-gauge sites. *Earth Future* **2014**, *2*, 383–406. [CrossRef]
11. Montaggioni, L.F. History of Indo-Pacific Coral Reef Systems since the Last Glaciation: Development Patterns and Controlling Factors. *Earth-Science Rev.* **2005**, *71*, 1–75. [CrossRef]
12. Nicholls, R.J.; Cazenave, A. Sea-Level Rise and Its Impact on Coastal Zones. *Science* **2010**, *328*, 1517–1520. [CrossRef] [PubMed]
13. Hauer, M.E.; Fussell, E.; Mueller, V.; Burkett, M.; Call, M.; Abel, K.; McLeman, R. Sea-Level Rise and Human Migration. *Nat. Rev. Earth Environ.* **2020**, *1*, 28–39. [CrossRef]
14. Duvat, V.K.E.; Magnan, A.K. Rapid Human-Driven Undermining of Atoll Island Capacity to adjust to Ocean Climate-Related Pressures. *Sci. Rep.* **2019**, *9*, 15129. [CrossRef] [PubMed]
15. Kench, P.S.; Liang, C.; Ford, M.R.; Owen, S.D.; Aslam, M.; Ryan, E.J.; Turner, T.; Beetham, E.; Dickson, M.E.; Stephenson, W.; et al. Reef Islands Have Continually adjusted to Environmental Change over the Past Two Millennia. *Nat. Commun.* **2023**, *14*, 508. [CrossRef] [PubMed]
16. Masselink, G.; Beetham, E.; Kench, P. Coral Reef Islands Can Accrete Vertically in Response to Sea Level Rise. *Sci. Adv.* **2020**, *6*, eaay3656. [CrossRef] [PubMed]
17. Chen, J.; Huang, Z.; Jiang, C.; Deng, B.; Long, Y. An Experimental Study of Changes of Beach Profile and Mean Grain Size Caused by Tsunami-Like Waves. *J. Coast. Res.* **2012**, *284*, 1303–1312. [CrossRef]
18. Chen, J.; Jiang, C.; Yang, W.; Xiao, G. Laboratory Study on Protection of Tsunami-Induced Scour by Offshore Breakwaters. *Nat. Hazards* **2015**, *81*, 1229–1247. [CrossRef]
19. Dean, R.G. Equilibrium Beach Profiles: Characteristics and Applications. *J. Coast. Res.* **1991**, *7*, 53–84.
20. Jara, M.; González, M.; Medina, R. Shoreline Evolution Model from a Dynamic Equilibrium Beach Profile. *Coast. Eng.* **2015**, *99*, 1–14. [CrossRef]
21. Riazi, A.; Slovinsky, P.A. Subaerial Beach Profiles Classification: An Unsupervised Deep Learning Approach. *Cont. Shelf Res.* **2021**, *226*, 104508. [CrossRef]
22. Larson, M.; Kraus, N.C.; Wise, R.A. Equilibrium Beach Profiles under Breaking and Non-Breaking Waves. *Coast. Eng.* **1999**, *36*, 59–85. [CrossRef]
23. Francesco, M.; Mancinelli, A.; Corvaro, S.; Rocchi, S.; Lorenzoni, C. Coastal Submerged Structures Adaptation to Sea Level Rise over Different Beach Profiles. *Ital. J. Eng. Geol. Environ.* **2020**, *1*, 87–98.
24. Marini, F.; Corvaro, S.; Rocchi, S.; Lorenzoni, C.; Mancinelli, A. Semi-Analytical Model for the Evaluation of Shoreline Recession Due to Waves and Sea Level Rise. *Water* **2022**, *14*, 1305. [CrossRef]
25. Aagaard, T.; Brian, G.; Michael, H. Sediment Transport on Dissipative, Intermediate and Reflective Beaches. *Earth-Sci. Rev.* **2013**, *124*, 32–50. [CrossRef]
26. Castelle, B.; Gerd, M. Morphodynamics of Wave-Dominated Beaches. *Camb. Prism. Coast. Futures* **2023**, *1*, e1. [CrossRef]
27. Rector, R.L. *Laboratory Study of Equilibrium Profiles of Beaches*; US Beach Erosion Board: Washington, DC, USA, 1954.
28. Sunamura, T.; Horikawa, K. Two Dimensional Beach Transformation Due to Waves. In Proceedings of the Coastal Engineering 1974, Copenhagen, Denmark, 24–28 June 1974; pp. 920–938.
29. Guedes, R.M.; Calliari, L.J.; Holland, K.T.; Plant, N.G.; Pereira, P.S.; Alves, F.N. Short-Term Sandbar Variability Based on Video Imagery: Comparison between Time–Average and Time–Variance Techniques. *Mar. Geol.* **2011**, *289*, 122–134. [CrossRef]

30. Günaydın, K.; Kabdaşlı, M.S. Investigation of Offshore Bar Geometry under Regular and Irregular Waves. *J. Coast. Res.* **2005**, *212*, 374–382. [CrossRef]
31. Luo, E.C.-R. Formation of Beach Profile with the Design Criteria of Seawalls. *Civ. Eng. Arch.* **2014**, *2*, 24–32. [CrossRef]
32. Walstra, D.; Reniers, A.; Ranasinghe, R.; Roelvink, J.; Ruessink, B. On Bar Growth and Decay during Interannual Net Offshore Migration. *Coast. Eng.* **2012**, *60*, 190–200. [CrossRef]
33. de Kruijf, M.; Slootman, A.; de Boer, R.A.; Reijmer, J.J. On the Settling of Marine Carbonate Grains: Review and Challenges. *Earth-Science Rev.* **2021**, *217*, 103532. [CrossRef]
34. Bian, C.; Chen, J.; Jiang, C.; Wu, Z.; Yao, Z. Threshold of Motion of Coral Reef Sediment under Currents in Flume Experiments. *Sedimentology* **2023**, *70*, 1723–1740. [CrossRef]
35. Bian, C.; Chen, J.; Jiang, C.; Wu, Z.; Yao, Z.; Liu, J. Experimental Study of the Incipient Motion Threshold of Coral Sediment for Oscillatory Flow. *Ocean Eng.* **2023**, *278*, 114375. [CrossRef]
36. Chen, J.; Yao, Z.; Jiang, C.-B.; Wu, Z.-Y.; Deng, B.; Long, Y.-N.; Bian, C. Experiment Study of the Evolution of Coral Sand Particle Clouds in Water. *China Ocean Eng.* **2022**, *36*, 720–733. [CrossRef]
37. Xu, X. Types of Two-Dimension Sandybeaches and Their Criterion. *Ocean. Eng.* **1988**, *4*, 51–62.
38. Jonsson, I.G. Wave Boundary Layers and Friction Factors. In Proceedings of the Coastal Engineering 1966, Tokyo, Japan, 5–8 September 1966; pp. 127–148.
39. Wu, Z. Response Rule of Sandy Beach on Change of Wave Dynamic Factors. Master's Thesis, Changsha University of Science & Technology, Changsha, China, 2014.
40. Silvester, R.; Hsu, J.R.C. *Coastal Stabilization*; World Scientific: Singapore, 1997; Volume 14.
41. Kench, P.S.; Brander, R.W. Wave Processes on Coral Reef Flats: Implications for Reef Geomorphology Using Australian Case Studies. *J. Coast. Res.* **2006**, *221*, 209–223. [CrossRef]
42. Hardy, T.A.; Young, I.R. Field Study of Wave Attenuation on an Offshore Coral Reef. *J. Geophys. Res. Ocean.* **1996**, *101*, 14311–14326. [CrossRef]
43. Nelson, R. Hydraulic Roughness of Coral Reef Platforms. *Appl. Ocean Res.* **1996**, *18*, 265–274. [CrossRef]
44. Lowe, R.J.; Falter, J.L.; Monismith, S.G.; Atkinson, M.J. Wave-Driven Circulation of a Coastal Reef–Lagoon System. *J. Phys. Oceanogr.* **2009**, *39*, 873–893. [CrossRef]
45. Vetter, O.; Becker, J.M.; Merrifield, M.A.; Pequignet, A.; Aucan, J.; Boc, S.J.; Pollock, C.E. Wave Setup over a Pacific Island Fringing Reef. *J. Geophys. Res. Oceans* **2010**, *115*, C12066. [CrossRef]
46. Taebi, S.; Lowe, R.J.; Pattiaratchi, C.B.; Ivey, G.N.; Symonds, G.; Brinkman, R. Nearshore Circulation in a Tropical Fringing Reef System. *J. Geophys. Res. Ocean.* **2011**, *116*, C02016. [CrossRef]
47. Becker, J.M.; Merrifield, M.A.; Ford, M. Water Level Effects on Breaking Wave Setup for Pacific Island Fringing Reefs. *J. Geophys. Res. Oceans* **2014**, *119*, 914–932. [CrossRef]
48. Pomeroy, A.; Lowe, R.; Symonds, G.; Van Dongeren, A.; Moore, C. The Dynamics of Infragravity Wave Transformation over a Fringing Reef. *J. Geophys. Res. Oceans* **2012**, *117*, C11022. [CrossRef]
49. Lentz, S.J.; Churchill, J.H.; Davis, K.A.; Farrar, J.T. Surface Gravity Wave Transformation across a Platform Coral Reef in the Red Sea. *J. Geophys. Res. Ocean.* **2016**, *121*, 693–705. [CrossRef]
50. Watts, G.M. *Laboratory Study of Effect of Varying Wave Periods on Beach Profiles*; US Beach Erosion Board: Washington, DC, USA, 1954.
51. Lin, C.; Hwung, H.H. External and Internal Flow Fields of Plunging Breakers. *Exp. Fluids* **1992**, *12*, 229–237. [CrossRef]
52. Chang, K.-A.; Liu, P.L.-F. Experimental Investigation of Turbulence Generated by Breaking Waves in Water of Intermediate Depth. *Phys. Fluids* **1999**, *11*, 3390–3400. [CrossRef]
53. Celikoğlu, Y.; Yüksel, Y.; Kabdaşlı, M.S. Longshore Sorting on a Beach under Wave Action. *Ocean Eng.* **2004**, *31*, 1351–1375. [CrossRef]

Disclaimer/Publisher's Note: The statements, opinions and data contained in all publications are solely those of the individual author(s) and contributor(s) and not of MDPI and/or the editor(s). MDPI and/or the editor(s) disclaim responsibility for any injury to people or property resulting from any ideas, methods, instructions or products referred to in the content.

Journal of
Marine Science and Engineering

MDPI

Article

Tectonic Subsidence on the East China Coast Recorded by Magnetic Properties of Pliocene Red Clay in the Yangtze Delta

Xianbin Liu [1,2], Jing Chen [2], Liping Xu [3], Xiaoli Sun [4], Lei Tan [1], Minghao Lv [1] and Jian Song [1,*]

1 School of Resource and Environmental Engineering, Ludong University, Yantai 264025, China; liuxb_801@163.com (X.L.); tanl971220@163.com (L.T.); lvminghao0513@gmail.com (M.L.)
2 State Key Laboratory for Estuarine and Coastal Research, East China Normal University, Shanghai 200241, China; jchen@geo.ecnu.edu.cn
3 Pilot National Laboratory for Marine Science and Technology, Qingdao 266237, China; lpxu@qnlm.ac
4 Shandong Zhengyuan Digital City Construction Company, Yantai 264025, China; shdsunxiaoli@126.com
* Correspondence: songjian04@163.com

Abstract: Thick red clay in northern China contains rich information about the uplifting of the Qinghai–Tibet Plateau, the drying process of Asian inland, the East Asian monsoon changes, and global cooling since 22 Ma. In comparison, the red clay widely distributed in southern China is generally much younger (<1 million years), thus limiting the paleoclimate and paleoenvironment reconstruction over a longer geological time. We conducted a comprehensive magnetic investigation on the Pliocene red clay of the core LQ11, located in the Yangtze Delta, to reveal its paleoclimate and paleoenvironment implications for the eastern China coast. Our results revealed that the Pliocene red clay in the Yangtze Delta has higher S-ratio and lower HIRM (Hard isothermal remanent magnetizations) values than Quaternary vermiculate red clay of hot–humid climate origin in southern China. This indicates a weaker transformation from maghemite to hematite during the process of pedogenesis. The lack of net-like white veins in the Pliocene red clay also indicates a relatively low intensity of pedogenesis. We believe that the Pliocene red clay, which is presently 250 m below the mean sea level, was formed in high-altitude topography before the Quaternary period, where paedogenic intensity was remarkably low. This finding shows rapid tectonic subsidence occurring on the eastern China coast since the late Pliocene and enriches the theoretical research on paleoenvironment reconstruction based on red clay.

Keywords: Pliocene; magnetic properties; paleoenvironment; East Asian monsoon; Yangtze Delta

Citation: Liu, X.; Chen, J.; Xu, L.; Sun, X.; Tan, L.; Lv, M.; Song, J. Tectonic Subsidence on the East China Coast Recorded by Magnetic Properties of Pliocene Red Clay in the Yangtze Delta. *J. Mar. Sci. Eng.* **2024**, *12*, 66. https://doi.org/10.3390/jmse12010066

Academic Editor: Gemma Aiello

Received: 19 October 2023
Revised: 19 December 2023
Accepted: 23 December 2023
Published: 27 December 2023

Copyright: © 2023 by the authors. Licensee MDPI, Basel, Switzerland. This article is an open access article distributed under the terms and conditions of the Creative Commons Attribution (CC BY) license (https://creativecommons.org/licenses/by/4.0/).

1. Introduction

The red clay, also named red soil or red earth, is widely distributed in northern and southern China and is an ideal sediment archive for exploring the late Cenozoic climate and environmental evolution (Figure 1a). The red clay underlain by the Quaternary loess in northern China was formed in Miocene–Pliocene and provides key insights into the onset of East Asian monsoons, the uplift of the Tibetan Plateau, and related desertification of the Asian interior since 22 Ma [1–3]. In comparison, the red clay in southern China, covering an area of 20×10^5 km^2, is generally composed of—from the bottom to the top—yellow vermiculated earth, uniform red clay, and brown earth [4]. The red clay in the southern regions contains abundant Palaeolithic sites harbouring rich information on human evolution. However, its deposition began about one million years ago and is much younger than that in northern China, thus limiting, to a certain extent, the reconstruction of the paleoenvironment in the subtropical and tropical areas of Asia [5,6].

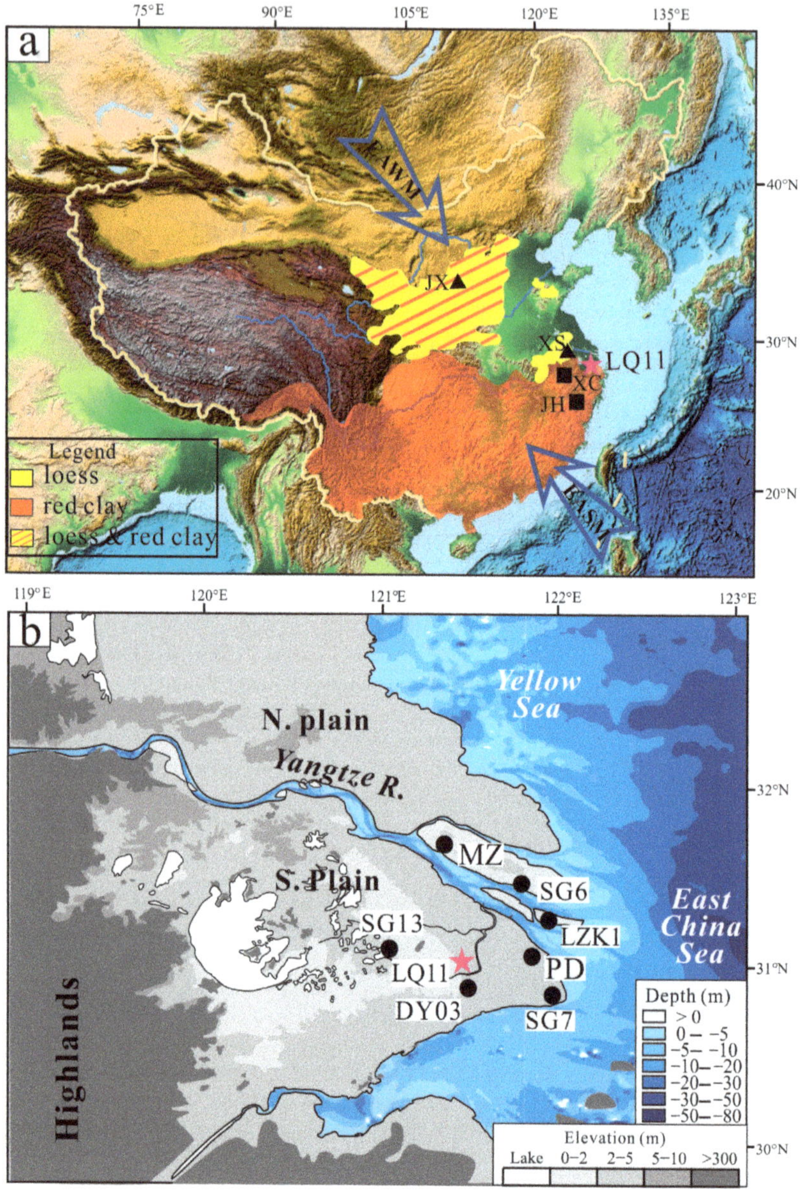

Figure 1. (**a**) Map showing the distributions of Neogene red clay and Quaternary loess in northern China and Quaternary red clay in southern China. EASM: East Asia summer monsoon; EAWM: East Asia winter monsoon. The location of the sampling site (star); Tertiary red clay sections (triangle; JX: Jiaxian) in northern China, and Quaternary red clay sections (dimand; XS: Xiashu; XC: Xuancheng; JH: Jinhua) in southern China. (**b**) The topography of the Yangtze Delta and its surrounding areas; the cores distributed in the Yangtze Delta.

The magnetic properties of rocks and sediments are sensitive to environmental changes and are described in different research fields, such as magnetostratigraphy, palaeomagnetism, and environmental magnetism. Environmental magnetism investigates the formation, transport, deposition, and post-depositional alteration processes of magnetic minerals under various environmental conditions. [7–11]. For Neogene red clay in northern China, ultrafine magnetic minerals (maghemite) produced during pedogenesis enhance the magnetic susceptibility (χ_{lf}) of sediments under warm and wet climates [12]. The red clay in southern China, especially vermiculate red clay, has experienced a much stronger pedogenesis with higher temperature and rainfall than that in northern China, and more paedogenic maghemite has been transformed into hematite [13,14]. As recorded by the variations in magnetic properties, the red clay from both northern and southern China documented the evolution of the paleoenvironment in Asia, especially the East Asian paleomonsoon [13–17].

We recently discovered red clay from the Pliocene age in a 301-m-long sediment core LQ11 in the Yangtze Delta, with a buried depth greater than 250 m, which is younger than that in northern China, but older than in southern China (Figure 1b). Here, we analyse the magnetic properties of the Pliocene red clay and further reveal its paleoclimate and paleoenvironment implications by comparing it with Neogene red clay in northern China and Quaternary red clay in southern China.

2. Geographic Background

The Yangtze River, one of the longest rivers (>6300 km) in the world, originates from the Tibetan Plateau and flows into the East China Sea [18]. The Yangtze River Delta is located between 30°20′ to 32°30′ N and 119°24′ to 122°30′ E. It borders the Yellow Sea and East China Sea to the east and is bounded by the Tianmu Mountain and Maoshan to the west. The northern boundary is marked by the line connecting Yangzhou City, Taizhou City, and Rudong City, while the southern boundary is defined by the northern shore of Hangzhou Bay.

The predominant bedrock in the Yangtze Delta region is composed of igneous rocks that were formed from the Neoarchean to the Late Cenozoic. Among them, the Late Jurassic and Cretaceous igneous rocks have the widest distributions, accounting for more than two-thirds of the total bedrock area. The residual hills exposed in the southern region are also of igneous origin. Sedimentary rocks are scattered and include sandstone, tuffaceous limestone, and volcanic breccia. Metamorphic rocks have the smallest distribution area in the region and are only found in the southern areas.

The extent of marine transgression in the Yangtze River estuary region reached its maximum 8000–7000 years ago [18]. The sea level then remained relatively stable, resulting in the accumulation of a large amount of sediment carried by the Yangtze River into the estuarine area. This caused the coastline to steadily advance seaward, gradually forming the modern delta.

The delta has low-lying topography, with an average elevation of about 4 m. Apart from a few exposed bedrock hills in the southern region, the majority of the Yangtze Delta is characterized by a cover of thick and loosely deposited sediments that are in unconformable contact with the underlying Mesozoic intermediate-acidic igneous bedrock. Pliocene strata are found only in sporadic depressions, typically with a thickness of less than 40 m. Moving into the Quaternary period, the strata reach their maximum thickness along the east coast, exceeding 300 m, and gradually thin out and disappear towards the west.

3. Materials and Methods

A 301 metre long and continuous sediment core (LQ11) was taken from the Yangtze delta (Figure 1b). A total of 277 samples from the whole core, including 19 samples in the Pliocene strata, were collected for measurement (Table 1).

Table 1. Sampling depth and sediment granularity.

Sample No.	Depth (m)	Clay (%)	Silt (%)	Sand (%)	Md (μm)
LQ11-1	301	27.40	61.90	10.70	24.34
LQ11-2	300	44.60	54.50	0.90	14.75
LQ11-3	297	40.10	59.89	0.01	8.08
LQ11-4	295	33.60	61.00	5.40	21.13
LQ11-5	292.5	38.30	60.00	1.70	11.14
LQ11-6	288	57.40	42.50	0.10	6.88
LQ11-7	285	56.60	43.20	0.20	7.20
LQ11-8	284	25.60	68.90	5.50	19.77
LQ11-9	282.5	30.50	63.80	5.70	19.43
LQ11-10	280	60.00	40.00	0.00	6.64
LQ11-11	276.5	30.10	62.30	7.60	21.13
LQ11-12	273.5	35.20	58.70	6.10	20.38
LQ11-13	270.8	36.90	57.50	5.60	19.11
LQ11-14	268	44.70	54.70	0.60	12.17
LQ11-15	265.5	26.70	64.90	8.40	24.46
LQ11-16	262	36.20	56.00	7.80	20.71
LQ11-17	257.5	24.20	71.00	4.80	18.32
LQ11-18	255	32.60	64.20	3.20	14.39
LQ11-19	253.2	35.40	59.00	5.60	17.83

Magnetic susceptibility (χ_{lf} and χ_{hf}) was measured using a Bartington MS2. After being demagnetized with a Dtech2000 (maximum alternating magnetic field: 100 mT, direct current magnetic field: 0.04 mT) to obtain non-hysteretic remanent magnetization, the sample was placed in a Minispin magnetometer to measure its remanent magnetization. Saturation isothermal remanent magnetization (IRM) was obtained using an MMPM10 pulse magnetometer at a 1 T magnetic field. S-ratio was calculated as $-$IRM $-$ 300 mT/SIRM, and HIRM = (SIRM $-$ IRM $-$ 300 mT)/2.

Based on the above parameters of the samples at room temperature, typical samples were selected for hysteresis loop and thermal demagnetization curve measurements using a variable field magnetic balance (VFTB). First-order reversal curve (FORC) was performed with a MicroMag3902 gradient magnetometer.

Four typical samples were selected to extract magnetic minerals, perform mineral morphology scanning, and measure the energy spectrum. The scanning electron microscope used was a JEOL JSM4800F field emission scanning electron microscope, and the energy spectrum testing instrument used was an Oxford Instruments X-ray energy spectrum analyser. Particle size testing was performed using a Coulter LS-100Q laser particle size analyser (measuring range: 0.02–2000 μm).

4. Results

4.1. Lithology of the Core LQ11

The LQ11 borehole recorded the Gauss, Matuyama, and Bruhess polarities based on a comparison of the magnetic inclination changes with a standard polarity column [19]. The boundaries of the three polarities were marked at 252 m and 112 m, where the section above 252 m is characterized by Gauss normal polarity, the strata between 252 m and 112 m exhibits Matuyama reverse polarity, and the interval from 112 m to the top of the borehole corresponds to Bruhess normal polarity.

The regional bedrock present below the red clay was Mesozoic basalt. The Pliocene red clay, found at depths between 301 and 252.7 m, is unconformably positioned on the regional bedrock (Figures 2 and 3). Within the layers of the Early Pleistocene period (252.7–112 m) are noticeable occurrences of four sedimentary cyclothems. These are characterized by a base of grey gravelly sand and coarse sand, which are subsequently covered by layers of brown silt. Moving into the Middle Pleistocene strata (112–67.6 m), there is an observable pattern of alternating strata. These comprise layers of gravelly sand along with various grades of sand, ranging from coarse to fine. Transitioning to the late Pleistocene strata (67.6–41.3 m), distinct thick cyclic units become apparent and are predominantly composed of greyish coarse sand and silt. The Holocene strata, spanning from 41.3 to 0 m, are characterized by deposits of fine sand and clayey silt.

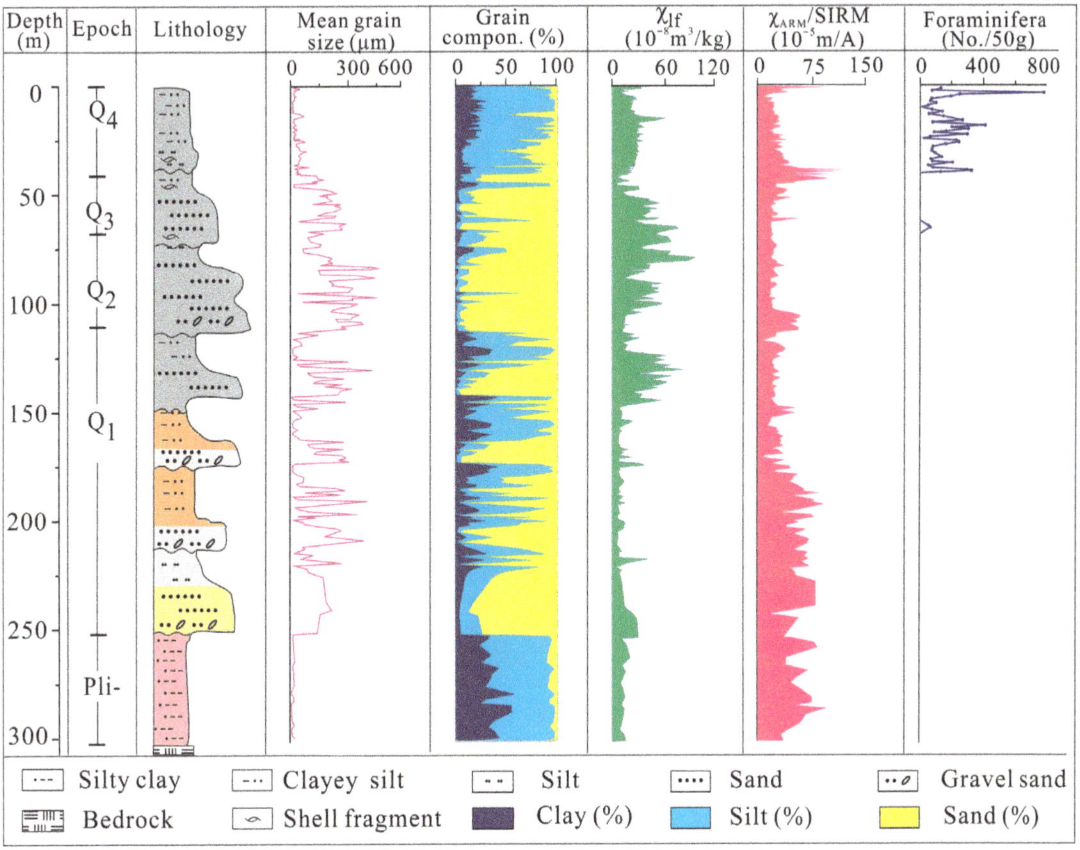

Figure 2. Lithology, mean grain size, abundance of foraminifera, and magnetic properties of the core LQ11.

4.2. Magnetic Properties of Pliocene Red Clay in Core LQ11

Generally, the magnetic parameters of Pliocene red clay in core LQ11 exhibit small down-core fluctuations (Figure 4). The χ_{lf} is lower than 21.18×10^{-8} m^{3}kg^{-1}, and SIRM ranges from 615.09 to 2623.45×10^{-5} Am^{2}kg^{-1}, which means that there are fewer ferrimagnetic minerals in the clay [7]. χ_{fd}% is generally lower than 5%, indicating the small contribution of viscous superparamagnetic particles (VSP, ~0.02 μm) of magnetic grains to the susceptibility signal. Low values of χ_{ARM} and χ_{ARM}/χ_{lf} imply low concentrations of single-domain magnetic minerals [8]. The S-ratios are all less than 0.8, indicating a high proportion of hard magnetic minerals.

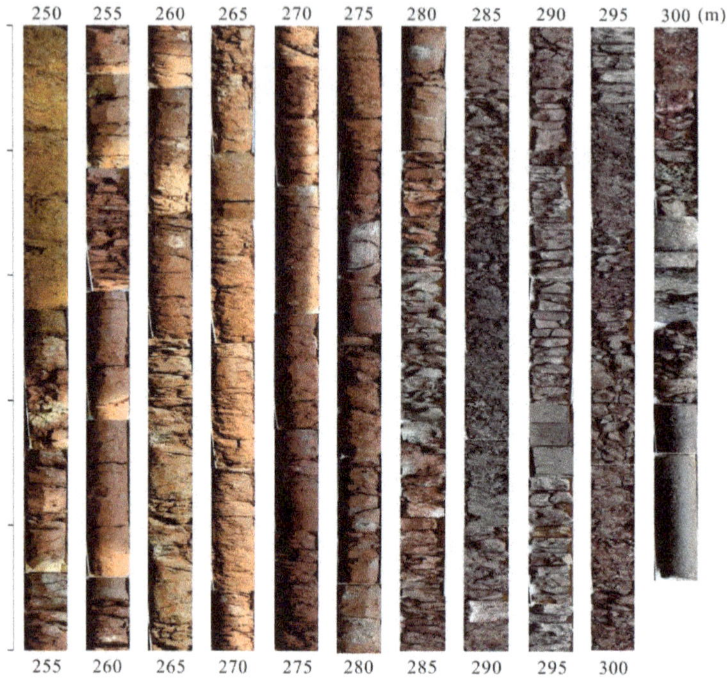

Figure 3. Lithology photos of the Pliocene strata in the core LQ11. Below 301 m, consists of a bedrock of black basalt; 301–252.7 m represents the Pliocene strata; and the lower stratum (301–283.5 m) is mostly dark red, while the upper section (283.5–252.7 m) is bright red, with the presence of calcareous nodules; above 252.7 m, consists of the Quaternary strata, yellow-brown coarse sand, and a mixture of gravel and clay.

The thermomagnetic curves, hysteresis loops, and FORC diagrams obtained from the two representative samples exhibit similar patterns, indicating the presence of the same magnetic minerals in the red clay (Figure 5). A small convex shape observed in the heating curve between 150 °C and 300 °C suggests the formation of maghemite from Fe-hydroxides such as goethite. A significant decrease in magnetization occurs before reaching 580 °C, indicating the widespread occurrence of magnetite (Figure 5a,d). Slight decreases in magnetization up to 680 °C suggest the presence of hematite (Figure 5a,d). The raw and dia/para-corrected hysteresis loops display noticeable differences, which can be attributed to contributions from paramagnetic or diamagnetic components (Figure 5b,e). The corrected loops exhibit a characteristic wasp-waisted shape and do not close until a field strength of 1000 mT is reached. This behaviour is often associated with the coexistence of two magnetic components with distinctly different coercivities (Figure 5b,e). The FORC diagrams provide valuable insights into the magnetic properties of the samples. Along the Hu axis (>60 mT), the contours show a greater degree of divergence, indicating a wider range of coercivity values. On the other hand, along the Hc axis (<80 mT) is a distribution of low coercivity values. These features suggest that the samples are dominated by multi-domain (MD) magnetite grains (Figure 5c,f).

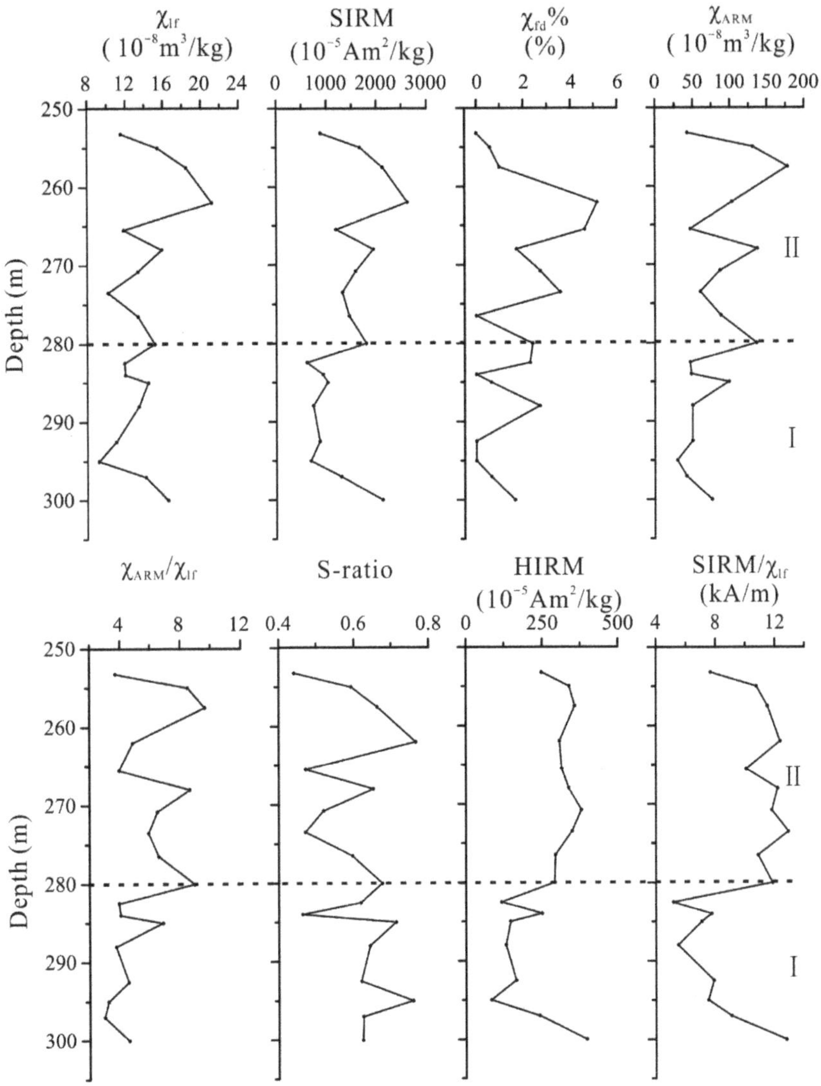

Figure 4. Magnetic parameters, including χ_{lf}, SIRM, $\chi_{fd}\%$, χ_{ARM}, χ_{ARM}/χ_{lf}, S-ratios, HIRM, and SIRM/χ_{lf}, of the Pliocene red clay.

To determine the specific magnetic minerals present, we conducted SEM observations and EDS analyses on representative samples' magnetic extracts. The SEM images revealed that the magnetic particles consist of irregularly shaped iron oxides, such as magnetite and hematite, which is also revealed by the high peaks of O and Fe on the EDS spectra (Figure 6).

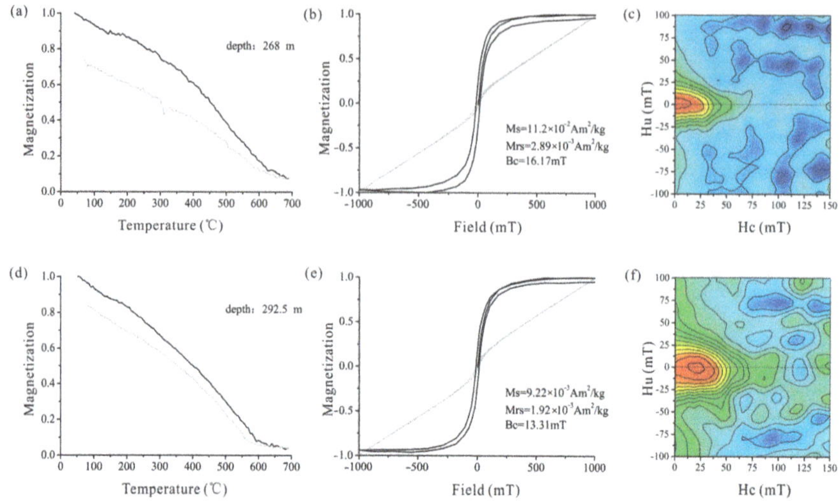

Figure 5. Magnetic properties of representative samples at depths of 268 and 292.5 m. (**a**,**d**), thermomagnetic curves; (**b**,**e**), black: heating curves; grey: cooling curves), hysteresis loops; (**c**,**f**), FORC diagrams.

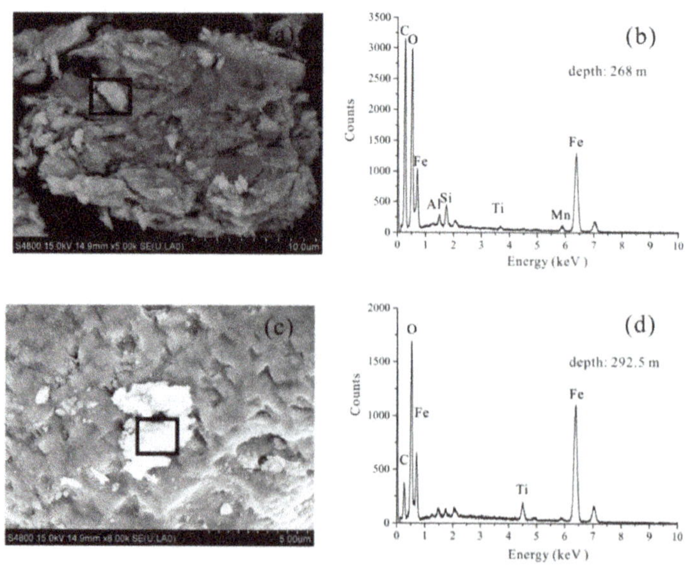

Figure 6. SEM images (**a**,**c**) and EDS spectra (**b**,**d**) of representative samples at depths of 268 and 292.5 m, respectively. Under the scanning electron microscope, magnetic minerals exhibit irregular shapes, smooth surfaces, and relatively large particles.

5. Discussion

5.1. Main Controlling Role of the Magnetic Properties of Pliocene Red Clay

The magnetic minerals may originate from detrital, biogenic, or authigenic sources [8]. Biogenic magnetite or greigite, typically formed through magnetotactic bacteria's biomineralization process, usually exhibits a single domain size (SD). In the FORC diagram, this type of mineral would be characterized by a narrow vertical spread along the Hu axis (<20 mT) and a wide horizontal spread along the Hc axis [19]. However, no iron sulphides were detected in the Pliocene red clay, which is supported by their low SIRM/χ values, the presence of divergent contours in the FORC diagram, and the absence of an S peak in the EDS spectra (Figure 5). Authigenic iron sulphides are formed through post-depositional reductive diagenesis. However, no iron sulphides are detected in the Pliocene red clay, which is supported by their low SIRM/χ values, a FORC diagram characterized by more divergent contours along the Hu axis, and no S peak in the EDS spectra (Figures 4–6). The total organic matter content is generally low, likely minimizing the vulnerability of the sediment to reductive dissolution (Table 2). Therefore, the magnetic minerals in Pliocene red clay were mainly of detrital origin.

Table 2. Magnetic susceptibility, mean grain size, TiO_2/Al_2O_3, and TOC values of representative samples.

Depth (m)	χ_{lf} (10^{-8} m^3kg^{-1})	Md (μm)	TiO_2/Al_2O_3	TOC (%)
295	9.18	21.13	0.070	0.33
285	14.45	7.20	0.063	0.30
270.8	13.36	19.11	0.068	0.27
255	15.45	14.39	0.066	0.25

Either wind-blown dust or terrigenous materials (soil) from local catchments could be potential sources of detrital sediments. The magnetic properties of the sediments are different from those in the northern Chinese Loess Plateau (Figure 7). Therefore, the magnetic minerals of sediments are probably derived from the soil of local catchments, which is also supported by the surrounding cores in terms of their magnetic, heavy minerals, and geochemical proxies [20,21].

Although there were slight differences in the magnetic properties in the upper part of the core, the provenance of red clay did not change significantly during the Pliocene era (Figure 4). This is supported by its stable Ti/Al ratios, which are widely used as a provenance indicator for various sediments [22,23]. Therefore, the Pliocene red clay can shed light on our understanding of the local paleoclimate and paleoenvironmental changes in the Yangtze Delta.

5.2. High Altitude Formation of Pliocene Red Clay

Maghemite is a transitional mineral, from ferrihydrite to haematite, during pedogenesis processes in soil [24,25]. As the degree of pedogenesis increases, more paedogenic maghemite is transformed into haematite [26]. The transformation of maghemite to haematite during a strong pedogenesis process played a crucial role in the magnetic depletion of the Quaternary vermiculate red clay in the Xuancheng (XC), Jinhua (JH), and Qiyang (QY) sections of southern China [13,27,28]. Therefore, magnetic parameters such as S-ratios and HIRM, which reflect the relative content of haematite, can be used to estimate the degree of paedogenic processes. Compared with Quaternary vermiculate red clay in southern China, Pliocene red clay generally has higher S-ratios and lower HIRM values, suggesting lower concentrations of haematite and intensity of pedogenesis (Figure 7d).

The Quaternary red clay in southern China generally consists of—from the bottom to the top—yellow vermiculated earth, uniform red clay, and brown earth. The net-like white veins in the Vermiculated earth resulted from iron depletion under high temperatures and abundant rainfall conditions, implying an extreme East Asian summer monsoon [29]. However, the net-like white veins were not observed in the Pliocene red clay of the core LQ11 (Figure 3), probably reflecting a temperate and slightly dry climate in the Yangtze Delta during the Pliocene, and that the Pliocene red clay had undergone a relatively low intensity of pedogenesis.

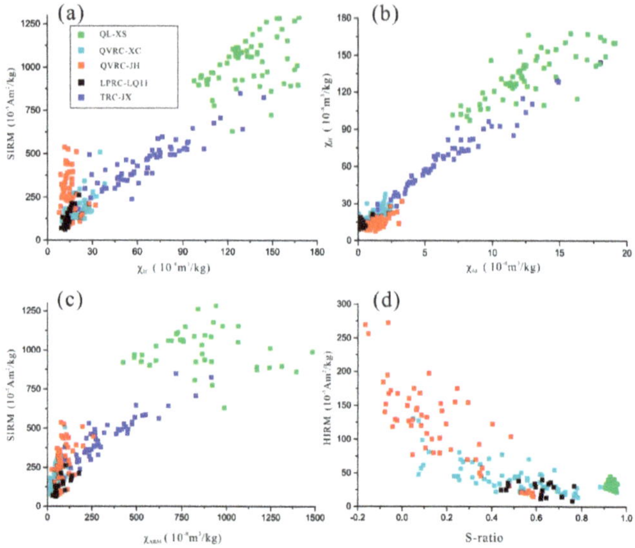

Figure 7. Magnetic properties of sediments from different strata, including Tertiary red clay from Jiaxian (TRC-JX) in northern China [30], and Quaternary loess from Xiashu (QL-XS) [31], Quaternary vermiculate red clay from Jinhua (QVRC-JH) [28], Xuancheng (QVRC-XC) [32], and Pliocene red clay of the core LQ11 (LPRC-LQ11) in southern China.

Pedogenesis, the process of soil formation, often influences the magnetic properties of the soil through the transformation of magnetic iron oxides [7]. The degree of chemical weathering and pedogenesis in East Asia are generally controlled by the local climate and are closely associated with monsoon intensity [33]. Thick and continuous Quaternary Loess–Neogene red clay sequences on the Chinese Loess Plateau preserve high-resolution archives for fluctuations of the East Asian monsoon during the late Cenozoic. Paedogenic ultrafine (nanoscale) magnetite/maghemite in warm and humid conditions enhanced the magnetic properties of sediments significantly, and thus magnetic parameters such as χ_{lf} were widely used to interpret the effect of environmental and climatic change on the Chinese Loess Plateau. Magnetic records combined with geochemical data from a series of sections indicate that the summer monsoon intensity has been enhanced since 8 Ma and reached a maximum in 4.5–2.6 Ma, possibly due to tectonic events such as the phased uplift of the Tibetan Plateau [34]. Due to global cooling, the summer monsoon initiated a weakening trend in East Asia after 2.6 Ma [1]. Consistent trends of summer monsoon are also recorded in the depositional sequences of the South China Sea, as indicated by clay minerals and terrigenous mass accumulation rate at Ocean Drilling Program (ODP) Site 1146, 1148 [35,36]. Although the strength and time of the East Asian summer monsoon has varied over the long geologic time, it has commonly been accepted that the Pliocene era was warmer and wetter than the Quaternary era [37–41]. Correspondingly, stronger chemical weathering and pedogenesis in both the Chinese Loess Plateau and South China Sea generally occurred in Pliocene sediments than in Quaternary sediments [41,42]. However, the pedogenesis of sediments in the Yangtze Delta differed, and the Pliocene red clay underwent a relatively low intensity of pedogenesis. The Yangtze Delta was probably at a high altitude during the Pliocene, resulting in weak weathering of sediments, which further limited the transformation from maghemite to haematite during pedogenesis.

5.3. Tectonic Subsidence of the Yangtze Delta during the Quaternary Period

The evolution of the sedimentary environment of the Yangtze Delta during the late Cenozoic period goes together with continuous Neo-tectonic subsidence, and climate and sea level changes [43]. Detailed sedimentary research from hundreds of boreholes revealed that either alluvial fan or lacustrine facies occurred in Pliocene stratigraphy in some areas, and some boreholes even lack Pliocene deposition, indicating a great difference in altitude in the region, and consequently, depositional and erosion areas coexisted in the south plain of Yangtze delta during the Pliocene period (Figure 8). *Picea* and *Abies* pollen, which are indicators of cold climate, remained at relatively stable proportions in Pliocene stratigraphy, implying high-elevation landforms and a slightly dry climate in the lower Yangtze regions [44]. In the early Pleistocene, the stratigraphic layers did not contain marine biotic fossils, and the depositional environment was characterized by alluvial and fluvial facies. The sediment consists mainly of coarse sand with gravel (angular and sub-angular) and silty sand, gradually transitioning upwards to fine silt and silty clay, exhibiting poor sorting. The C–M (Cailleux–Middleton) diagram indicates an overwhelming dominance of the rolling component and sediment accumulates rapidly, reflecting strong tectonic subsidence control in the depositional basin. In the middle Pleistocene, the sediment in the stratigraphic sequence shifts from gravelly sand to upper clayey silty sand, with sporadic occurrences of foraminifera and ostracods in marine facies. The C–M diagram reflects sedimentation processes involving rolling, jumping, and suspension mixing, showing improved sorting. The sediment transport distance between the source and sink increases, reflecting a fluvial environment formed during the processes of peneplanation. From the late Pleistocene to the Holocene, there was a significant increase in the thickness of fine-grained strata, with improved sorting. The C–M diagram shows dominance of jumping and suspension, with a gradual disappearance of the rolling component. The sediment transport distance between the source and sink continued to increase. The strata reveal abundant marine foraminifera and marine ostracods, indicating the development of estuarine, shallow marine, and deltaic facies. This suggests frequent marine influence in the deltaic region during this period, with the topography significantly lower than in the earlier stages, approaching present-day elevation [43,44].

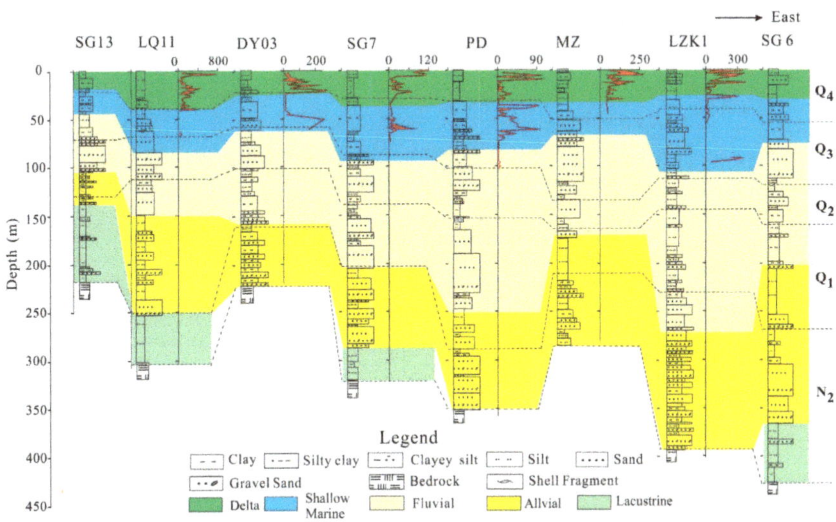

Figure 8. Sedimentary facies changes and the abundance of foraminifera (No./50 g) of sediment cores in the Yangtze Delta.

Gradual tectonic subsidence also occurred in the nearby Zhe-Min uplift on the East China Sea shelf during the Quaternary period [45,46]. With the subsidence of those regions, the main river channel shifted gradually from north to south, thus inducing a southward migration of depocenter beginning from the early Pleistocene (Figure 9). The mega-chronostratigraphic cycle of cores implies not only the local sedimentary evolution of the Yangtze coast but also a hint of the channelization time of the Yangtze River.

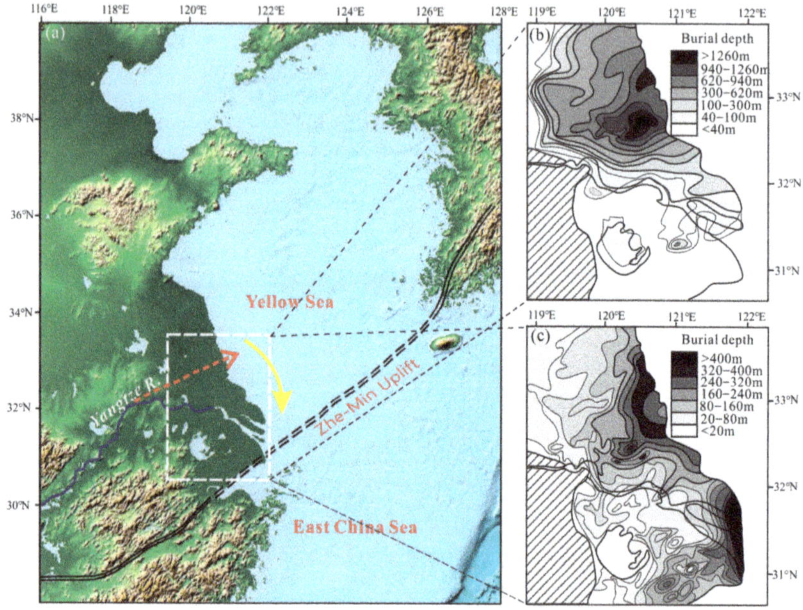

Figure 9. (**a**) Geomorphology in the Eastern Coastal Area of China and a southward migration of river channel (yellow arrow). (**b**) Neogene sedimentary thickness of the Yangtze Delta and Subei basin, modified after [18]. (**c**) Quaternary sedimentary thickness of the Yangtze Delta and Subei basin, modified after [18].

6. Conclusions

Magnetic properties combined with particle size, SEM observations, and EDS of the Pliocene red clay in the Yangtze Delta were analysed to reveal the paleoclimate and paleoenvironment of the East China coast. Based on the results of this study, we can reach the following conclusions: (1) the magnetic minerals of sediments are derived from the soil in local catchments, and the magnetic properties are generally controlled by the paedogenic intensity. (2) The Pliocene red clay generally has high S-ratios and low HIRM values, suggesting low pedogenesis intensity in the Pliocene. (3) High altitude constrained the chemical weathering and pedogenesis of Pliocene red clay and limited the transformation from paedogenic maghemite to haematite. Our results demonstrate a rapid tectonic subsidence occurring on the eastern China coast during the Quaternary era.

Author Contributions: X.L.: designed the research study and wrote the manuscript; J.C.: designed the research study; L.T.: conducted the fieldwork and measurements; M.L.: conducted the fieldwork and measurements; L.X.: analysed the data; X.S.: analysed the magnetic data; J.S.: designed the research study and reviewed the manuscript. All authors have read and agreed to the published version of the manuscript.

Funding: This research was funded by the National Science Foundation of China (Grant No. 41901102) and the Natural Science Foundation of Shandong Province, China (No. ZR2019PD013, ZR2019MD009).

Institutional Review Board Statement: Not applicable.

Informed Consent Statement: Not applicable.

Data Availability Statement: The data presented in this study are available on request from the corresponding author.

Acknowledgments: We would like to sincerely thank Baocheng Zhao for their help with field sampling, and Zhongyuan Chen for his helpful suggestions on the manuscript.

Conflicts of Interest: The authors declare no conflicts of interest.

References

1. An, Z.; Kutzbach, J.E.; Prell, W.L.; Porter, S.C. Evolution of Asian monsoons and phased uplift of the Himalaya-Tibetan plateau since Late Miocene times. *Nature* **2001**, *411*, 62–66.
2. Guo, Z.; Ruddiman, W.F.; Hao, Q.; Wu, H.; Qiao, Y.; Zhu, R.; Peng, S.; Wei, J.; Yuan, B.; Liu, T. Onset of Asian desertification by 22 Myr ago inferred from loess deposits in China. *Nature* **2002**, *416*, 159–163. [CrossRef] [PubMed]
3. Peng, W.; Zhang, H.; Pullen, A.; Li, M.; Pan, B.; Xiao, W.; Nie, J. Stepwise increased spatial provenance contrast on the Chinese loess plateau over late Miocene-Pleistocene. *Commun. Earth Environ.* **2023**, *4*, 60. [CrossRef]
4. Hu, X.; Zhao, J.; Zhang, P.; Xue, Y.; Liu, X. Fe isotopic composition of the Quaternary red clay in subtropical Southeast China: Redoxic Fe mobility and its paleoenvironmental implications. *Chem. Geol.* **2019**, *524*, 356–367. [CrossRef]
5. Qiao, Y.; Guo, Z.; Hao, Q.; Wu, W.; Jiang, W.; Yuan, B.; Zhang, Z.; Wei, J.; Zhao, H. Loess-soil sequences in south Anhui Province: Magnetostratigraphy and paleoclimatic significance. *China. Sci. Bull.* **2003**, *48*, 2088–2093. [CrossRef]
6. Liu, C.; Zheng, L.; Wei, W.; Xu, X.; Sheng, H.; Yuan, B. Mineral magnetism to probe into the nature of palaeomagnetic signals of subtropical red soil sequences in southern China. *Geophys. J. Int.* **2010**, *18*, 1395–1410. [CrossRef]
7. Thompson, R. *Environmental Magnetism*; Springer Science & Business Media: Berlin/Heidelberg, Germany, 1986.
8. Liu, Q.; Roberts, A.P.; Larrasoaña, J.C.; Banerjee, S.K.; Guyodo, Y.; Tauxe, L.; Oldfield, F. Environmental magnetism: Principles and applications. *Rev. Geophys.* **2012**, *50*, 1–124. [CrossRef]
9. Kasbohm, J.; Schoene, B.; Montanari, A.; Coccioni, R. High-precision U-Pb zircon geochronology of the Miocene Bisciaro Formation, Contessa Section, Italy: A case study for requisite radioisotopic calibration of bio- and magnetostratigraphy. *Palaeogeogr. Palaeoclimatol. Palaeocol.* **2021**, *576*, 110487. [CrossRef]
10. Zheng, Y.; Qiu, Z.; Qiu, Z.; Li, L.; Wei, X.; Zhang, R.; Yue, L.; Deng, T. Revised magnetostratigraphy of the Linxia Basin in the northeast Tibetan Plateau, constrained by micromammalian fossils. *Palaeogeogr. Palaeoclimatol. Palaeocol.* **2023**, *623*, 111620. [CrossRef]
11. Lazos, I.; Sboras, S.; Chousianitis, K.; Kondopoulou, D.; Pikridas, C.; Bitharis, S.; Pavlides, S. Temporal evolution of crustal rotation in the Aegean region based on primary geodetically-derived results and palaeomagnetism. *Acta Geod. Geophys.* **2022**, *57*, 317–334. [CrossRef]
12. Nie, J.; Song, Y.; King, J. A review of recent advances in red-clay environmental magnetism and paleoclimate history on the Chinese Loess Plateau. *Front. Earth Sci.* **2016**, *4*, 1–13. [CrossRef]
13. Wang, S.; Lin, S.; Lu, S. Rock magnetism, iron oxide mineralogy and geochemistry of Quaternary red earth in central China and their paleopedogenic implication. Palaeogeography, Palaeoclimatology. *Palaeoecology* **2013**, *379*, 95–103. [CrossRef]
14. Liu, C.; Wei, W.; Deng, C. A new weathering indicator from high-temperature magnetic susceptibility measurements in an Argon atmosphere. *Geophys. J. Int.* **2020**, *221*, 2010–2025. [CrossRef]
15. Nie, J.; Zhang, R.; Necula, C.; Heslop, D.; Liu, Q.; Gong, L.; Banerjee, S. Late Miocene–early Pleistocene paleoclimate history of the Chinese Loess Plateau revealed by remanence unmixing. *Geophys. Res. Lett.* **2014**, *41*, 2163–2168. [CrossRef]
16. Wang, S.; Lu, S. A rock magnetic study on red palaeosols in Yun-Gui Plateau (Southwestern China) and evidence for uplift of Plateau. *Geophys. J. Int.* **2014**, *196*, 736–747. [CrossRef]
17. Hong, H.; Fang, Q.; Churchman, G.J.; Zhao, L.; Ji, K.; Jin, X.; Yin, K.; Wang, C.; Liu, C. Study of organic-matter intercalated vermiculitic phases in the hydromorphic red clay sediments. *Appl. Clay Sci.* **2020**, *196*, 105744. [CrossRef]
18. Chen, Z.; Zong, Y.; Wang, Z.; Wang, H.; Chen, J. Migration patterns of Neolithic settlements on the abandoned Yellow and Yangtze River deltas of China. *Quat. Res.* **2008**, *70*, 301–314. [CrossRef]
19. Liu, X.; Chen, J.; Maher, B.A.; Zhao, B.; Yue, W.; Sun, Q.; Chen, Z. Connection of the proto-Yangtze River to the East China Sea traced by sediment magnetic properties. *Geomorphology* **2018**, *303*, 162–171. [CrossRef]
20. Roberts, A.P.; Pike, C.R.; Verosub, K.L. First-order reversal curve diagrams: A new tool for characterizing the magnetic properties of natural samples. *J. Geophys. Res.* **2000**, *105*, 28461–28475. [CrossRef]
21. Chen, J.; Wang, Z.; Chen, Z.; Wei, Z.; Wei, T.; Wei, W. Diagnostic heavy minerals in Plio–Pleistocene sediments of the Yangtze delta, China with special reference to the Yangtze River connection into the sea. *Geomorphology* **2009**, *113*, 129–136. [CrossRef]
22. Yue, W.; Liu, J.; Zhang, D.; Wang, Z.; Zhao, B.; Chen, Z.; Chen, J. Magnetite with anomalously high Cr2O3 as a fingerprint to trace upper Yangtze sediments to the sea. *Geomorphology* **2016**, *268*, 14–20. [CrossRef]
23. Sheldon, N.D.; Tabor, N.J. Quantitative paleoenvironmental and paleoclimatic reconstruction using paleosols. *Earth Sci. Rev.* **2009**, *95*, 1–52. [CrossRef]
24. Hao, Q.; Guo, Z.; Qiao, Y.; Xu, B.; Oldfield, F. Geochemical evidence for the provenance of middle Pleistocene loess deposits in southern China. *Quat. Sci. Rev.* **2010**, *29*, 3317–3326. [CrossRef]

25. Barrón, V.; Torrent, J.; De Grave, E. Hydromaghemite, an intermediate in the hydrothermal transformation of 2-line ferrihydrite into hematite. *Am. Mineral.* **2003**, *88*, 1679–1688. [CrossRef]
26. Hu, P.; Liu, Q.; Torrent, J.; Barrón, V.; Jin, C. Characterizing and quantifying iron oxides in Chinese loess/paleosols: Implications for pedogenesis. *Earth Planet. Sci. Lett.* **2013**, *369*, 271–283. [CrossRef]
27. Torrent, J.; Barrón, V.; Liu, Q. Magnetic enhancement is linked to and precedes hematite formation in aerobic soil. *Geophys. Res. Lett.* **2006**, *33*, 140–165. [CrossRef]
28. Liu, C.; Deng, C.; Liu, Q. Mineral magnetic studies of the vermiculated red soils in southeast China and their paleoclimatic significance. *Palaeogeogr. Palaeoclimatol. Palaeocol.* **2012**, *329*, 173–183. [CrossRef]
29. Ji, R.; Hu, Z.; Zhang, W. Magnetic properties of the Jieshou red clay sequence in the Jinhua-Quzhou Basin, Southeastern China and its paleoenvironmental implications. *Quat. Sci.* **2015**, *35*, 1020–1029. (In Chinese)
30. Yin, Q.; Guo, Z. Mid-Pleistocene vermiculated red soils in southern China as an indication of unusually strengthened East Asian monsoon. *China Sci. Bull.* **2006**, *51*, 213–220. [CrossRef]
31. Zhao, Z.; Qiao, Y.; Wang, Y.; Fu, J.; Wang, S.; Li, C.; Yao, H.; Jiang, F. Can the magnetic susceptibility record of Chinese red clay sequence be used for palaeomonsoon reconstructions? Geophys. *J. Int.* **2016**, *204*, 1421–1429.
32. Zhang, W.; Yu, L.; Lu, M.; Zheng, X.; Shi, Y. Magnetic properties and geochemistry of the Xiashu Loess in the present subtropical area of China, and their implications for pedogenic intensity. *Earth Planet. Sci. Lett.* **2007**, *260*, 86–97. [CrossRef]
33. Liu, C.; Dupont-Nivet, G.; Wang, W.; Deng, C. Magnetic response to pedogenesis in aerobic soils of different weathering degree. *Palaeogeogr. Palaeoclimatol. Palaeocol.* **2021**, *567*, 1978–2012. [CrossRef]
34. Jenny, H. *Factors of Soil Formation: A System of Quantitative Pedology*; Courier Corporation: North Chelmsford, MA, USA, 1994.
35. Song, Y.; Fang, X.; Chen, X.; Torii, M.; Ishikawa, N.; Zhang, M.; Yang, S.; Chan, H. Rock magnetic record of late Neogene red clay sediments from the Chinese Loess Plateau and its implications for East Asian monsoon evolution. *Palaeogeogr. Palaeoclimatol. Palaeoecol.* **2018**, *510*, 109–123. [CrossRef]
36. Wan, S.; Li, A.; Clift, P.D.; Jiang, H. Development of the East Asian sum-mer monsoon: Evidence from the sediment record in the South China Sea since 8.5 Ma. *Palaeogeogr. Palaeoclimatol. Palaeoecol.* **2006**, *241*, 139–159. [CrossRef]
37. Gai, C.; Liu, Q.; Roberts, A.P.; Chou, Y.; Zhao, X.; Jiang, Z.; Liu, J. East Asian monsoon evolution since the late Miocene from the South China Sea. *Earth Planet. Sci. Lett.* **2019**, *530*, 115960. [CrossRef]
38. Wang, H.; Lu, H. Climate controls on evolution of grassland ecosystems since late Cenozoic: A phytolith perspective. *Earth Sci. Rev.* **2022**, 104059. [CrossRef]
39. Zachos, J.; Pagani, M.; Sloan, L.; Thomas, E.; Billups, K. Trends, rhythms, and aberrations in global climate 65 Ma to present. *Science* **2001**, *292*, 686–693. [CrossRef] [PubMed]
40. Clemens, S.C.; Prell, W.L.; Sun, Y.; Liu, Z.; Chen, G. South Hemisphere forcing of Pliocene δ18O and the evolution of Indo-Asian monsoons. *Paleoceanography* **2008**, *23*, 1–15. [CrossRef]
41. Sun, Y.; An, Z.; Clemens, S.C.; Bloemendal, J.; Vandenberghe, J. Seven million years of wind and precipitation variability on the Chinese Loess Plateau. *Earth Planet. Sci. Lett.* **2010**, *297*, 525–535. [CrossRef]
42. Tian, J.; Xie, X.; Ma, W.; Jin, H.; Wang, P. X-ray fluorescence core scanning records of chemical weathering and monsoon evolution over the past 5 Myr in the south Southern China Sea. *Paleoceanography* **2011**, *26*, 1–17. [CrossRef]
43. Xiong, S.; Ding, Z.; Zhu, Y.; Zhou, R.; Lu, H. A∼6Ma chemical weathering history, the grain size dependence of chemical weathering intensity, and its implications for provenance change of the Chinese loess–red clay deposit. *Quat. Sci. Rev.* **2010**, *29*, 1911–1922. [CrossRef]
44. Qiu, J.; Li, X. *Quaternary Stratigraphy and Sedimentary Environment of Shanghai*; Shanghai Science and Technology Press: Shanghai, China, 2007; pp. 1–230. (In Chinese)
45. Liu, J.; Zhang, X.; Mei, X.; Zhao, Q.; Guo, X.; Zhao, W.; Liu, J.; Saito, Y.; Wu, Z.; Li, J.; et al. The sedimentary succession of the last ~ 3.50 Myr in the western south Yellow Sea: Paleoenvironmental and tectonic implications. *Mar. Geol.* **2018**, *399*, 47–65. [CrossRef]
46. Yi, L.; Ye, X.; Chen, J.; Li, Y.; Long, H.; Wang, X.; Du, J.; Zhao, S.; Deng, C. Magnetostratigraphy and luminescence dating on a sedimentary sequence from northern East China Sea: Constraints on evolutionary history of eastern marginal seas of China since the Early Pleistocene. *Quat. Int.* **2014**, *349*, 316–326. [CrossRef]

Disclaimer/Publisher's Note: The statements, opinions and data contained in all publications are solely those of the individual author(s) and contributor(s) and not of MDPI and/or the editor(s). MDPI and/or the editor(s) disclaim responsibility for any injury to people or property resulting from any ideas, methods, instructions or products referred to in the content.

Journal of
Marine Science and Engineering

MDPI

Article

New Seismoacoustic Data on Shallow Gas in Holocene Marine Shelf Sediments, Offshore from the Cilento Promontory (Southern Tyrrhenian Sea, Italy)

Gemma Aiello [1,*] and Mauro Caccavale [1,2]

1 Institute of Marine Sciences (ISMAR), National Research Council of Italy (CNR), 80133 Naples, Italy
2 Istituto Nazionale di Geofisica e Vulcanologia (INGV), Sezione Osservatorio Vesuviano (OV), Via Diocleziano 328, 80125 Naples, Italy
* Correspondence: gemma.aiello@cnr.it

Abstract: High-resolution seismoacoustic data represent a useful tool for the investigations of gas-charged sediments occurring beneath the seabed through the identification of the diagnostic intrasedimentary features associated with them. Acoustic blanking revealed shallow gas pockets in the seismostratigraphic units of the inner shelf off the Northern Cilento promontory. Six main seismostratigraphic units were recognized based on the geological interpretation of the seismic profiles. Large shallow gas pockets, reaching a lateral extension of 1 km, are concentrated at the depocenter of Late Pleistocene–Holocene marine sediments that are limited northwards by the Solofrone River mouth and southwards by the Licosa Cape promontory. A morphobathymetric interpretation, reported in a GIS environment, was constructed in order to show the main morphological lineaments and to link them with the acoustic anomalies interpreted through the Sub-bottom chirp profiles. A newly constructed workflow was assessed to perform data elaboration with Seismic Unix software by comparing and improving the seismic data of the previously processed profiles that used Seisprho software. The identification of these anomalies and the corresponding units from the offshore Cilento promontory represent a useful basis for an assessment of marine geohazards and could help to plan for the mitigation of geohazards in the Cilento region.

Keywords: high-resolution seismic profiles; shallow gas; seismic processing; acoustic anomalies; Cilento promontory; Southern Tyrrhenian Sea

Citation: Aiello, G.; Caccavale, M. New Seismoacoustic Data on Shallow Gas in Holocene Marine Shelf Sediments, Offshore from the Cilento Promontory (Southern Tyrrhenian Sea, Italy). *J. Mar. Sci. Eng.* **2022**, *10*, 1992. https://doi.org/10.3390/jmse10121992

Academic Editor: Antoni Calafat

Received: 28 November 2022
Accepted: 13 December 2022
Published: 14 December 2022

Publisher's Note: MDPI stays neutral with regard to jurisdictional claims in published maps and institutional affiliations.

Copyright: © 2022 by the authors. Licensee MDPI, Basel, Switzerland. This article is an open access article distributed under the terms and conditions of the Creative Commons Attribution (CC BY) license (https://creativecommons.org/licenses/by/4.0/).

1. Introduction

Reflection seismics is an important tool for shallow-gas investigations as it provides a rapid mean for detecting and mapping gas accumulations [1–4]. Over the last decades, high-resolution seismic exploration data (sub-bottom and Sparker) have been very useful for shallow-gas evaluations [5–7] (among others).

Shallow gas pockets are often limited upwards by a strong seismic reflector, totally masking the underlying sediments, which look absolutely transparent. Acoustically transparent zones that are well shown by high-resolution reflection profiles are often related to gas-charged sediments [8,9]. Hovland and Judd [8] have described in detail the acoustic anomalies related to gas occurrence. Various seismic signatures of gas within sediments have been recognized, including acoustic blanking, bright spots, and enhanced reflections. For the aims of our paper, the most significant acoustic anomaly is acoustic blanking. The reflection or absorption of acoustic energy by gas prevents the recording of reflections from lower sediment layers. Although acoustic blanking is widespread on several of the world's continental margins, it is not always found in close association with pockmarks [8].

The acoustic anomalies related to the occurrence of gas within the sedimentary record have been studied in detail, both in the central Adriatic Sea [10] and in the central Mediterranean Sea [11]. Geletti et al. [10] have highlighted gas seepages, fracture systems, and deep salt structures based on the interpretation of Sub-bottom chirp profiles. Spatola et al. [11] have shown the acoustic anomalies related to gas occurrence in the Sicily Channel based on the geological interpretation of Sub-bottom profiles, consisting of vertical blanketed areas characterized by acoustically transparent seismic facies.

The acoustic anomalies associated with gas seepage have been observed both within the sediments and at the seabed [12]. Cold seeps are a seafloor expression of the migration of fluids through sediments from the subsurface to the sea bottom and into the water column until they arrive in the atmosphere. They may be generated by microbial activity in the sediments or by thermal processes involving deeper layers. These geological processes mainly ensue on portions of the seafloor, where the expulsion of free and hydrated gas, including methane and other hydrocarbons, occurs. These structures have been recognized globally on active and passive continental margins. On the active continental margins, they have been found in subduction zones and in deltaic environments, where the rapid deposition of sediments and high subsidence have been found. Cold seeps are widespread features of passive continental margins, including the northern US Atlantic Margin. Methane seepage is expected to intensify at these relatively shallow seeps as bottom waters warm and underlying methane hydrates dissociate.

Missiaen et al. [6] have shown seismic evidence of shallow gas. The most evident acoustic anomaly observed on the seismic profiles is acoustic blanking, appearing as diffuse and chaotic seismic facies masking all the other reflections. Acoustic blanking has been observed in the upper sequence and occasionally reaching up to the seabed. Although the acoustic turbidity reaches the surface locally, the sea-floor morphology does not reveal any pockmark.

Andreassen et al. [7] have analyzed shallow gas and fluid migration on the Barents Sea continental margin based on three-dimensional seismic data. The occurrence of gas is inferred from bright spots, while the fluid migration is inferred from vertical zones of acoustic masking and acoustic pipes. The most important indicator of fluid migration is acoustic masking, represented by an area with low seismic reflectivity. It indicates a scattering of the acoustic energy caused by interstitial gas bubbles in the sediments.

The occurrence of gas in Holocene sedimentary successions is of interest both because of shallow-gas accumulations, which may reduce the shear strength of the sediments and pose a hazard to hydrocarbon exploration and development, and because the occurrence of shallow-gas accumulations, and the underlying indications of fluid flow may point towards deeper prospective reservoirs [5,13].

Marine geohazards include fluid flows [13]. Fluid flow through marine sediment is an ordinary process appearing when water from the underlying sediment is removed, when the sediments have been buried and fossilized, biogeochemical processes, hydrothermal activity, or the migration of deep-seated fluid from the acoustic basement. In most cases, fluid accumulates under the sediment when there is rapid sedimentation. Secondary fluids (water and gases) will be added by several geological processes, including the decay of organic matter.

The aim of this paper is to show the occurrence of shallow gas pockets and acoustic anomalies in the inner shelf of the Cilento offshore area based on the geological interpretation of Sub-bottom chirp profiles. A processing sequence has been constructed by reading the seismic sections through Seismic Unix software [14]. Then, this sequence was applied to the seismic record in order to highlight the acoustic anomalies, indicating the occurrence of Quaternary marine deposits enriched by gas. These anomalies include acoustic blanking and the occurrence of wide, acoustically transparent seismic units corresponding to seismic units filled by gas. The location of the study area is shown in Figure 1.

New geologic, seismostratigraphic, and sedimentological data have been recently shown, highlighting the depositional environments in the Cilento offshore are based on marine geological data [15]. An onshore–offshore DEM off the Cilento promontory has been constructed by superimposing the geological results of the marine geological survey at the 1:25.000 scale. Moreover, the geological interpretation of Sub-bottom chirp profiles has shown the occurrence of shallow gas pockets. However, in a previous paper, the acoustic anomalies related to the occurrence of gas have not yet been discussed in detail; this will be the subject of this paper. Herein, we have improved the quality of previously processed seismic profiles (Seisprho) [16] by assessing a new workflow for the seismic processing of Sub-bottom data with the Seismic Unix software. This has allowed us to improve the previously obtained results on seismic stratigraphy [15] and to focus on the identification of acoustic anomalies related to gas occurrence by constructing new thematic maps in a GIS environment showing the distribution of the shallow gas pockets in the Cilento offshore area.

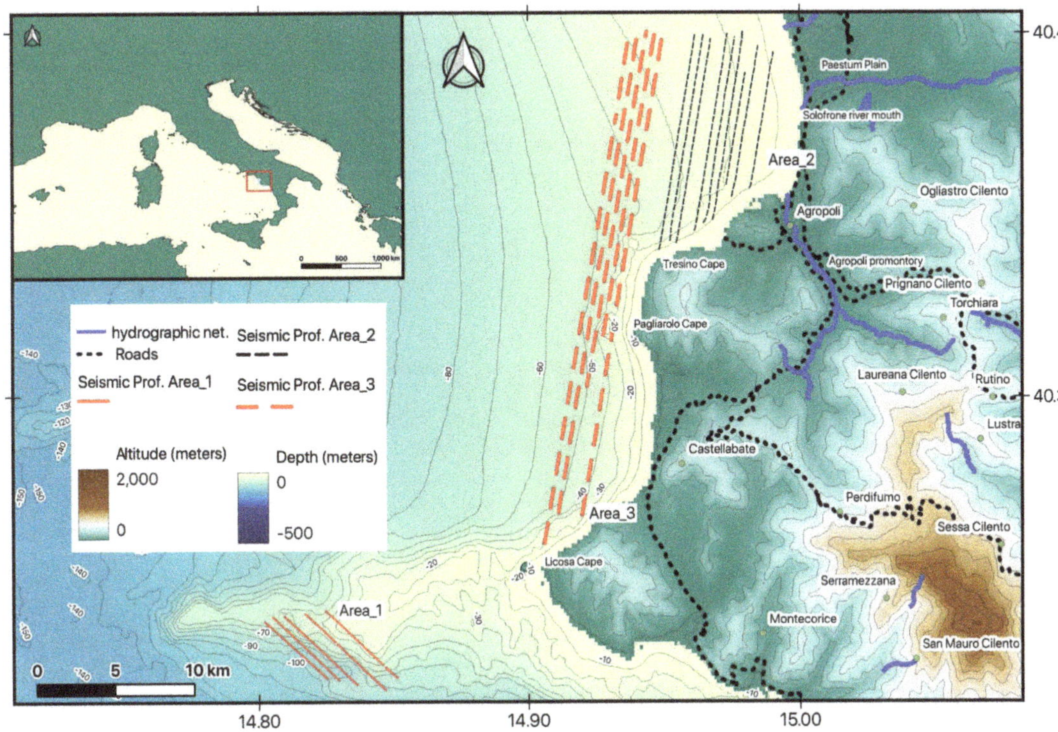

Figure 1. Sketch map showing the location of the seismic profiles analyzed in this paper. Based on the location of the seismic profiles, three main areas have been distinguished, characterized by distinctive acoustic anomalies that are genetically related to the occurrence of gas. The inset on the upper left side shows the general location map of the study area. GIS base was modified after Aiello and Caccavale [15].

Our study will allow us to solve the contrasting hypotheses on the occurrence of shallow gas in the study area, which was not previously singled out based on the bathymetric and geomorphological data recently published on the Cilento offshore area [17]. Violante [17] has carried out a computer-aided geomorphologic seabed classification and habitat mapping of the Punta Licosa Marine Protected Area. The Punta Licosa habitat map shows several marine landforms, including spurs of coralligenous bioconstructions, wave-cut terraces with a mixed organogenic cover, slopes with mixed organogenic sediments, depositional terraces, rocks, deep terraces with bioclastic covers, ledges with coralligenous shelf muddy plains, and sandy fringes with bioclasts and offshore transitions [17]. The pockmarks or other seabed structures that are genetically related to gas occurrence have not been highlighted. Since basic geophysical studies on gas and related seabed structures have suggested that pockmarks are not necessarily associated with the shallow gas occurring in the subsurface [6,7] (among others), our study will integrate the existing knowledge on shallow gas and the corresponding acoustic anomalies from the offshore Cilento promontory.

2. Geologic Setting

The origin of the sedimentary basins and their subsidence along the Campania sector of the eastern Tyrrhenian margin has been the subject of various studies based on geological field observations and on seismic and well data in the subsurface on land and at sea [18–30]. In this geological context, an important lineament is represented by the 41st parallel line, separating the Northern Tyrrhenian Sea from the Southern Tyrrhenian Sea [31–35].

Along the Campania margin, south of the Aurunci Mountains, the coastal basins are represented by the Garigliano Plain, the Campania Plain, and the Sele Plain. In this sector of the Tyrrhenian margin, the extensional phases were active along the NE-SW trending normal faults, ranging in age from the Late Pliocene to the Quaternary [25,26,36–38]. A chronostratigraphic calibration of the five seismic profiles has been carried out by using data provided by some exploration wells located along the Latium-Campania margin (Michela 1; Mara 1) [25,27].

The continental shelf between the Gulf of Gaeta and the Gulf of Policastro is the seaward prolongation of the alluvial coastal plains (Campania Plain, Sarno Plain, and Sele Plain), bounding the Tyrrhenian sector of the Apennines, which is controlled by high rates of tectonic subsidence [37].

These coastal plains, whose continuity is interrupted by structural highs with a NE-SW trend and by volcanic complexes (Phlegrean Fields and Somma Vesuvius), are limited northeastwards by the inner reliefs of the Apennines. Their sedimentary fill consists of marine and continental clastic deposits, alternating with abundant volcanic products. The acoustic basement is represented by Meso-Cenozoic carbonates ("Campania-Lucania carbonate platform") [39–41] and by Cenozoic-deformed sequences and the related flysch deposits (Cilento Flysch; Sicilide Units; Liguride Units: Figure 2) [42–45].

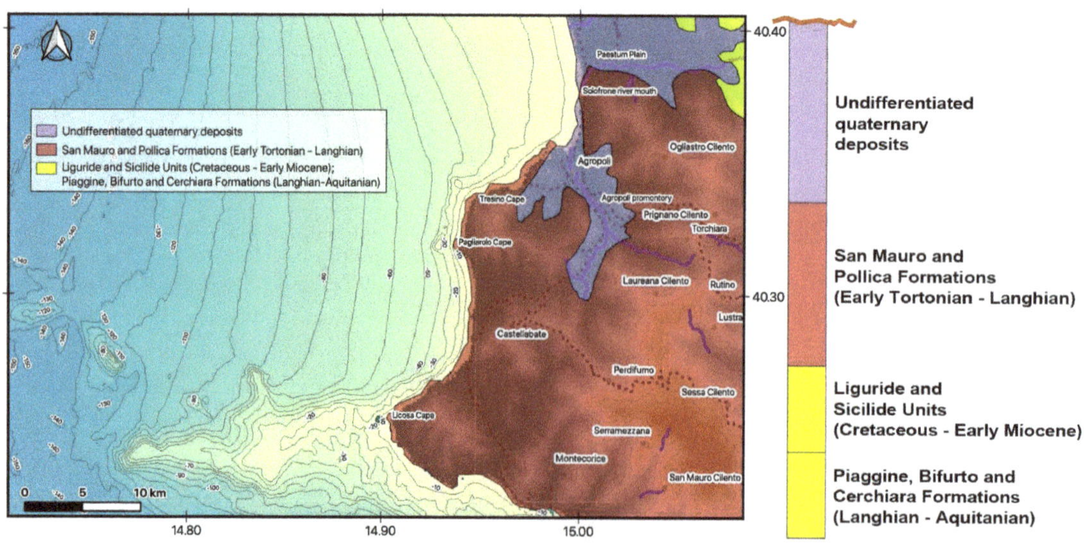

Figure 2. Geologic sketch map of the northern Cilento promontory. Note: the occurrence of main geological units, including the undifferentiated Quaternary deposits, the Cilento Flysch deposits (S. Mauro and Pollica Formations), the Liguride and Sicilide Units, the Trentinara Formation, and the Mesozoic shallow-water limestones. GIS base was modified after Aiello and Caccavale [15].

The Tyrrhenian offshore area between the Sorrento Peninsula and the Cilento promontory, including the half-graben of the Salerno Valley [46,47], is characterized by a wide continental shelf, extending up to 25 km westwards of the Cilento coast, with thin sedimentary cover that becomes thick from a few tens of meters up to a few hundreds of meters.

Southwards of the Sorrento Peninsula, a narrow continental shelf occurs, which is associated with steep slopes, bounding the structural depression of the Salerno Basin. The individuation and the tectonic setting of the Salerno Basin have been mainly controlled by the regional fault Capri-Sorrento Peninsula, with a WSW-ENE trend and an average throw of 1500 m, bounding southwards to the structural high of the Sorrento Peninsula-Capri Island. This fault is composed of several coalescent segments and controls high submarine hazards in the Gulf of Naples [48].

A geological map of the northern Cilento promontory was constructed and imported into a GIS environment (Figure 2).

In this map, several geological units were mapped, including the Quaternary undifferentiated deposits, the S. Mauro and Pollica Formations, which are Langhian–Early Tortonian in age, the Liguride and the Sicilide Units, which are Cretaceous–Early Miocene in age, the Piaggine, Bifurto and Cerchiara Formations, which are Langhian–Aquitanian in age, the Trentinara Formation (limestones and marls), which is Langhian–Aquitanian in age, and the shallow water limestones, which are Late Cretaceous–Liassic in age (Figure 2).

3. Materials and Methods

A scientific theme was developed to carry out the acquisition, processing, and geologic interpretation of a densely spaced grid of high-resolution seismic profiles (Sub-bottom chirp) collected by the CNR-ISMAR onboard the R/V Urania (National Research Council of Italy). The grid of the Sub-bottom chirp seismic sections is reported in Figure 1, which has been constructed in a GIS environment. The chirp linear frequency modulated signals of 2–7 kHz or 8–23 kHz and 4 kW provide high-resolution soundings, reaching up to a 10–30 cm resolution within the uppermost marine sediments, according to the resolution values reported for chirp data by Gasperini and Stanghellini [16]. The obtained data, recorded onboard with a SEG Y format, have been processed by using the software Seismic Unix [14].

The CWP/SU: Seismic Unix is an open-source seismic utilities package that was originally created by the Center for Wave Phenomena at the Colorado School of Mines [14]. This software has been used to convert the seismograms between the SU standard file and the SEG Y files.

Simple processing has been applied to the seismic sections of the Cilento offshore area, consisting of the application of a linear vertical gain along the seismic section, which has allowed for a significant improvement in the quality of the seismic data. The file recorded in the field by the oceanographic cruise was converted from the SEG Y format to the SU format, used by the Seismic Unix software.

The workflow of the processing of the seismic sections consisted of the following steps:

(1) Conversion of the seismic traces from SEG Y to SU;
(2) Spectral analysis of the frequencies of the seismograms, in order to reconstruct the distribution of the frequencies;
(3) Application of fast Fourier transform (FFT) for the visualization and the analysis of the frequencies of the seismic signal. In particular, the Fourier transform converts waveform data in the time domain into the frequency domain. The Fourier transform breaks down the original time-based waveform into a series of sinusoidal terms, each with a single magnitude, frequency, and phase. This process converts a waveform in the time domain that is difficult to describe mathematically into a more manageable series of sinusoidal functions that, when added together, exactly reproduce the original waveform;
(4) Application of a high-pass filter with a low-cut frequency at 150 Hz in order to eliminate the seismic noise and the dark signal. A significant improvement in the

visualization of the seismic reflector of the sea bottom was obtained by using the frequency of 150 Hz;

(5) Application of a uniform gain on each seismic trace;

(6) Application of a time-variant gain (TVG) in order to improve both the seismic signal of the deeper seismic reflectors and the whole visualization of all the seismic profiles;

(7) Output of the seismic profiles with Seismic Unix in order to obtain the final files, which were plotted using a graphic interface.

The geological interpretation of the Sub-bottom chirp data was carried out based on the seismostratigraphic criteria [49–51]. The basic principle of seismic stratigraphy is that the seismic reflectors, determined by the contrasts of acoustic impedance, correspond to the strata plans. Perhaps, the geometry of the seismic reflectors corresponds to the depositional geometry. The contrasts of acoustic impedance, which are the product of the seismic velocity for the density of the crossed grounds, control the location of the seismic reflectors and are located along the strata surfaces or other discontinuities, having a chronostratigraphic meaning. The strata surfaces represent the old depositional surfaces and perhaps are coeval in the depositional area.

The discontinuities are old erosional or nondepositional surfaces, corresponding with significant stratigraphic gaps. Even though they represent events that vary over geological time, the discontinuities are chronostratigraphic surfaces because the layers overlying the discontinuity are younger than the underlying layers [49–51].

4. Results

The study area represents a wide continental shelf, for which the outer margin is located at a water depth of 250 m. While the continental shelf northwards of the Licosa Cape has uniform gradients, the marine area surrounding the Licosa Cape represents an E–W trending structural high, characterized by the remnants of the terraced surfaces, which erode the rocky basement. A break in slope at water depths ranging from 60–80 m marks the passage from the structural high of the Licosa Cape to the outer shelf through a steep slope. Towards the Policastro Gulf, the continental slope is characterized by an articulated morphology and high gradients at water depths ranging from 10–110 m [15] (Figure 3).

The relict morphologies that are NNW-SSE to E-W trending occur both northwestwards and southwestwards of the Licosa Cape and are genetically related to the Middle-Late Pleistocene continental shelf [15] (Figure 3). The palimpsest deposits are of an emerged–submerged beach form of elongated dunes, composed of well-sorted sands and gravels that are often cemented, with abundant bioclastic fragments that are sometimes overlain by thin drapes of bioclastic sands of the outer shelf. On the Sub-bottom chirp lines, the relict sands appear as deaf units with slightly inclined clinoforms, with offsets truncated at the sea bottom by polycyclic erosional surfaces [15].

The sedimentological analyses show the grain-size distribution of the sediments on the seafloor, including sandy gravels, gravelly sands, sands, silty sands, muddy sands, sandy silts, silts, and muds [15]. Ternary plots of the sea bottom samples were constructed (shales-sands-silts and gravels-sands-silts) in order to improve the processing of the sedimentological data. Fine-grained sands are widespread along the coast in the northern region and in the offshore area of the Licosa Cape. Coarse-grained sands are located in the offshore area of the Licosa Cape.

A sketch of a morphobathymetric map and a geologic map was constructed in a GIS environment, aimed at showing the main morphostructural features of the studied area (Figure 3) [15]. The main geomorphological features are represented by channels, abrasion terraces, an acoustic substratum, and sandy bodies. Important channels start from the 20 m isobaths on the inner shelf from the Pagliarolo Cape towards the Licosa Cape, while other channels start from the outcrops of the acoustic substratum surrounding the Licosa Cape (Figure 3) [15]. A wide outcrop of the acoustic basement that is elongated in shape and E–W trending is located offshore to the Licosa Cape promontory, reaching water depths exceeding 100 m (Figure 3). Northeastwards of the main outcrop, two minor outcrops occur that are, respectively, NE-SW and NW-SE trending (Figure 3) [15]. Otherwise, southwards, another outcrop has been detected (Figure 3). For a detailed description of the Quaternary marine deposits, refer to the paper of Aiello and Caccavale [15].

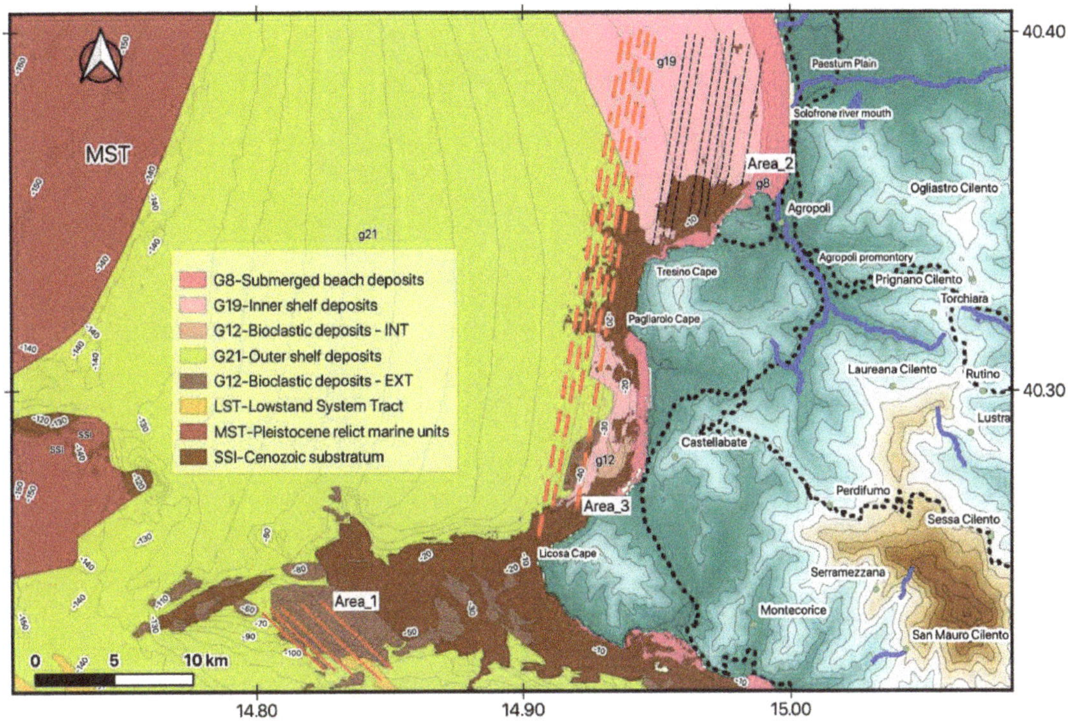

Figure 3. Sketch of a morphobathymetric and geologic map of the analyzed area. The main geomorphological lineaments occurring in this area have been reported, coupled with Quaternary marine deposits (modified after [15]).

Three areas were distinguished based on seismic interpretation in order to highlight the different acoustic features recognized in the Cilento offshore area (Figures 1 and 3).

A sketch table has been constructed in order to improve the understanding of the areas involved by gas and other corresponding acoustic features (Table 1).

Based on the identified features, the first area is located offshore from the Licosa Cape promontory at water depths ranging between 30 and 90 m (Figure 1). It is composed of seven seismic sections (B33_1, B33a_1, B35_1, B35a_1, B37_1, B38, and B39), whose distinctive acoustic features are represented by acoustic blanking, controlled by the occurrence of gas. The wide occurrence of rocky acoustic basements composed of the ssi unit corresponding with the seaward prolongation of the Cilento Flysch, widely cropping out in the adjacent coastal sector, can be singled out. In fact, the seismic lines appear to be located on the flank of a wide outcrop of the rocky acoustic basement, distinguishing the offshore sector (Figure 3) [15].

Table 1. Sketch table showing the areas of the acoustic features, the corresponding seismic sections and the related shot points, and the acoustic features recognized based on seismic interpretation.

	Seismic Sections	Shot Points	Acoustic Features
1	B33_1	400–550; 800–1100; 1450–1750	acoustic blanking
1	B33a_1	350–490	acoustic blanking
1	B35_1	1250–1400 1900–2100 2350–3100	acoustic blanking
1	B35a_1	0–380 750–900 1600–1850 2600–2800 3200–3400	acoustic blanking
1	B37_1	1000 2500 3500–4000	acoustic blanking
1	B38	700–800 1200–1500 2300–2500 3200–3600 4550–5100	acoustic blanking
1	B39	100–500 700–750 1100–1400 2400–2550 3100–3400 3900–4000 4400–4600 5000–5400	acoustic blanking
2	B43	2900–3200	shallow gas pockets
2	B45		no feature
2	B46		no feature
2	B47		no feature
2	B48		no feature
2	B49	2000–2500	shallow gas pockets
2	B51	0–0.1 ($\times 10^4$)	shallow gas pockets
2	B52	0–0.18 ($\times 10^4$) 0.25–0.45 ($\times 10^4$) 0.7–0.85 ($\times 10^4$)	seismic units impregnated of gas
2	B53	0.4–0.7 ($\times 10^4$) 0.75–1.1 ($\times 10^4$)	seismic units impregnated gas
3	B61	0.3–0.45 ($\times 10^4$) 0.45–1.2 ($\times 10^4$)	seismic units impregnated gas
3	B62a	1.15–1.2 ($\times 10^4$) 1.3–1.45 ($\times 10^4$)	shallow gas pockets
3	B63	0.1–0.48 ($\times 10^4$) 0.55–0.7 ($\times 10^4$)	seismic units impregnated gas

Table 1. *Cont.*

	Seismic Sections	Shot Points	Acoustic Features
3	B64	0.2–0.5 ($\times 10^4$) 0.5–0.7 ($\times 10^4$)	shallow gas pockets seismic units impregnated gas
3	B65	180–400 500 600–800	shallow gas pockets
3	B65_1	0–0.8 ($\times 10^4$)	seismic units impregnated gas

The second area is, instead, located in the northern Cilento promontory from the seaward prolongation of the Paestum Plain, proceeding southwards up to the Tresino Cape (Figure 1). It is composed of nine seismic sections (B43, B45, B46, B47, B48, B49, B51, B52, and B53), whose distinctive acoustic features are represented by shallow gas pockets (Table 1). The seismic sections have highlighted the occurrence of thick sedimentary cover, organized in several seismostratigraphic units, mainly Holocene in age [15].

The third area is located on the northern Cilento promontory, starting from the offshore prolongation of the Paestum Plain up to the Licosa Cape promontory, at water depths ranging between 10 and 60 m (Figure 1). It is composed of six seismic sections (B61, B62a, B63, B64, B65, and B65_1; Table 1), whose distinctive features are represented by shallow gas pockets and by the seismic units impregnated by gas ("gassy sediments").

A seismic analysis of the Sub-bottom chirp profiles has allowed us to study the seismostratigraphic characteristics of the Cilento offshore area between the Solofrone River mouth and the Licosa Cape. The main outcrops at the sea bottom of the rocky acoustic basement have been bounded relative to the sedimentary covers, which are composed of coarse-grained sands, grading towards middle-fine-grained sands and fine-grained sands. The sandy facies are prevalent in the sector of the continental shelf located between the Solofrone River mouth and Agropoli, where they form a wide system of coarse-grained sandy belts with an N–S trend parallel to the isobaths and are located at water depths ranging between 10 m and 17 m. In the same area, the rocky outcrops have a limited extension and occur at water depths ranging between 15 m and 20 m.

The rocky acoustic basement (ssi) widely crops out next to the shoreline from Agropoli to the Tresino Cape and from the Tresino Cape to the Pagliarolo Cape (Figure 3), where it represents the seaward prolongation of the coastal cliffs incised in the deposits of the Cilento Flysch [41,43,45].

The seismic lines B51, B52, and B53 have been taken as an example among the existing seismic lines in order to highlight the occurrence of wide shallow gas pockets and to show the significant improvement in the seismic lines obtained by introducing the Seismic Unix processing with respect to the processing carried out with the Seisprho [15].

In the northern Cilento offshore area, between the Solofrone River mouth and the Licosa Cape, the seismostratigraphic analysis showed that the recent sedimentary cover, ranging in age between the Late Pleistocene and the Holocene, is organized into three main seismostratigraphic units, overlying the undifferentiated acoustic basement (ssi).

The seismic interpretation (Figure 4) shows that the first unit (seismostratigraphic unit 1) is characterized by acoustically transparent seismic facies and ranges in thickness between 7 m and 10 m. Unit 1, probably composed of sands, unconformably overlies the undifferentiated acoustic basement (ssi). Its top is strongly eroded, forming palaeochannels in which the seismostratigraphic unit 2 was deposited.

The second unit (seismostratigraphic unit 2; Figure 4) is distinguished from alternating acoustically transparent intervals and continuous intervals, probably corresponding with alternating sands and shales, for a whole thickness of about 10 m. The unit forms the filling of the palaeochannels individuated at the top of unit 1 and the erosional depressions located at the top of the ssi substratum.

The third unit (seismostratigraphic unit 3; Figure 4) is characterized by acoustic facies with parallel and discontinuous reflectors of high amplitude, and represents the recent filling, Holocene in age.

The seismostratigraphic setting of the first area, as shown in Table 1 and in Figure 4, is exemplified by the seismic interpretation of the seismic sections shown in Figures 5–7.

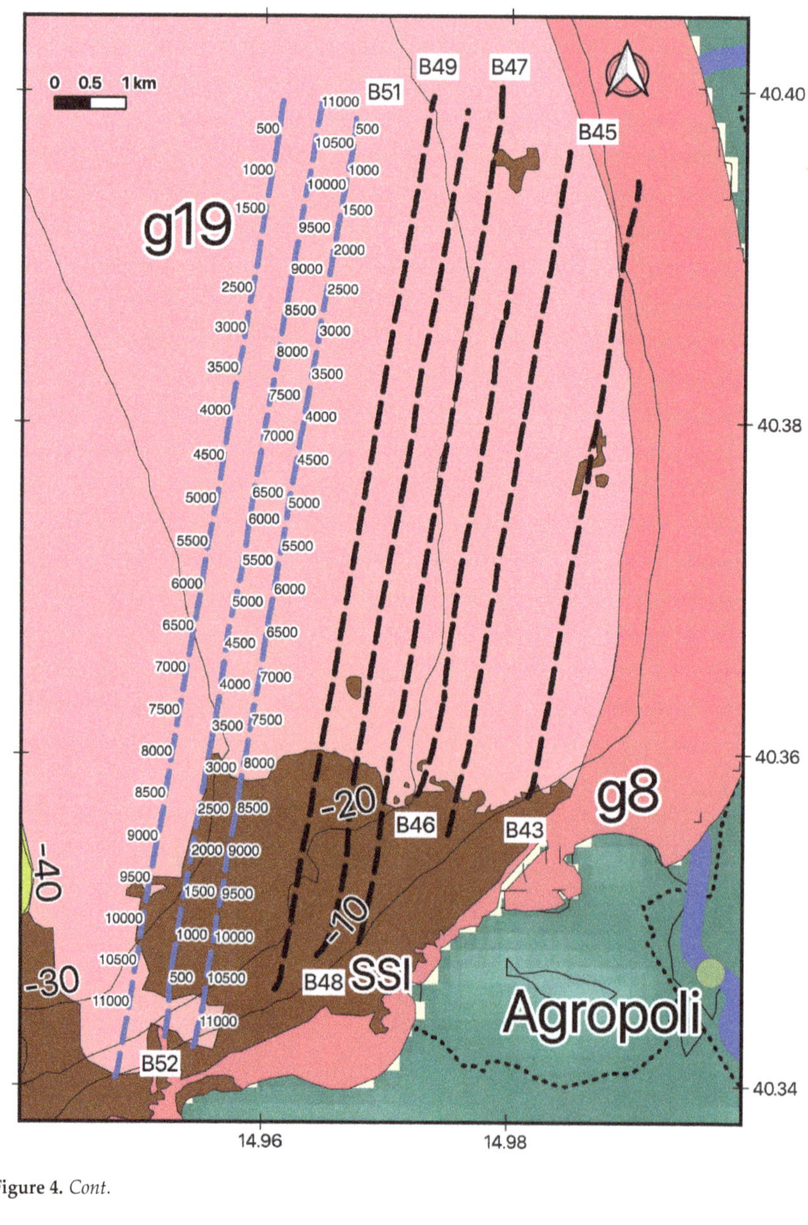

Figure 4. *Cont.*

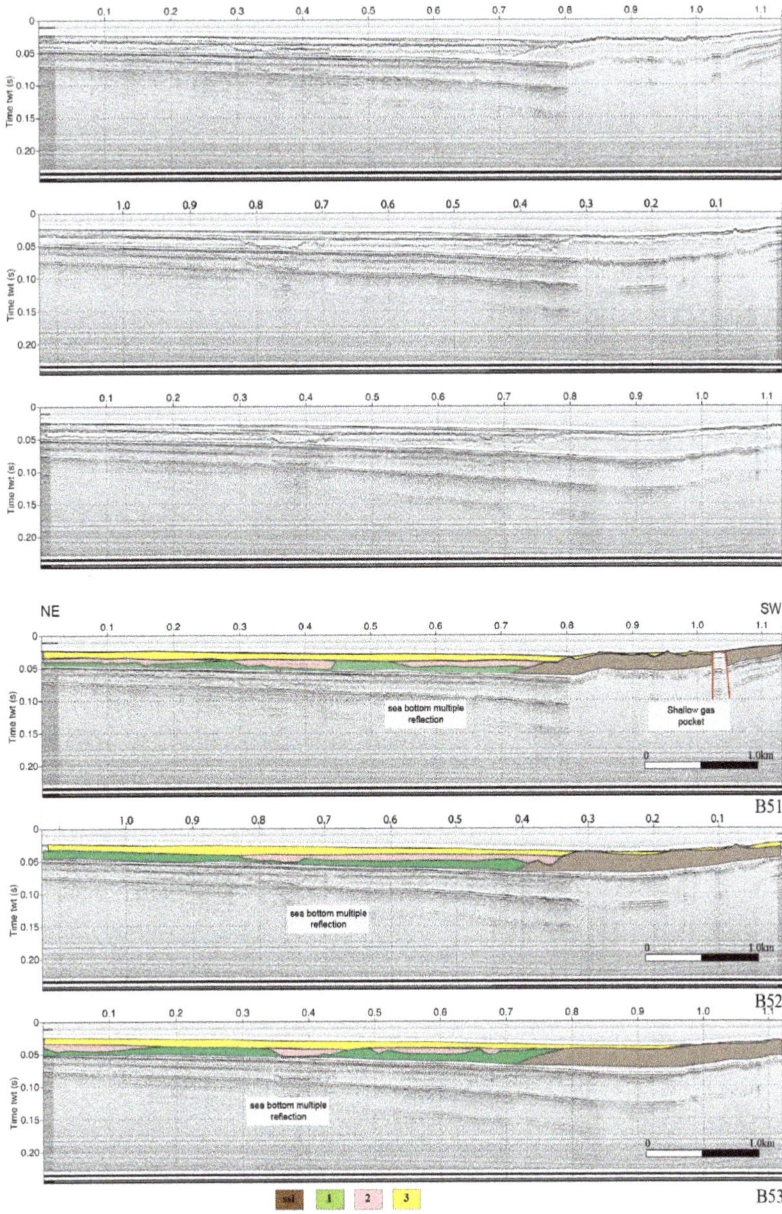

Figure 4. (**top**) Sketch location map of the Sub-bottom chirp profiles; (**middle**) Sub-bottom chirp profiles B51, B52, and B53 and (**bottom**) corresponding geologic interpretation. Note the occurrence of the three main seismostratigraphic units (1, 2, 3) overlying the undifferentiated acoustic basement (ssi; see the text for further description).

The seismic profile B33_1 crosses a wide outcrop of the acoustic basement (ssi unit), which is genetically related to the Cilento Flysch (Figure 5). On the north-western side of the seismic section, three wide areas of acoustic blanking have been identified, suggesting the occurrence of gas impregnations (Figure 5). The seismic profile B33a_1 (area 1) has shown the wide outcrop of the acoustic basement (ssi unit), overlain by a recent sedimentary cover (Figure 5). The interpretation of the seismic section has suggested the occurrence of a wide area of acoustic blanking between the shot points 350 and 480, highlighting the occurrence of gas impregnations (Figure 5).

The seismic profiles B35_1 and B35a_1 show a similar stratigraphic setting, with wide outcrops for the acoustic basement (ssi unit), overlain by the recent sedimentary cover (Figure 6). The interpretation of the seismic sections has shown the occurrence of several areas of acoustic blanking (Table 1) with gas impregnations.

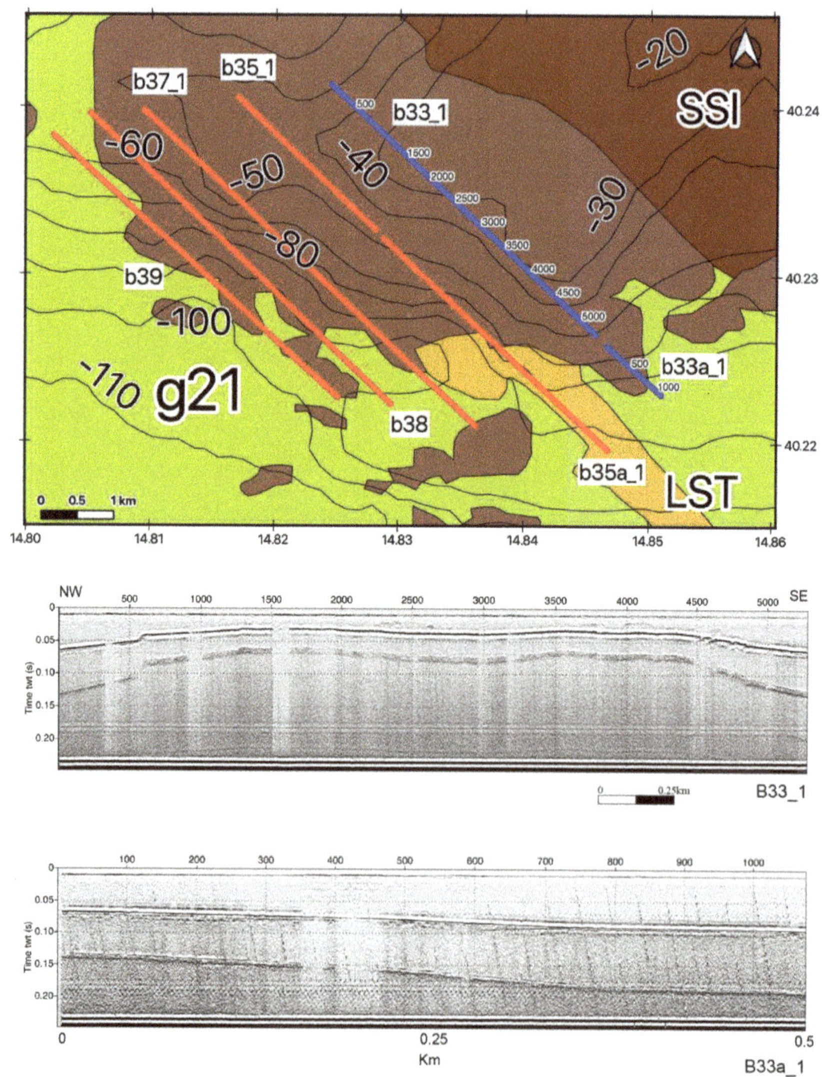

Figure 5. *Cont.*

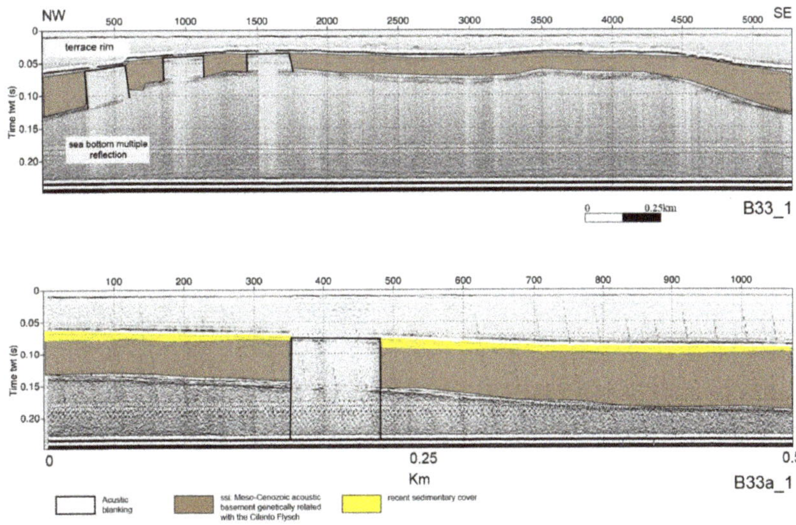

Figure 5. (**top**) Sketch location map of the Sub-bottom chirp profiles of area 1; (**middle**) Sub-bottom chirp profiles B33_1 and B33a_1 and corresponding geologic interpretation (**bottom**).

The seismic profiles B37_1, B38, and B39 show wide areas of acoustic basement (ssi unit) overlain by a thin recent sedimentary cover. All three seismic sections have highlighted the occurrence of shallow gas pockets, as shown by the seismic interpretation (Figure 7).

The seismic profiles B43 and B45 (area 2: Figure 8-top) show a stratigraphic setting characterized by three main seismostratigraphic units, corresponding with Holocene marine deposits, Holocene coastal and marine sands, and coastal sands, which are Late Holocene in age (Figure 8-middle). On the seismic profile B43, a wide shallow gas pocket was identified (Figure 8-middle).

The seismic profile B46 shows three main seismostratigraphic units, respectively, corresponding to the Holocene marine deposits, with coastal and marine sands that are Holocene in age, and coastal sands, which are Late Holocene in age (Figure 9). The coastal and marine sands are interpreted as a wide seismic unit of gassy sediments due to their seismic facies. The interpretation of the seismic profile B47 shows the occurrence of four main seismostratigraphic units, respectively corresponding with the ssi unit, interpreted as the Meso-Cenozoic acoustic basement that is genetically related to the Cilento Flysch, with the marine deposits (Holocene in age), and with the coastal sands. The coastal sands represent the filling of a wide palaeodepression since they thicken toward the center of the depression and thin toward both the sides (Figure 9).

The seismic profiles B48 and B49 (Figure 10) show the occurrence of five main seismostratigraphic units, respectively, represented by the ssi unit (Meso–Cenozoic acoustic basement, genetically related with the Cilento Flysch), marine deposits, the coastal and marine sands (both Holocene in age), a seismic sequence filling the palaeochannels, and by the coastal sands (Late Holocene in age) (Figure 10).

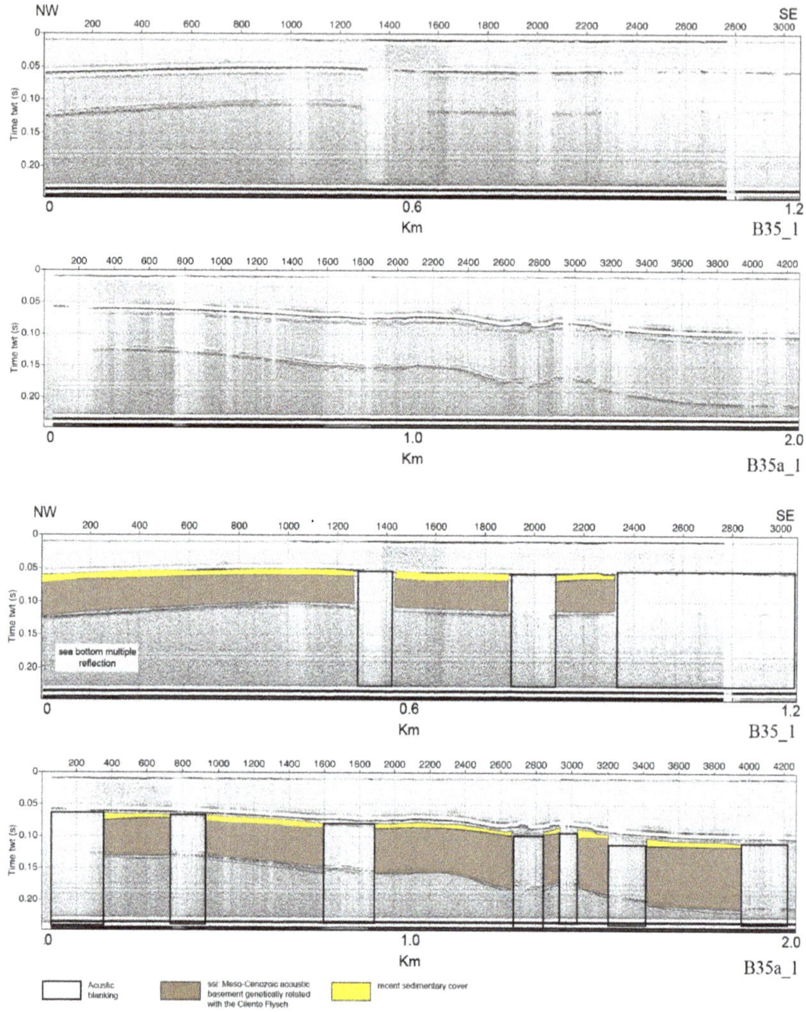

Figure 6. (**top**) Sketch location map of the Sub-bottom chirp profiles of area 1; (**middle**) Sub-bottom chirp profiles B35_1 and B35a_1 and the corresponding geologic interpretation (**bottom**).

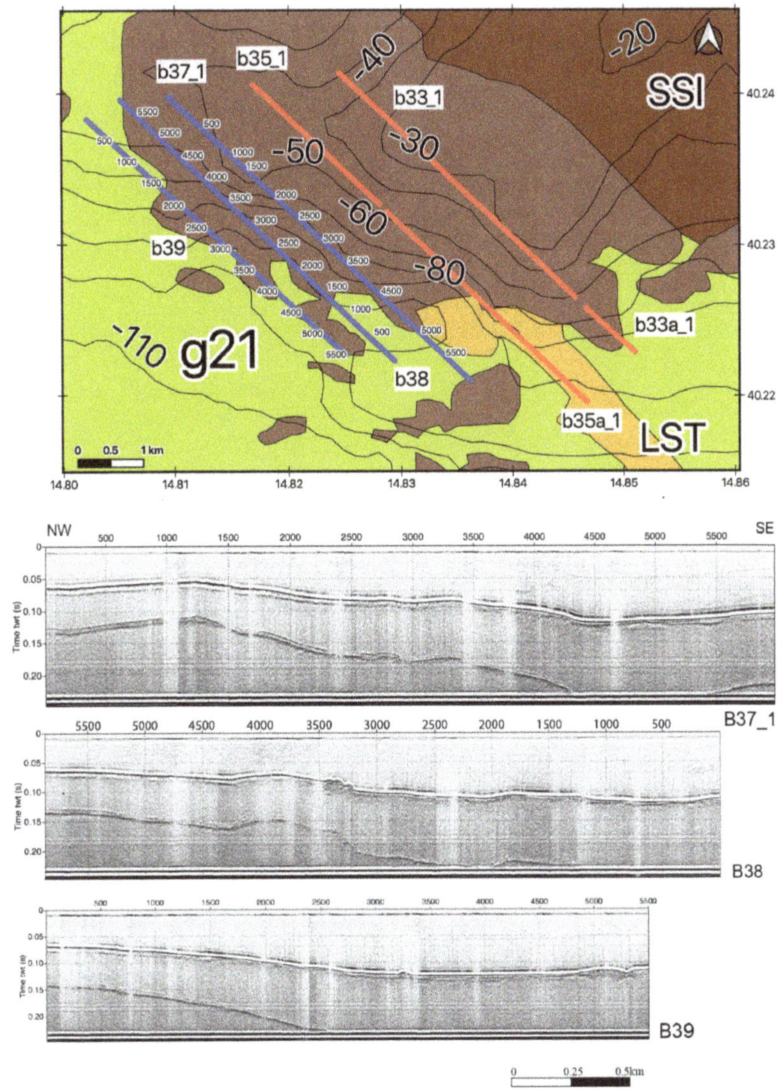

Figure 7. *Cont.*

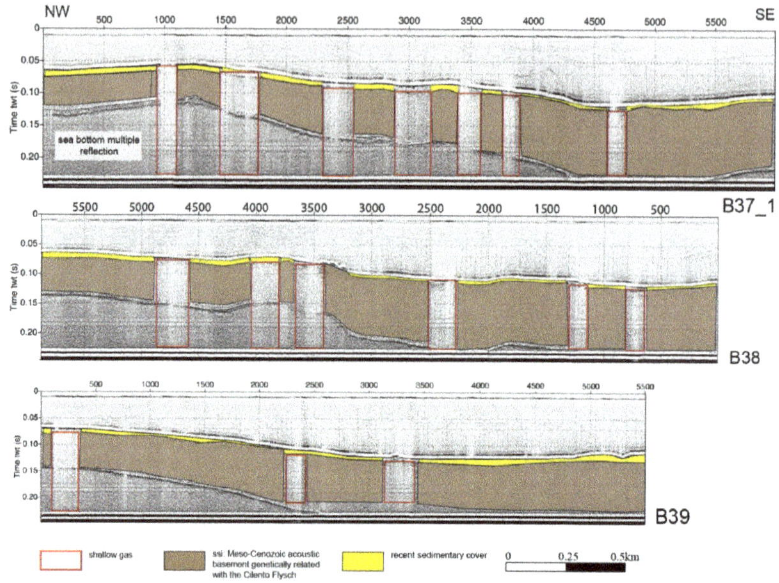

Figure 7. (**top**) Detailed location map; (**middle**) Sub-bottom chirp profiles B37_1, B38, and B39 and the corresponding geologic interpretation (**bottom**).

The seismic profiles B61 and B62_a show the occurrence of five main seismostratigraphic units (Figure 11). The ssi unit constitutes a small morphostructural high occurring in the central part of the seismic section from the shot point 1.2 to 1.8. Proceeding onshore, this seismic unit prolongates on the side of the morphostructural high, below the seismo-stratigraphic unit of the coastal and marine sands, representing gassy sediments below the coastal sands (Late Holocene in age). On the seismic profile B62_a, the same unit composes a small outcrop. The coastal and marine sands (Holocene in age) represent a wide, acoustically transparent seismic unit, interpreted as gassy sediments (Figure 11). Moreover, the coralligenous deposits are interpreted as facies hetheropy with coastal sands (Holocene in age) (Figure 11).

The seismic profiles B63 and B64 show the wide seismostratigraphic units of coastal sands, interpreted as gassy sediments, seismostratigraphic units for channel filling, and seismostratigraphic units for coralligenous deposits in facies hetheropy with coastal sands (Figure 12).

The seismic profiles B65 and B65_1 show the seismostratigraphic units of coastal and marine sands (Holocene in age) overlain by coastal sands (Late Holocene in age) (Figure 13). Two wide areas of acoustic blanking were detected at the shot points between 200 and 800 (Table 1). Moreover, the seismic profile B65_1 shows wide outcrops of acoustic basement (ssi unit). The seismostratigraphic units of channel filling are characterized by disrupted seismic reflectors controlled by the occurrence of gas (Figure 13).

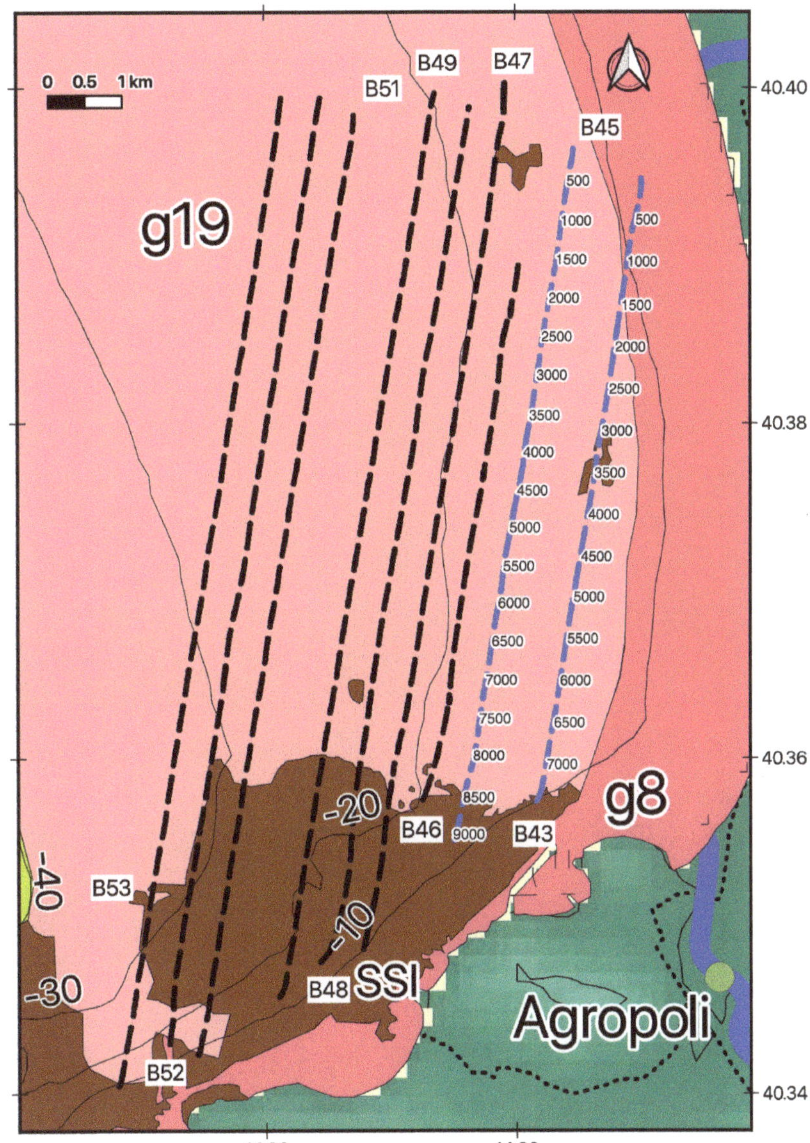

Figure 8. *Cont.*

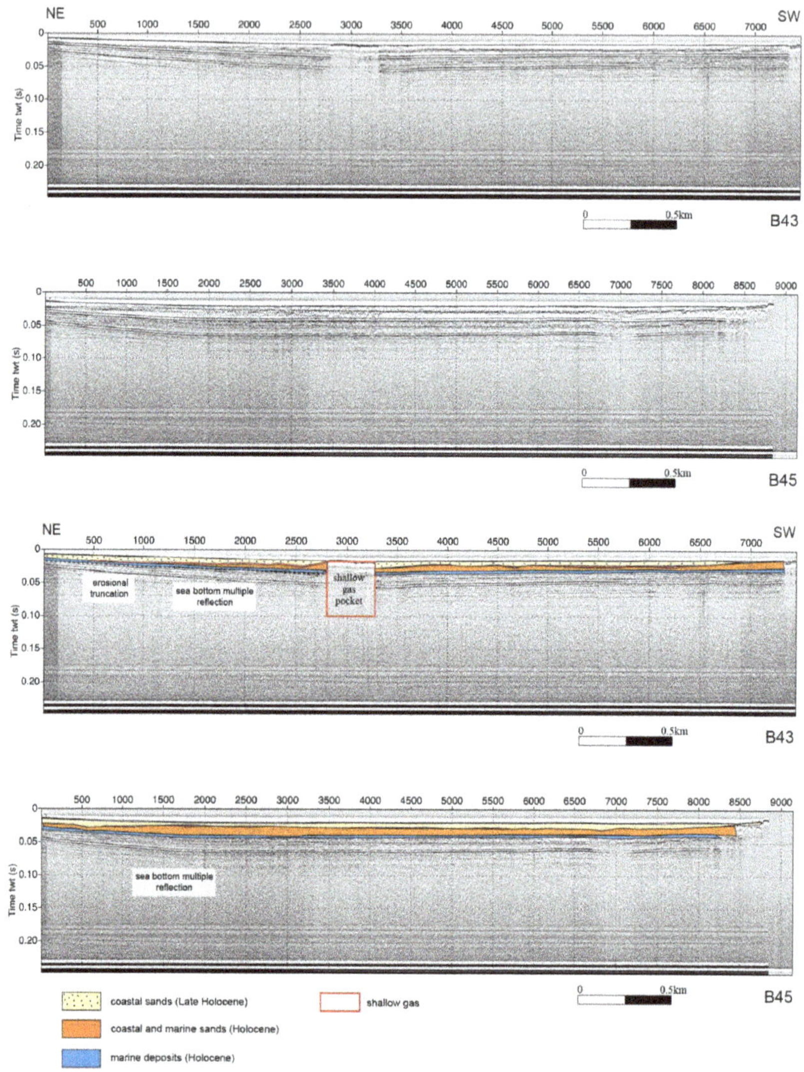

Figure 8. (**top**) Sketch location map of the Sub-bottom chirp profiles of area 2; (**middle**) Sub-bottom chirp profiles B43 and B45 and the corresponding geologic interpretation (**bottom**).

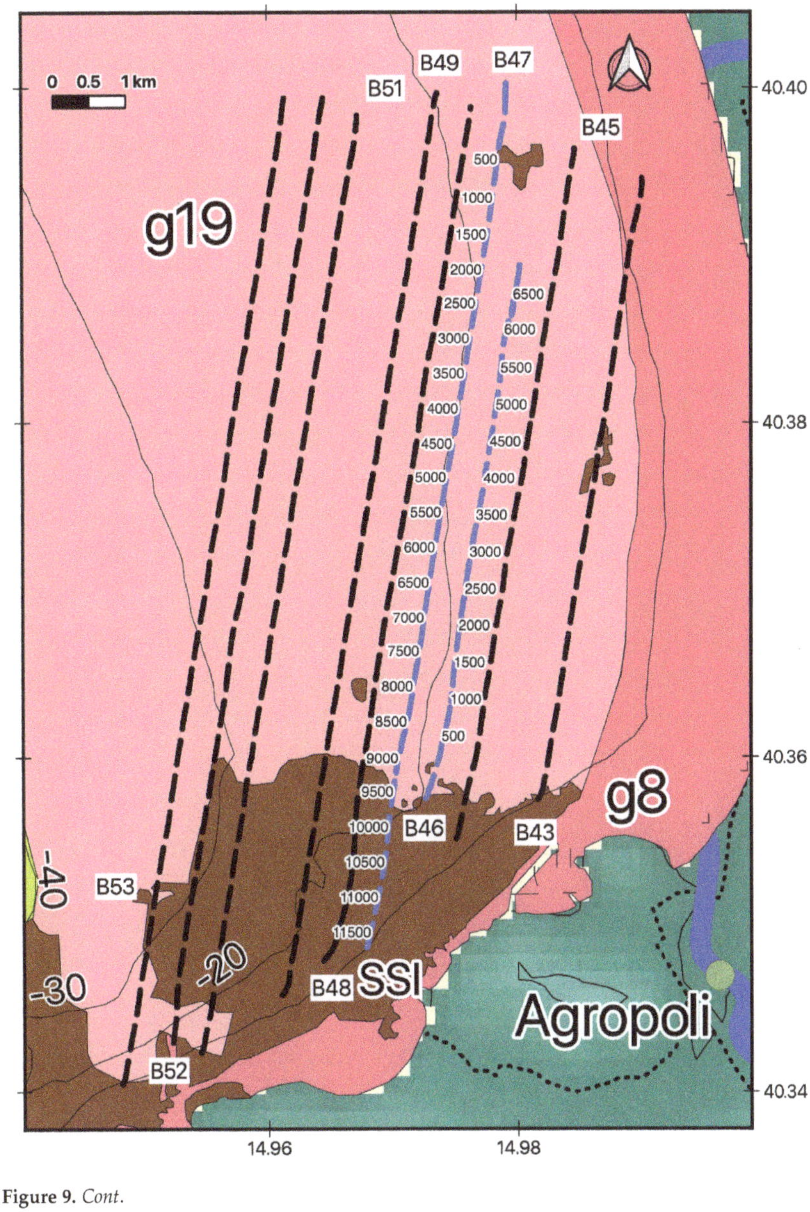

Figure 9. *Cont.*

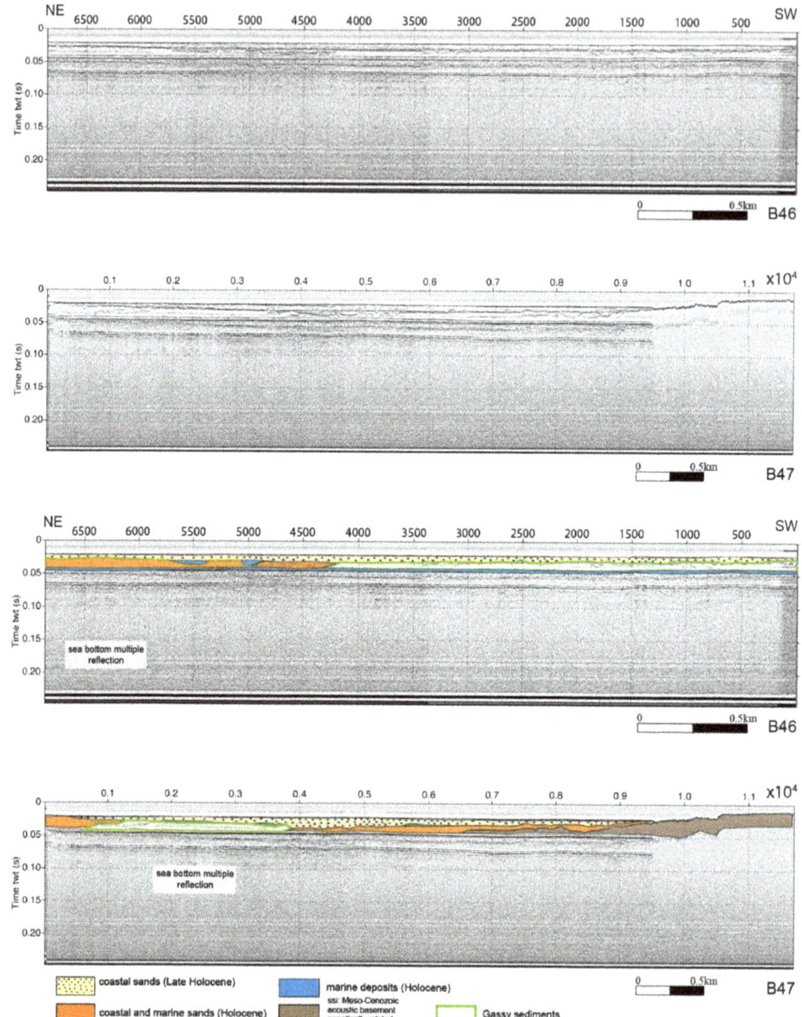

Figure 9. (**top**) Detailed location map; (**middle**) Sub-bottom chirp profiles B46 and B47 and the corresponding geologic interpretation (**bottom**).

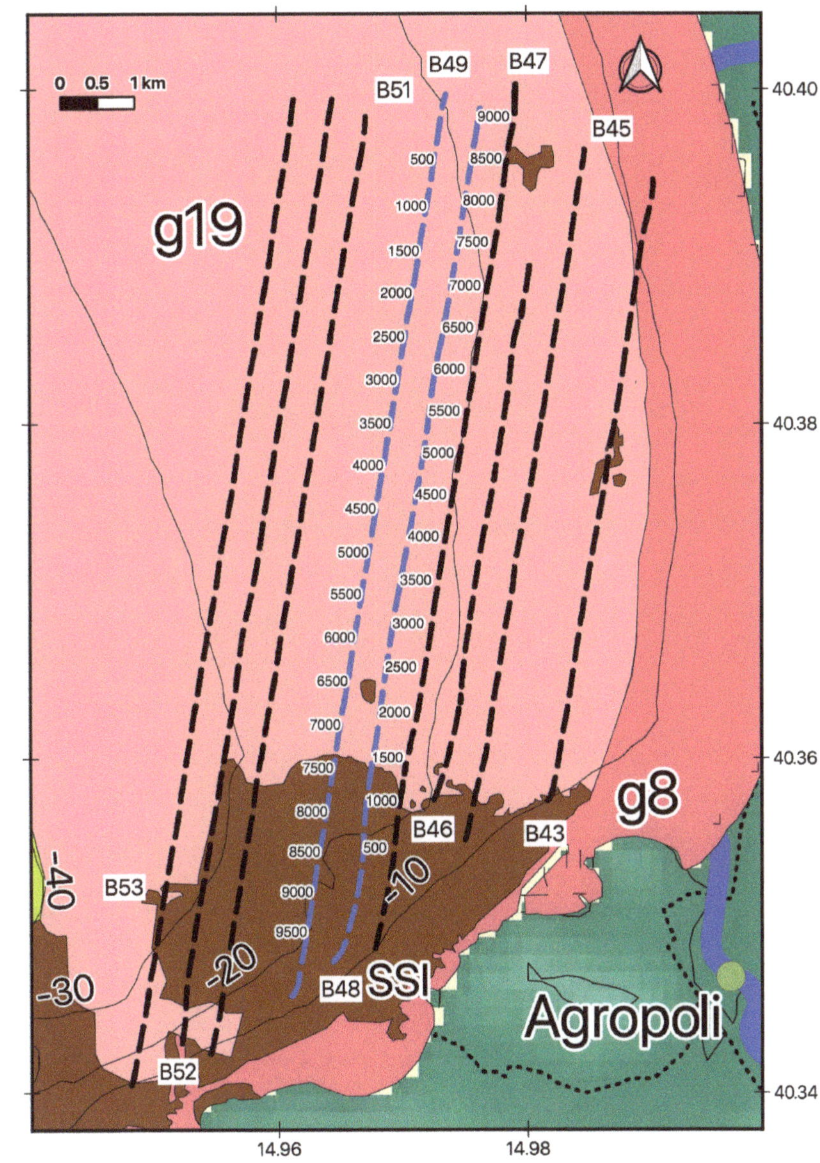

Figure 10. *Cont.*

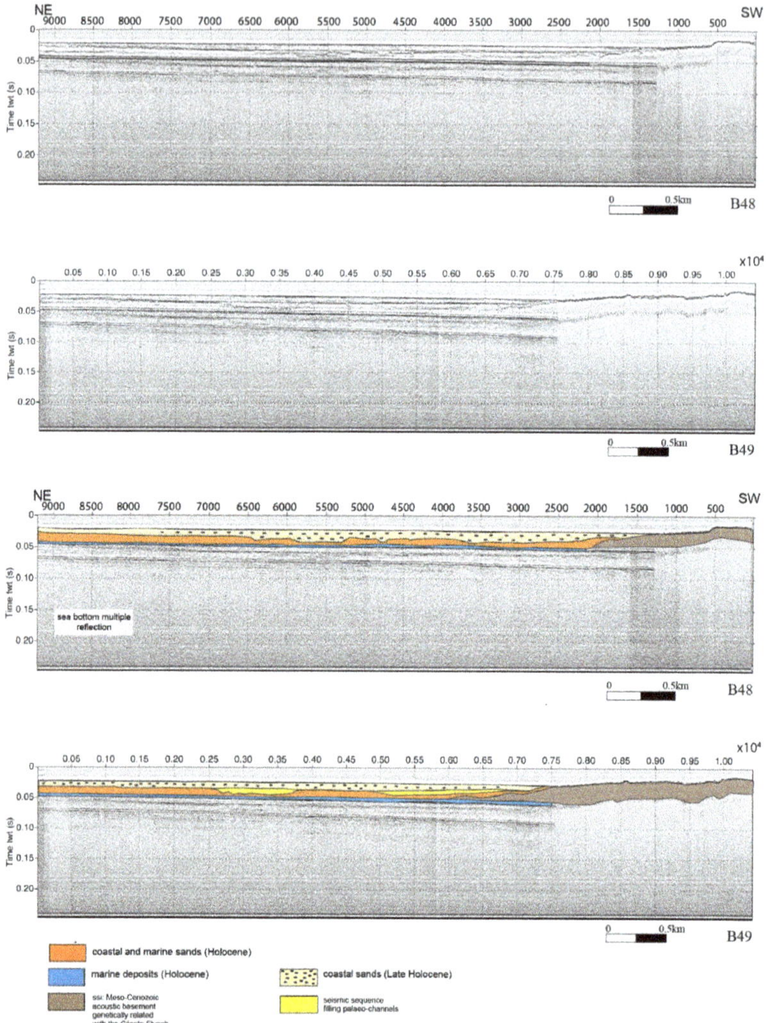

Figure 10. (**top**) Detailed location map; (**middle**) Sub-bottom chirp profiles B48 and B49 and the corresponding geologic interpretation (**bottom**).

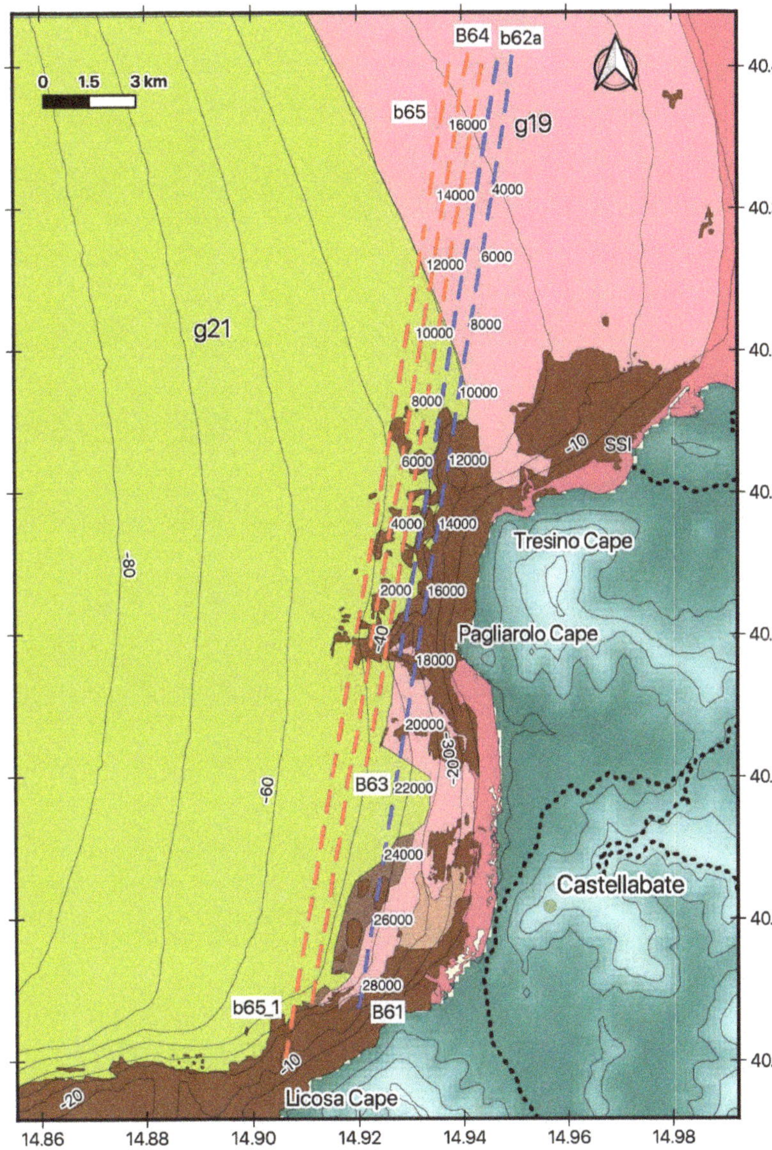

Figure 11. *Cont.*

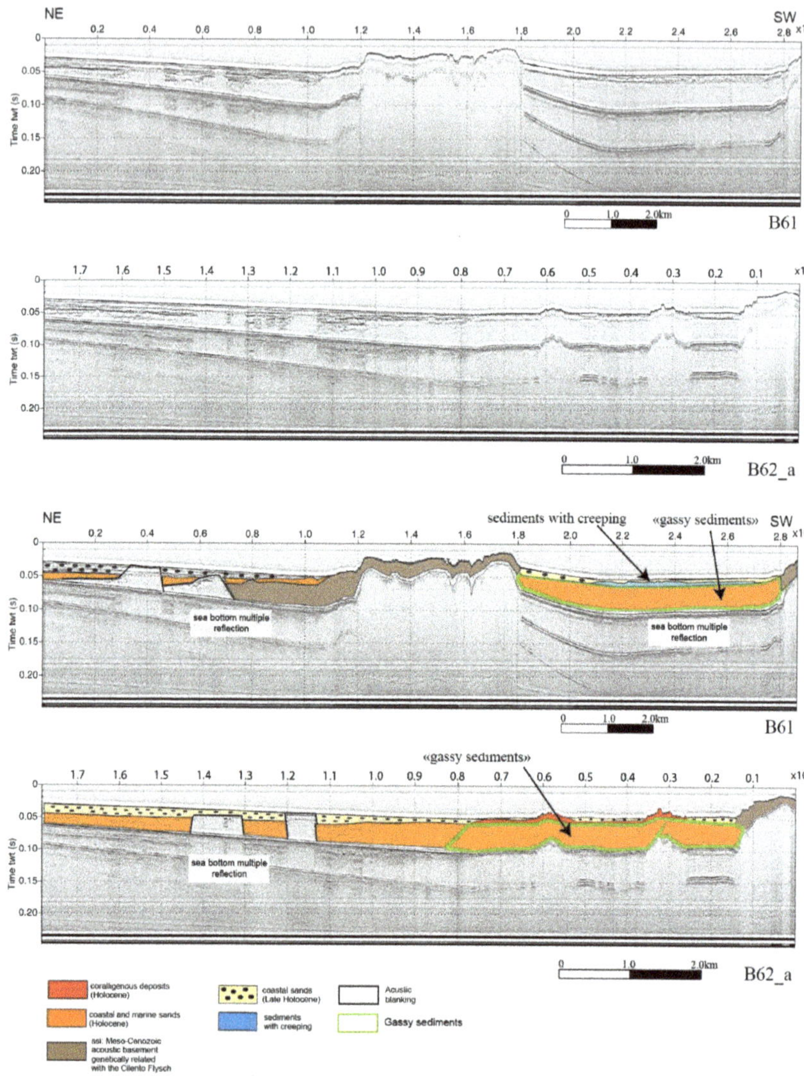

Figure 11. (**top**) Sketch location map of the Sub-bottom chirp profiles of area 3; (**middle**) Sub-bottom chirp profiles B61 and B62_a and the corresponding geologic interpretation (**bottom**).

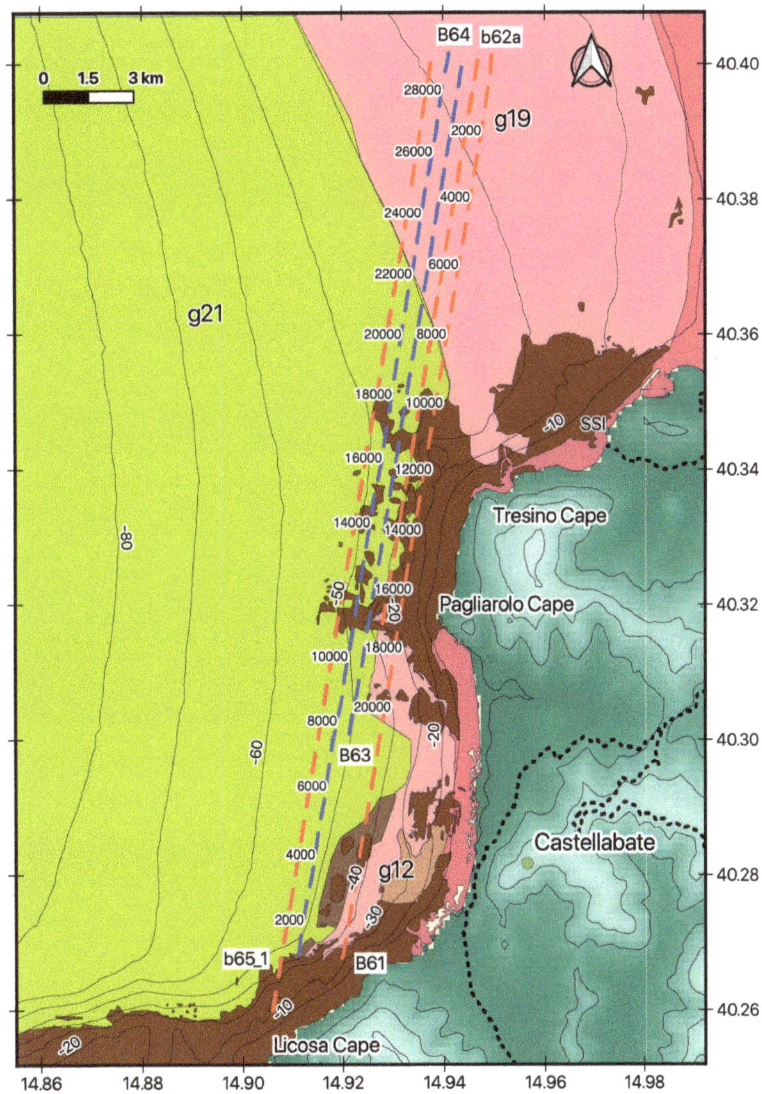

Figure 12. *Cont.*

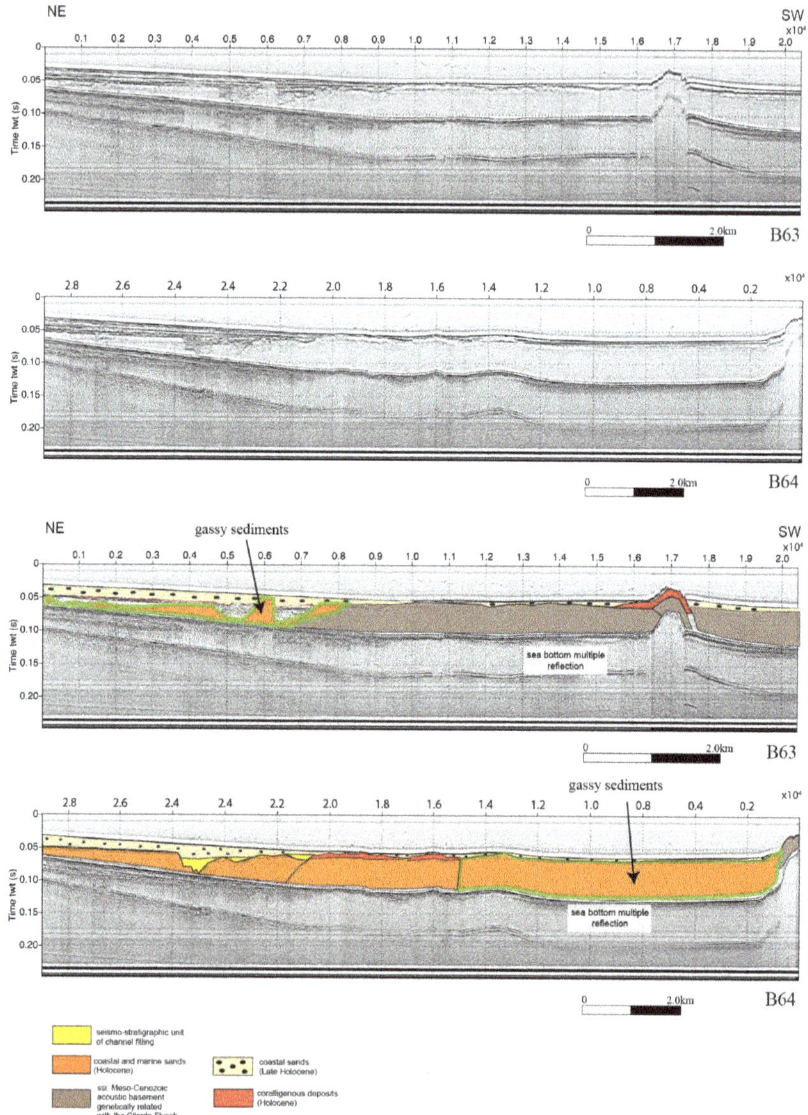

Figure 12. (**top**) Detailed location map; (**middle**) Sub-bottom chirp profiles B63 and B64 and the corresponding geologic interpretation (**bottom**).

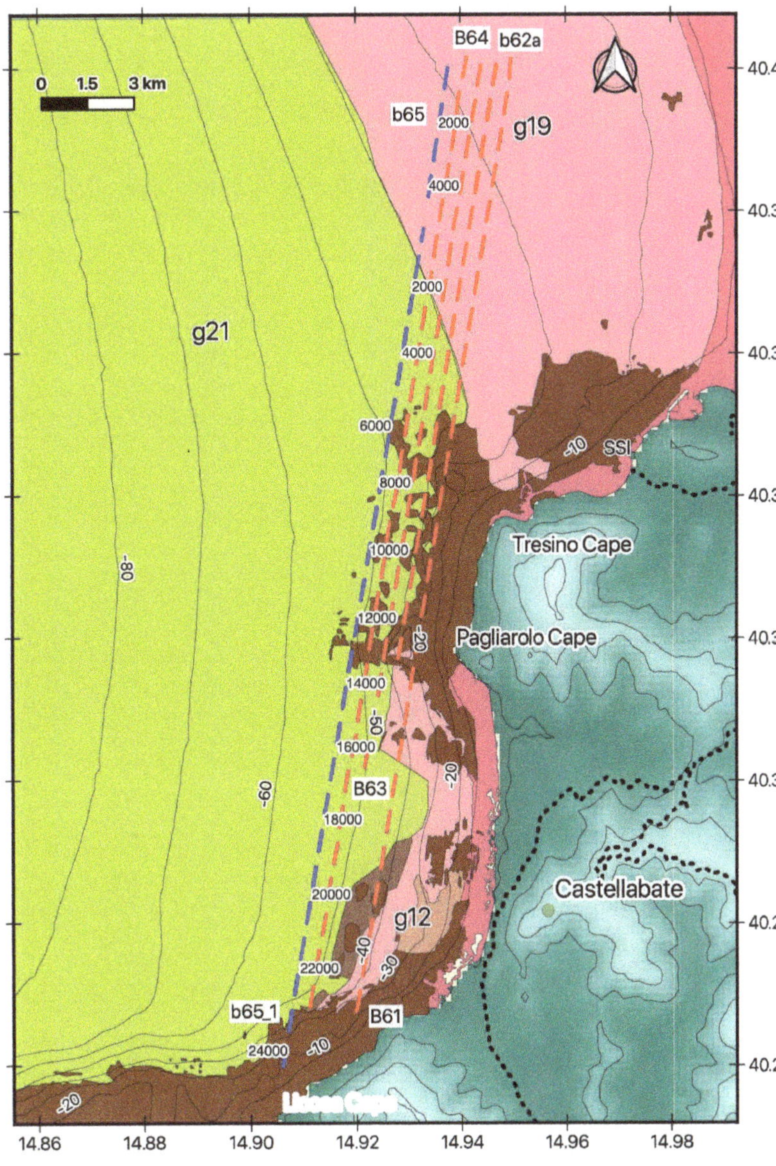

Figure 13. *Cont.*

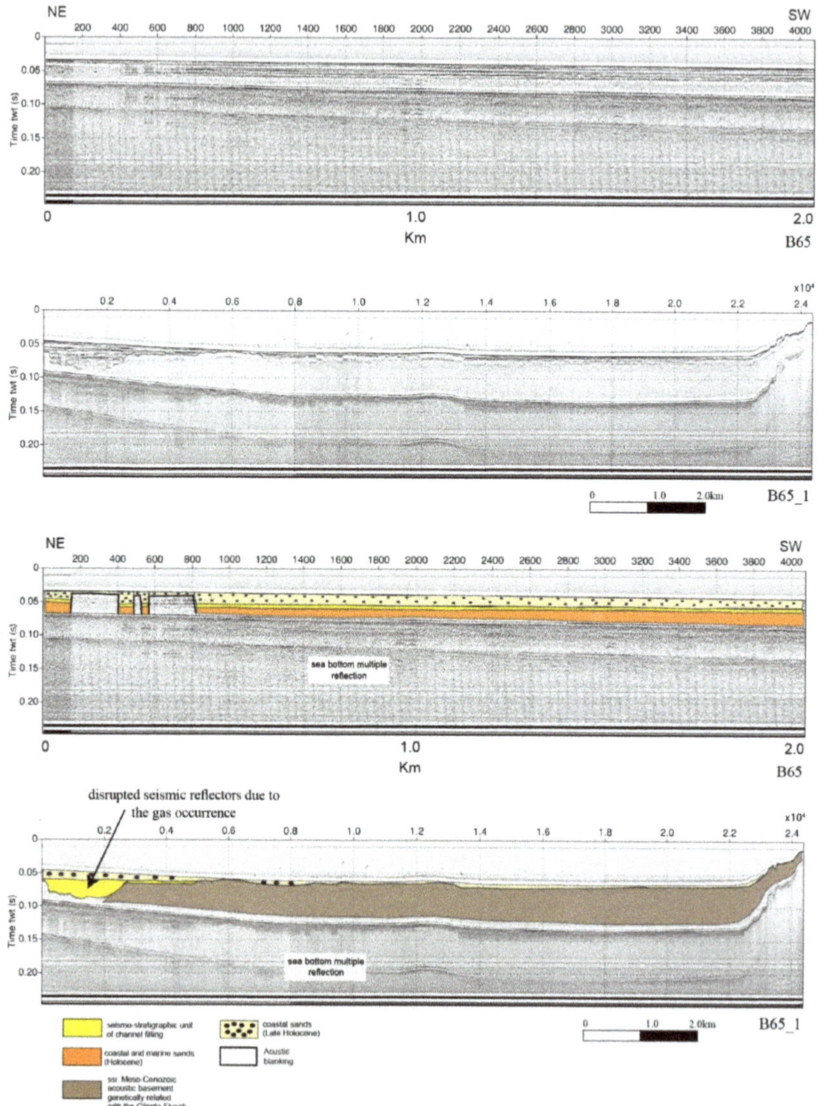

Figure 13. (**top**) Detailed location map; (**middle**) Sub-bottom chirp profiles B65 and B65_1 and the corresponding geologic interpretation (**bottom**).

5. Discussion

The interpretation of the seismoacoustic data has shown that significant gas impregnation affects the Holocene marine deposits, whose stratigraphic architecture is characterized by four or more main seismostratigraphic units overlying the ssi unit, which represents the acoustic basement (Figures 4–13). The number of seismostratigraphic units varies in the three areas distinguished in this paper (Table 1), namely area 1 (Figures 5–7), area 2 (Figures 8–10), and the area 3 (Figures 11–13).

Seismostratigraphic unit 1 is probably composed of sands and overlies the rocky acoustic basement through erosional unconformity. Significant variations in thickness have been observed in correspondence with this unit. These variations could have been controlled by the occurrence of a rough palaeomorphology, in which unit 1 was deposited. Moreover, strong erosional processes were probably active after the deposition of unit 1, as suggested by the erosional unconformity located at its top, triggering the individuation of the palaeochannels, in which seismostratigraphic 2 was deposited (Figure 4). These filled palaeochannels were observed (Figure 4), from which seismostratigraphic unit 2 is characterized by bidirectional onlaps. The thickness of this unit seems to be quite constant and is in the order of tens of meters. Unit 2 is characterized by acoustically transparent intervals alternating with parallel and continuous seismic reflectors and deposits in the palaeochannels (Figure 4). Seismostratigraphic unit 3 is distinguished by parallel and discontinuous seismic reflectors, interpreted as the first phase of the Holocene basin filling. Seismostratigraphic unit 4 is characterized by a seismic facies with parallel and continuous–discontinuous seismic reflectors, interpreted as the second phase of the Holocene basin filling (Figure 4).

The seismostratigraphic units of area 1 were recognized based on the geological interpretation of the corresponding seismic sections (Figures 5–7). The two main seismostratigraphic units were distinguished, including the ssi unit, which represents the rocky acoustic substratum that is genetically related to the Cilento Flysch, overlain by the recent sedimentary cover. Based on this stratigraphic architecture, it may be suggested that this area represents a structural high whose stratigraphic setting was controlled by the occurrence of the Cilento structural high onshore.

The seismostratigraphic units of area 2 are shown by the corresponding seismic sections (Figures 8–10). Five seismostratigraphic units were distinguished, including the ssi unit, representing the rocky acoustic substratum that is genetically related to the Cilento Flysch, the marine deposits (Holocene in age), the coastal and marine sands (Holocene in age), the seismic sequence filling the palaeochannels, which is correlated with seismostratigraphic unit 2 (previously described), and coastal sands (Holocene in age). Based on this stratigraphic setting, it may be supposed that this area represents a shallow coastal environment, hosting coastal and marine sandy sedimentation spanning the Holocene and Late Holocene. The seismic unit composed of coastal and marine sands, Holocenic in age, is interpreted as gassy sediments.

The seismostratigraphic units of area 3 were interpreted based on the corresponding seismic profiles (Figures 11–13). The six main seismostratigraphic units were distinguished based on seismic interpretation. They include the five seismostratigraphic units of the area 2, while the sixth seismic unit is represented by coralligenous deposits (Holocene in age). Based on this seismostratigraphic setting, this area can be interpreted as a coastal and marine area in which the coralligenous deposits locally grew.

The observed stratigraphic architecture suggests that the inner Cilento continental shelf from the Solofrone River mouth to Agropoli represents a depocenter of marine deposits, showing mainly as sandy grain in size. From Agropoli to the Licosa Cape, the continental shelf clearly represents a structural high. Both from a physiographic and depositional point of view, it is included in the domain of the Licosa Cape morphostructural high. This is highlighted by the wide outcrops of rocky acoustic basement at the sea bottom from Agropoli to the Tresino Cape and from the Tresino Cape to the Pagliarolo Cape. The occurrence of the rocky outcrops has favored the growth of coralligenous deposits, appearing as tabular seismic units cropping out at the sea bottom and overlying the rocky acoustic basement.

The seismic interpretation shows a significant improvement over the seismic data, as processed with the Seismic Unix software, with respect to the same seismic dataset previously processed with the Seisprho software [15]. In fact, the seismic lines processed with Seismic Unix have allowed for the identification of new seismostratigraphic units, including the "gassy sediments" and the coralligenous deposits, among others. These units were not previously distinguished within the seismic data processed using the Seisprho software [15] due to the lower resolution of the seismic data.

The gassy sediments were interpreted as occurring in a wide, acoustically transparent seismic unit (Figures 4–13), represented by coastal and marine sands that are Holocene in age. The characteristics of the gassy sediments have been deeply studied in several areas of the world [11,52–55]. Yuan et al. [52] studied the acoustic and physical characteristics of the gassy sediments in the western Irish Sea based on seismic interpretations of Sub-bottom and Uniboom seismic profiles. Broad areas of acoustic blanking have been highlighted, which were probably controlled by a high content of gas in the fine-grained Holocene sediments. This kind of situation can be compared with the gas-charged sediments occurring in the northern Cilento promontory. These gassy sediments have been basically highlighted in two large areas, with a few smaller pockets in between. Similar to our seismic sections and located in three areas distributed between the northern Cilento Promontory and the Licosa Cape, one of the most predominant acoustic features is represented by acoustic blanking. Seismic and well log evidence have suggested that the gas in the western Irish Sea is probably biogenic in origin and has accumulated in situ in the fine-grained sediments.

Coralligenous deposits were observed in several of the seismic sections in the third area (Figures 11 and 12), outcropping at the sea bottom as localized seismic units and in facies hetheropy with the recent coastal deposits. These kinds of deposits have been recently singled out offshore to the Licosa Cape promontory based on habitat mapping and multibeam bathymetric analysis [17]. It can be suggested that the seismic units observed in Sub-bottom chirp profiles correspond to several coalescent coralligenous bioconstructions (Figures 11 and 12).

In the Cilento offshore area, several acoustic anomalies were identified based on the seismic interpretations. The acoustic features are mainly represented by acoustic blanking, which was mainly detected in the first area, located offshore to the Licosa Cape promontory at water depths ranging between 30 and 90 m (Figure 1), by the shallow gas pockets, which were mainly observed in the second and in the third areas, and by the seismic units impregnated with gas, namely the "gassy" sediments, which were observed both in the second and in the third areas (Table 1). The second and the third areas are located in the northern Cilento promontory from the seaward prolongation of the Paestum Plain, proceeding southwards up to the Tresino Cape, and the third area is located on the northern Cilento promontory, starting from the offshore prolongation of the Paestum Plain up to the Licosa Cape promontory, at water depths ranging between 10 and 60 m (Figure 1).

A sketch map was constructed in order to show the distribution of the acoustic anomalies based on the seismic interpretations (Figure 14). This map was constructed in a GIS environment, mapping the shot points where the main acoustic anomalies were detected based on the seismic interpretations (acoustic blanking, shallow gas pockets, and gassy sediments, see also Table 1). Based on this map, significant acoustic anomalies occur in the depocenter located from the offshore of the Paestum Plain to the Tresino Cape (northern Cilento offshore) and on the outer shelf offshore area to the Licosa Cape. Scattered anomalies occur on the inner shelf offshore to the S. Maria di Castellabate Plain (Figure 14).

The origin of gas in the Cilento offshore area also needs to be discussed. Gas in shallow marine sediments has two main potential sources, including the biogenic gas produced by the bacterial degradation of organic matter at low temperatures and the thermogenic gas produced by the high-temperature degradation and cracking of organic compounds at considerable burial depths [6]. The geology of the study area does not suggest any possible deep thermogenic sources, and it can be hypothesized that the gas has accumulated in situ in the shallow organic-rich sediments. Further core and well log data are required to test this hypothesis since only one core is available from the literature in the study area (Licosa core) [15], but this core has mainly shown relict and palimpsest deposits and organogenic deposits [15]. It is possible that shallow marine and coastal deposits hosted a layer that was rich in organic matter, favoring the development of biogenic gas. In addition, the abundant presence of fine-grained muddy sediments and the high sedimentation rates (resulting in fast burial of the organic matter) also formed an ideal basis for the generation of biogenic gas, but further studies are necessary.

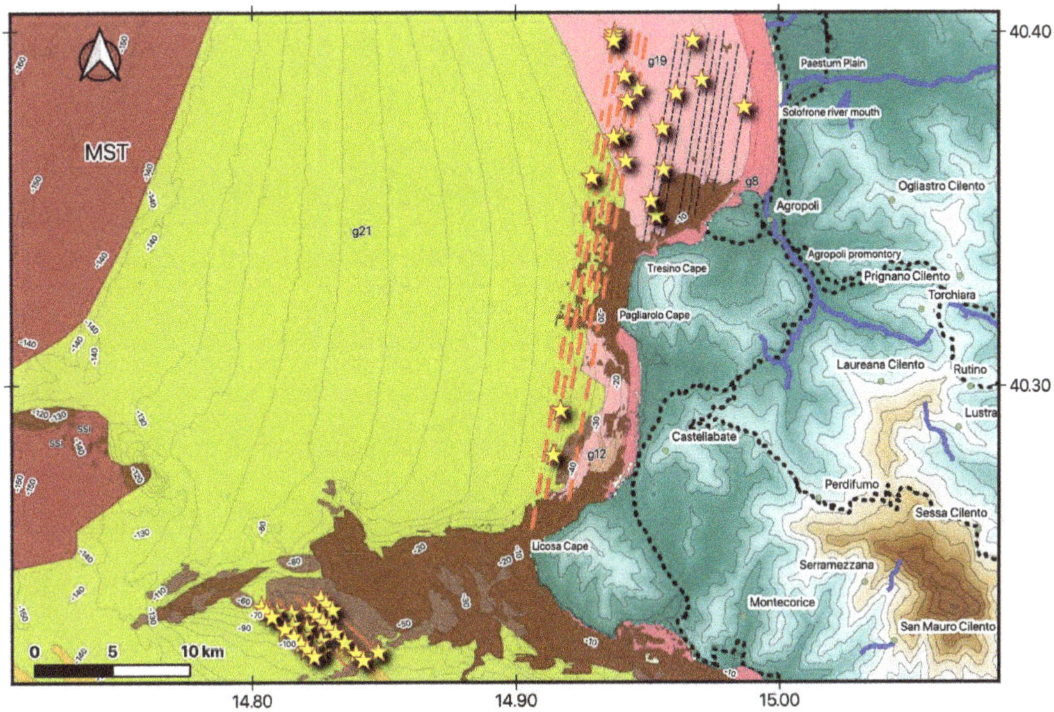

Figure 14. Sketch map showing the distribution of the acoustic anomalies (yellow stars) in the study area based on seismic interpretation.

Literature studies have shown several cases in which the pockmarks are not necessarily associated with the occurrence of shallow gas in the subsurface, as shown by the significant acoustic anomalies identified in the seismic profiles. Missiaen et al. [6] have highlighted that although acoustic blanking locally reaches the surface for the seismic profiles, the seabed morphology does not reveal any clear gas-related features. Hovland and Curzi [9] have shown that acoustic blanking, assumed to represent gas-charged sediments, is not associated with pockmarks but with a locally elevated seabed (mounds or ridges), as controlled by mud diapirism. Hovland and Judd [8] have stated that although acoustic blanking is widespread in the North Sea, including the pockmark areas, it is not ubiquitous, neither is it always found in close association with pockmarks. In the present work, the whitened band on the seismic profiles does not arrive at the seafloor; it arrives at the seabed only locally. This should mean that the gas does not escape into the water column that generates the pockmarks; these features are not present in the high-resolution bathymetry [17]. In that case, the pockmarks are not associated with the occurrence of gas in the Cilento offshore area, but buried gassy sediments have been detected. A number of studies on the subject under consideration using modern software exist. Among them, a significant study has been recently carried out by Bosikov and Klyuev [56], highlighting that the use of modern computer-aided methods, in particular the use of the Micromine software, is an important part of the integrated research for the determination of deposit prospects for various ores. This study has analyzed the prospects and estimation of reserves for open-pit and underground mining in the Berezkinskoye ore field, demonstrating the high potentiality of modern software, as it was in this paper.

The occurrence of gas in the Campania offshore area is not novel. Important structures linked to gas occurrence have been previously described in the Campania region, in particular, the offshore of the Volturno River mouth. Distinct sites of vertically focused flows, marked by columnar zones of amplitude that mask and disrupt reflectors, reached the seabed at the outer shelf and corresponded with the alignment of elongated pockmarks [57]. In that case, acoustic anomalies are associated with pockmarks.

6. Conclusions

High-resolution seismoacoustic data provide an appropriate technique for studying gas-charged sediments through the recognition of the diagnostic intrasedimentary features associated with the occurrence of gas in marine deposits.

The acoustic anomalies of the Cilento offshore area were studied based on the geological interpretation of Sub-bottom chirp profiles. Acoustic blanking, shallow gas pockets, and gassy sediments, occurring as wide, acoustically transparent seismic units, were identified in the study area. The occurrence of these features, however, suggests that the northern Cilento promontory represents an interesting area whose features need to be studied in more detail in subsequent research in order to relate these relationships with marine geohazards.

A processing sequence was constructed by using the Seismic Unix software and was applied to the seismic record in order to highlight acoustic anomalies, indicating the occurrence of Quaternary marine deposits that are impregnated with gas. A new processing sequence for high-resolution seismic data was applied to Sub-bottom seismic profiles located in the northern Cilento offshore area by using Seismic Unix software [14]. The workflow allowed us to plot the Sub-bottom profiles using Seismic Unix, therefore obtaining high-resolution seismic sections. This processing sequence has allowed for a significant improvement in the seismic profiles with respect to the previous versions obtained through Seisprho (Figure 4) [15].

A sketch morphobathymetric map was constructed through GIS (Geographic Information System) in order to show the morphostructural features of the study area. These features consist of the outcrops of the rocky acoustic basement (ssi unit), sandy ridges–representing relict and palimpsest deposits occurring both northwestwards and southwestwards of the Licosa Cape promontory–channels, and abrasion terraces. This map has been integrated into another map and constructed in the GIS environment, reporting the location of the main acoustic anomalies of the Cilento offshore area based on seismic interpretation.

In the Cilento offshore area, three areas were distinguished, typified by different acoustic features (Figure 1; Table 1). In the first area, located offshore from the Licosa Cape at between 30 and 90 m of water depth, the distinctive acoustic feature is represented by acoustic blanking controlled by the occurrence of gas [6,7]. The second area is in the northern Cilento promontory, from the seaward prolongation of the Paestum Plain to the Tresino Cape, and the distinctive acoustic features are represented by shallow gas pockets. The third area is located on the northern Cilento promontory, from the offshore prolongation of the Paestum Plain up to the Licosa Cape at water depths ranging between 10 and 60 m; the distinctive acoustic features are represented by shallow gas pockets and by the seismic units impregnated by gas ("gassy sediments").

The seismostratigraphic setting of the Cilento offshore area was reconstructed based on the geological interpretation of Sub-bottom chirp profiles, highlighting the different seismostratigraphic units of the three areas. While the first area represents a structural high, the second and the third areas represent shallow coastal and marine environments, where coralligenous deposits locally grew. In this geologic setting, and taking into account similar case histories, we may suppose that the gas is biogenic in origin, but further studies are necessary.

Author Contributions: Conceptualization, G.A. and M.C.; methodology, M.C.; software, M.C.; formal analysis, G.A.; investigation, G.A.; data curation, M.C.; writing—original draft preparation, G.A.; writing—review and editing, G.A. All authors have read and agreed to the published version of the manuscript.

Funding: This research received no external funding.

Institutional Review Board Statement: Not applicable.

Informed Consent Statement: Not applicable.

Data Availability Statement: Not applicable.

Conflicts of Interest: The authors declare no conflict of interest.

References

1. Carlson, P.R.; Golan-Bac, M.; Karl, H.A.; Kvenvolden, K.A. Seismic and geochemical evidence for shallow gas in sediment of Navarin continental margin, Bering Sea. *AAPG Bull.* **1985**, *69*, 422–436.
2. Judd, A.G.; Hovland, M. The evidence of shallow gas in marine sediments. *Continent. Shelf Res.* **1992**, *12*, 1081–1095. [CrossRef]
3. Lee, S.H.; Chough, S.K. Distribution and origin of shallow gas in deep-sea sediments of the Ulleung Basin, East Sea (Sea of Japan). *Geo-Mar. Lett.* **2003**, *22*, 204–209. [CrossRef]
4. Hovland, M. The Geomorphology and Nature of Seabed Seepage Processes. In *Bathymetry and Its Applications*; Blondel, P., Ed.; IntechOpen: Rjeka, Croatia, 2012; Chapter 4; pp. 79–104.
5. Heggland, R. Gas seepage as an indicator of deeper prospective reservoirs. A study based on exploration 3D seismic data. *Mar. Petrol. Geol.* **1998**, *15*, 1–9. [CrossRef]
6. Missiaen, T.; Murphy, S.; Loncke, L.; Henriet, J.P. Very high-resolution seismic mapping of shallow gas in the Belgian coastal zone. *Cont. Shelf Res.* **2002**, *22*, 2291–2301. [CrossRef]
7. Andreassen, K.; Espen, G.; Nilssen, E.G.; Ødegaard, C.M. Analysis of shallow gas and fluid migration within the Plio-Pleistocene sedimentary succession of the SW Barents Sea continental margin using 3D seismic data. *Geo-Mar. Lett.* **2007**, *27*, 155–171. [CrossRef]
8. Hovland, M.; Judd, A.G. *Seabed Pockmarks and Seepages: Impact on Geology, Biology and the Marine Environment*; Graham and Trotman: London, UK, 1988; p. 293.
9. Hovland, M.; Curzi, P.V. Gas seepage and assumed mud diapirism in the Italian central Adriatic Sea. *Mar. Petrol. Geol.* **1989**, *6*, 161–169. [CrossRef]
10. Geletti, R.; Del Ben, A.; Busetti, M.; Ramella, R.; Volpi, V. Gas seeps linked to salt structures in the Central Adriatic Sea. *Basin Res.* **2008**, *20*, 473–487. [CrossRef]
11. Spatola, D.; Micallef, A.; Sulli, A.; Basilone, L.; Basilone, G. Evidence of active fluid seepage (AFS) in the southern region of the central Mediterranean sea. *Measurement* **2018**, *128*, 247–253. [CrossRef]
12. Ceramicola, S.; Dupré, S.; Somoza, L.; Woodside, J. Cold seep systems. In *Submarine Geomorphology*; Micallef, A., Krastel, S., Savini, A., Eds.; Springer International Publishing: Cham, Switzerland, 2018; pp. 367–387.
13. Kopp, H.; Chiocci, F.L.; Berndt, C.; Çağatay, M.N.; Ferreira, T.; Fortes, C.J.E.M.; Gràcia, E.; González Vega, A.; Kopf, A.J.; Sørensen, M.B.; et al. Marine geohazards: Safeguarding society and the Blue Economy from a hidden threat. In *Position Paper 26 of the European Marine Board*; Muñiz Piniella, A., Kellett, P., van den Brand, R., Alexander, B., Rodríguez Perez, A., Van Elslander, J., Heymans, J.J., Eds.; European Marine Board: Ostend, Belgium, 2021; p. 100.
14. Colorado School of Mines. Seismic Unix. 2000. Available online: https://wiki.seismic-unix.org (accessed on 30 October 2022).
15. Aiello, G.; Caccavale, M. The Depositional Environments in the Cilento Offshore (Southern Tyrrhenian Sea, Italy) Based on Marine Geological Data. *J. Mar. Sci. Eng.* **2021**, *9*, 1083. [CrossRef]
16. Gasperini, L.; Stanghellini, G. SEISPRHO: An interactive computer program for processing and interpretation of high-resolution seismic reflection profiles. *Comput. Geosci.* **2009**, *35*, 1497–1507. [CrossRef]
17. Violante, C. Computer-Aided Geomorphic Seabed Classification and Habitat Mapping at Punta Licosa MPA, Southern Italy. In *ICCSA 2020: Computational Science and Its Applications*; Gervasi, O., Morgante, B., Misra, S., Garau, C., Blecic, I., Tanion, D., Apduhan, B.O., Rocha, A., Tarantino, E., Torre, C., et al., Eds.; Springer Nature: Bern, Switzerland, 2020; pp. 681–695.
18. Ortolani, F.; Aprile, F. Nuovi dati sulla struttura profonda della Piana Campana a sud-est del fiume Volturno. *Boll. Soc. Geol. Ital.* **1978**, *97*, 591–608.
19. Fabbri, A.; Gallignani, P.; Zitellini, N. Geologic evolution of the peri-Tyrrhenian sedimentary basins. In *Sedimentary Basins of Mediterranean Margins*; Wezel, F.C., Ed.; Tecnoprint: Bologna, Italy, 1981; pp. 101–126.
20. Bartole, R. Tectonic structure of the Latian-Campanian shelf (Tyrrhenian Sea). *Boll. Oceanol. Teor. Appl.* **1984**, *2*, 197–230.
21. Bartole, R.; Savelli, D.; Tramontana, M.; Wezel, F.C. Structural and sedimentary features in the Tyrrhenian margin off Campania, Southern Italy. *Mar. Geol.* **1983**, *55*, 163–180. [CrossRef]
22. Finetti, I.; Morelli, C. Esplorazione sismica per riflessione dei Golfi di Napoli e Pozzuoli. *Boll. Geof. Teor. Appl.* **1973**, *16*, 175–222.
23. Trincardi, F.; Zitellini, N. The rifting of the Tyrrhenian Basin. *Geo-Mar. Lett.* **1987**, *7*, 1–6. [CrossRef]
24. Mariani, M.; Prato, R. I bacini neogenici del margine tirrenico: Approccio sismico-stratigrafico. *Mem. Soc. Geol. Ital.* **1988**, *41*, 519–531.
25. Aiello, G.; Marsella, E.; Sacchi, M. Quaternary structural evolution of Terracina and Gaeta basins (Eastern Tyrrhenian margin, Italy). *Rend. Lincei* **2000**, *11*, 41–58. [CrossRef]

26. Aiello, G.; Marsella, E.; Cicchella, A.G.; Di Fiore, V. New insights on morpho-structures and seismic stratigraphy along the Campania continental margin (Southern Italy) based on deep multichannel seismic profiles. *Rend. Lincei* **2011**, *22*, 349–373. [CrossRef]
27. Aiello, G.; Cicchella, A.G.; Di Fiore, V.; Marsella, E. New seismo- stratigraphic data of the Volturno Basin (northern Campania, Tyrrhenian margin, southern Italy): Implications for tectono-stratigraphy of the Campania and Latium sedimentary basins. *Ann. Geophys.* **2011**, *54*, 265–283.
28. Aiello, G.; Iorio, M.; Molisso, F.; Sacchi, M. Integrated morpho-bathymetric, seismic stratigraphic and sedimentological data on the Dohrn canyon (Naples Bay, Southern Tyrrhenian Sea): Relationships with volcanism and tectonics. *Geosciences* **2020**, *10*, 319. [CrossRef]
29. Conti, A.; Bigi, S.; Cuffaro, M.; Doglioni, C.; Scrocca, D.; Muccini, F.; Cocchi, L.; Ligi, M.; Bortoluzzi, G. Transfer zones in an oblique back-arc basin setting: Insights from the Latium-Campania segmented margin (Tyrrhenian Sea). *Tectonics* **2017**, *36*, 78–107. [CrossRef]
30. Zitellini, N.; Ranero, C.R.; Loreto, M.F.; Ligi, M.; Pastore, M.; D'Oriano, F.; Sallares, V.; Grevemeyer, I.; Moeller, S.; Prada, M. Recent inversion of the Tyrrhenian Basin. *Geology* **2020**, *48*, 123–127. [CrossRef]
31. Doglioni, C. A proposal for the kinematic modelling of the Tyrrhenian-Apennines system. *Terra Nova* **1991**, *3*, 423–432. [CrossRef]
32. Spadini, G.; Wezel, F.C. Structural evolution of the "41st parallel zone": Tyrrhenian Sea. *Terra Nova* **1994**, *6*, 552–562. [CrossRef]
33. Carminati, E.; Wortel, M.J.R.; Spakman, W.; Sabadini, R. The role of slab detachment processes in the opening of the western central Mediterranean basins: Some geological and geophysical evidence. *Earth Planet. Sci. Lett.* **1998**, *160*, 651–665. [CrossRef]
34. Bruno, P.P.G.; Di Fiore, V.; Ventura, G. Seismic study of the '41st Parallel' Fault System offshore the Campanian–Latial continental margin, Italy. *Tectonophysics* **2000**, *324*, 37–55. [CrossRef]
35. Faccenna, C.; Becker, T.W.; Lucente, F.P.; Jolivet, L.; Rossetti, F. History of subduction and back-arc extension in the Central Mediterranean. *Geophys. J. Intern.* **2001**, *145*, 809–820. [CrossRef]
36. Billi, A.; Bosi, V.; De Meo, A. Caratterizzazione strutturale del rilievo del Monte Massico nell'ambito dell'evoluzione quaternaria delle depressioni costiere dei fiumi Garigliano e Volturno. (*Campania settentrionale*). *IL Quaternario* **1997**, *10*, 15–26.
37. Santangelo, N.; Romano, P.; Ascione, A.; Russo Ermolli, E. Quaternary evolution of the Southern Apennines coastal plains: A review. *Geol. Carpathica* **2017**, *68*, 43–56. [CrossRef]
38. Aiello, G.; Cicchella, A.G. Dati sismostratigrafici sul margine continentale della Campania tra Ischia, Capri ed il Bacino del Volturno (Tirreno meridionale, Italia) in base al processing sismico ed all'interpretazione geologica di profili sismici a riflessione multicanale. *Quad. Geofis.* **2019**, *149*, 1–52. (In Italian)
39. D'Argenio, B.; Pescatore, T.; Scandone, P. *Schema Geologico dell'Appennino Meridionale*; Verlag: Berlin, Germany; Volume 183, pp. 49–72.
40. Mostardini, F.; Merlini, S. Appennino centro-meridionale. Sezioni Geologiche e Proposta di Modello Strutturale. *Mem. Soc. Geol. Ital.* **1986**, *35*, 177–202.
41. Vitale, S.; Ciarcia, S. Tectono-stratigraphic setting of the Campania region (southern Italy). *J. Maps* **2018**, *14*, 9–21. [CrossRef]
42. Amore, F.O.; Bonardi, G.; Ciampo, G.; De Capoa, P.; Perrone, V.; Sgrosso, I. Relazioni tra flysch interni e domini appenninici: Reinterpretazione delle formazioni di Pollica, S. Mauro e Albidona e l'evoluzione infra-medio-miocenica delle zone esterne sud appenniniche. *Mem. Soc. Geol. Ital.* **1988**, *41*, 285–297.
43. Bonardi, G.; Amore, F.O.; Ciampo, G.; De Capoa, P.; Miconnet, P.; Perrone, V. Il Complesso Liguride *Auct.*: Stato delle conoscenze e problemi aperti sulla sua evoluzione pre-appenninica ed i suoi rapporti con l'Arco Calabro. *Mem. Soc. Geol. Ital.* **1988**, *41*, 7–35.
44. Zuppetta, A.; Mazzoli, S. Deformation history of a synorogenic sedimentary wedge, northern Cilento area, southern Apennines thrust and fold belt, Italy. *GSA Bull.* **1997**, *109*, 698–708. [CrossRef]
45. Cammarosano, A.; Cavuoto, G.; Martelli, L.; Nardi, G.; Toccaceli, R.M.; Valente, A. Il Progetto CARG nell'area silentina (area interna Appennino meridionale): Il nuovo assetto stratigrafico-strutturale derivante dal rilevamento dei fogli 503, 502, e 519 (Vallo della Lucania, Agropoli e Capo Palinuro). *Rend. Online Soc. Geol. Ital.* **2011**, *12*, 19–21.
46. Aiello, G.; Di Fiore, V.; Marsella, E.; D'Isanto, C. Stratigrafia sismica e morfobatimetria della Valle di Salerno. In Proceedings of the 26th National Congress GNGTS (Gruppo Nazionale di Geofisica della Terra Solida), Rome, Italy, 24–27 November 2007; Extended Abstract. pp. 495–498.
47. Aiello, G.; Marsella, E.; Di Fiore, V.; D'Isanto, C. Stratigraphic and structural styles of half-graben offshore basins: Multichannel seismic and Multibeam morpho-bathymetric evidences on the Salerno Valley (Southern Campania continental margin, Italy). *Quad. Geofis.* **2009**, *77*, 1–34. (In English)
48. Aiello, G.; Sacchi, M. New morpho-bathymetric data on marine hazard in the offshore of Gulf of Naples (Southern Italy). *Nat. Hazards* **2022**, *111*, 2881–2908. [CrossRef]
49. Vail, P.R.; Mitchum, R.M.; Thompson, S. Seismic stratigraphy and global changes of sea level, part IV: Global cycles of relative changes of sea level. In *Seismic Stratigraphy—Applications to Hydrocarbon Exploration*; Memoir 26; Payton, C.E., Ed.; American Association of Petroleum Geologists: Tulsa, OK, USA, 1977; pp. 83–98.
50. Mitchum, R.M.; Vail, P.R.; Thompson, S. Seismic stratigraphy and global changes of sea-level, part 2: The depositional sequence as a basic unit for stratigraphic analysis. In *Seismic Stratigraphy—Applications to Hydrocarbon Exploration*; Memoir 26; Payton, C.E., Ed.; American Association of Petroleum Geologists: Tulsa, OK, USA, 1977; pp. 53–62.

51. Anstey, N.A. *Simple Seismics: For the Petroleum Geologist, the Reservoir Engineer, the Well-Log Analyst, the Processing Technician, and the Man in the Field*; Heathfield: East Sussex, England, 1982; p. 168.
52. Yuan, F.; Bennell, J.D.; Davis, A.M. Acoustic and physical characteristics of gassy sediments in the western Irish sea. *Continent. Shelf Res.* **1992**, *12*, 1121–1134. [CrossRef]
53. Fleischer, P.; Orsi, T.H.; Richardson, M.D.; Anderson, A.L. Distribution of free gas in marine sediments: A global overview. *Geo-Mar. Lett.* **2001**, *21*, 103–122.
54. Gorgas, T.J.; Kim, G.Y.; Park, S.C.; Wilkens, R.H.; Kim, D.C.; Lee, G.H.; Seo, Y.K. Evidence for gassy sediments on the inner shelf of SE Korea from geoacoustic properties. *Continent. Shelf Res.* **2003**, *23*, 821–834. [CrossRef]
55. Leighton, T.G.; Robb, G.B.N. Preliminary mapping of void fractions and sound speeds in gassy marine sediments from subbottom profiles. *J. Acoust. Soc. Am.* **2008**, *124*, EL313–EL320. [CrossRef] [PubMed]
56. Bosikov, I.I.; Klyuev, R.V. Assessment of Berezkinskoye ore field prospectivity using Micromine software. *Geol. Miner. Depos.* **2022**, *7*, 192–202. [CrossRef]
57. Misuraca, M.; Budillon, F.; Tonielli, R.; Di Martino, G.; Innangi, S.; Ferraro, L. Coastal Evolution, Hydrothermal Migration Pathways and Soft Deformation along the Campania Continental Shelf (Southern Tyrrhenian Sea): Insights from High-Resolution Seismic Profiles. *Geosciences* **2018**, *21*, 121. [CrossRef]

Journal of
Marine Science and Engineering

MDPI

Article

Coupling Relationship of Geomorphic Evolution and Marine Hydrodynamics in the Stage-Specific Development of Urban Bays: A Modelling Case Study in Quanzhou Bay (1954–2017), China

Xianbiao Xiao [1,2,3], Yunhai Li [1,2], Junjian Tang [1], Fusheng Luo [1], Fangfang Shu [1], Liang Wang [1], Jia He [1], Xiaochun Zou [1], Wenqi Chi [1], Yuting Lin [1] and Binxin Zheng [1,*]

1 Laboratory for Ocean & Coast Geology, Third Institute of Oceanography, Ministry of Natural Resources, Xiamen 361005, China
2 Laboratory for Marine Geology, Qingdao National Laboratory for Marine Science and Technology, Qingdao 266237, China
3 Electric Power Research Institute of State Grid Fujian Electric Power Co., Ltd., Fuzhou 350007, China
* Correspondence: zhengbinxin@tio.org.cn

Abstract: With the development of social economy and human activities, the geomorphology and hydrodynamic conditions of coasts have been dramatically changed, causing serious environmental pollution and resource depletion. Taking Quanzhou Bay as an example, this study combined geomorphologic change with a hydrodynamic model to simulate the change in tidal currents in different periods. The results show a change in the coastline was the main cause of hydrodynamic change during the industrialization reform. During the past 70 years, the tidal prism decreased year by year, and the average velocity of the tidal current in the channel decreased by 33.7% and 30.8% at flood and ebb tide, respectively. In the early stages of industrialization, reclamation land was used in a single way. The tidal prism decreased by 22.2% and 29.8% in the spring and neap tide, respectively. In the middle and later stages, the tidal current velocity increased, and reclamation land was used in a variety of ways. In modern society, the reclamation land-use type was unitary. Based on this research, we show the influence of human activities on the evolution of the bay's geomorphology and provide suggestions for the management of the bay.

Keywords: human activities; geomorphology; coupling relationship; hydrodynamics; bays urbanization

Citation: Xiao, X.; Li, Y.; Tang, J.; Luo, F.; Shu, F.; Wang, L.; He, J.; Zou, X.; Chi, W.; Lin, Y.; et al. Coupling Relationship of Geomorphic Evolution and Marine Hydrodynamics in the Stage-Specific Development of Urban Bays: A Modelling Case Study in Quanzhou Bay (1954–2017), China. *J. Mar. Sci. Eng.* **2022**, *10*, 1677. https://doi.org/10.3390/jmse10111677

Academic Editor: Gemma Aiello

Received: 18 October 2022
Accepted: 1 November 2022
Published: 7 November 2022

Publisher's Note: MDPI stays neutral with regard to jurisdictional claims in published maps and institutional affiliations.

Copyright: © 2022 by the authors. Licensee MDPI, Basel, Switzerland. This article is an open access article distributed under the terms and conditions of the Creative Commons Attribution (CC BY) license (https://creativecommons.org/licenses/by/4.0/).

1. Introduction

As a link between the sea and the land with abundant natural resources, the advantageous geographical position makes a bay play an important role in the national economic development and international cultural exchange [1]. With the increase in human activities in the past century, the main factors controlling the environmental changes in the bay have gradually changed from natural factors to human factors [2]. As the earth entered the Anthropocene, the geomorphologic evolution process of the bay has also undergone changes different from the natural conditions. Therefore, the study of the impact of human activities on geomorphologic evolution is beneficial to the sustainable development of the marine environment in the urban bay [3].

With the development of the economy, urbanization in the bay has changed the topography and hydrodynamic environment [4]. Continual human activities lead to the gradual seaward advance of the bay coastline, with decreasing bay area [5]. Previous research showed that the area of bays around the East China Sea has shrunk by 11.23% over the past 20 years [6]. Similarly, between 1987 and 2017, the areas of San Francisco Bay, the New York Bay, the Tokyo Bay, and the Guangdong–Hong Kong–Macao Bay were reduced to varying degrees [7]. These human activities (such as reclamation, channel dredging, coastal aquaculture, etc.) not only changed the topography, but also affected the evolution of the hydrodynamic environment in the bay [8]. According to the present study of semi-enclosed bays, reclamation weakens the hydrodynamic forces and results in prolonged stays of the water at the same location [9,10]. Coastal aquaculture and port engineering have changed the local tidal flow and water structure in the bay and estuary areas [11]. This has led to increased levels of pollutants in the bay, which destroyed natural wetlands, leading to ecological deterioration around the bay [12]. Coastal aquaculture and port engineering have changed the tidal currents and water structure [11]. This has led to increased levels of pollutants in the bay, leading to ecological deterioration around the bay [12].

The hydrodynamic conditions at the estuary of the bay are closely related to the sedimentary environment of the bay, which would change the transport mode of the terrestrial materials and influence the evolution of the topography in the bay [13]. In a tidal-controlled bay with large terrestrial material flux, the sediment moves back and forth with the flood and ebb tide, and fine-grained sediments are prone to resuspension and then flocculation and deposition [14]. The reduction of most of the tidal prisms in the bay alters the hydrodynamic conditions of the bay, which, in turn, causes the movement of suspended particles in the water [10]. In the future, the transformation of the surrounding environment of the bay will be increasingly complex, and the contradiction between economic development and environmental evolution will be further highlighted [15]. Therefore, it is necessary to study the relationship between social development and the evolution of a bay's geomorphology.

At present, research regarding hydrodynamics mainly focuses on the hydrodynamic changes over short time scales and the hydrodynamic prediction after construction [7,10,11]. However, there is still a lack of research on the hydrodynamic change over the long time scale and the mechanism of sedimentary dynamics in the process of geomorphology evolution in a bay. In the past 70 years, Quanzhou has undergone the whole process of industrialization, with many periods of reclamation, resulting in significant changes in the topography of the bay [16]. There is a certain coupling relationship between the changes in the geomorphology and hydrodynamics and the development of urbanization in a bay, and this coupling relationship is obviously manifested in Quanzhou Bay, showing stage-specific development of urban bays. Therefore, Quanzhou Bay is a natural laboratory to investigate the hydrodynamic change process of a bay under the influence of human activities, and can be used as a typical area to study the relationship between the industrialization development and hydrodynamic evolution of a bay.

In this study, the coupling relationship between the geomorphologic environment of an urban bay and the hydrodynamics under the influence of human activities were analyzed by hydrodynamic model simulation. The influence of the construction of the reclamation expansion of the urban bay on the hydrodynamic and sedimentary dynamics of the bay was explored.

2. Materials and Methods

2.1. Basic Data

2.1.1. Water Depth and Topographic Data

In this study, six large-scale marine maps (1954–2017) of Quanzhou Bay were selected as the data source of the bathymetry and coastline (Figure 1, Table 1); the lowest theoretical datum was used for bathymetric data [17]. All marine map data were converted to the WGS84 coordinate system and Mercator projection (Figure 2, Table 1). Due to the dredging in Quanzhou Bay, there were some differences between the actual channel topography and the topographic data of the marine map. The accuracy and integrality of the bay's geomorphological data can be ensured by adding the actual channel data. The survey time of the channel maintenance bathymetric data was 2013, 2015, and 2017, and the main survey areas were Houzhu, Dazhui, and Neigang channels (Table 2). The survey depth datum adopted the local theoretical lowest tide level (Huanghai Vertical Datum 1956).

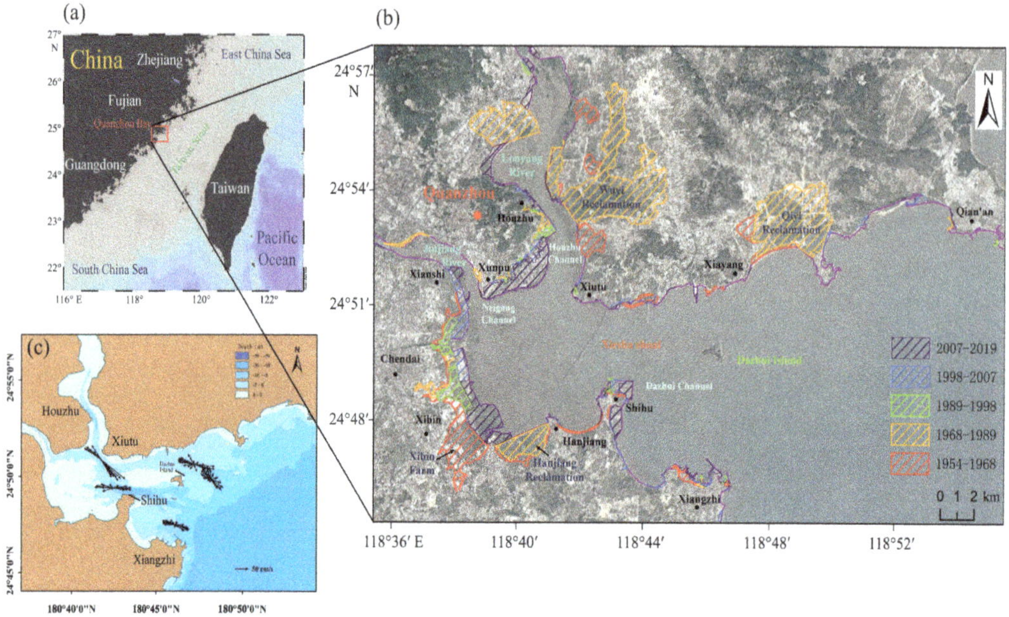

Figure 1. Overview of the study area. (**a**) The location of the study area. (**b**) The basic scenario of the study area, including cities, islands, channels, rivers, and reclamation areas at different stages. (**c**) Tidal currents during the Quanzhou Bay tidal cycle [18].

Table 1. Status of the historical marine map data.

Marine Map Name	Marine Map Number	Scale	Coordinate System	Time of Publication
Nanri Island to Quanzhou Bay	10–56	1:100,000	-	April 1973
Quanzhou Bay	5616	1:50,000	Beijing54	September 1975
Quanzhou Bay	14,181	1:35,000	Beijing54	April 2000
Quanzhou Bay	14,181	1:35,000	Beijing54	April 2003
Quanzhou Bay	14,181	1:35,000	WGS84	May 2008
Quanzhou Bay	14,181	1:35,000	CGCS2000	July 2019

The data comes from previous studies [16].

Table 2. Channel maintenance time and area information.

Measure Time	Measuring Area		
	Neigang Channel	Dazhui Channel	Houzhu Channel
2013	Up to the junction of Neigang channel and Houzhu channel, down to the downstream of Quanzhou Bridge. The measuring area is about 4.2 km^2.	Outside to the southern part of the Dazhui island, inside to the junction of the Shihu channel and the Dazhui island channel. The measuring area is about 3.59 km^2.	From the Dazhui island to the Houzhu Port.
2015	Up to the Quanzhou Bridge and down to the junction of Neigang channel and Houzhu channel.	From the anchorage outside Quanzhou Bay, follow the Dazhui island channel to the ports of Shihu and Xiutu. The measuring length was 11.04 km.	It is about 8.8 km from Shihu to Houzhu.
2017	/	From the anchorage outside Quanzhou Bay, follow the Dazhui island channel to the ports of Shihu and Xiutu. The measuring length was 11.04 km.	It is about 8.8 km from Shihu to Houzhu.

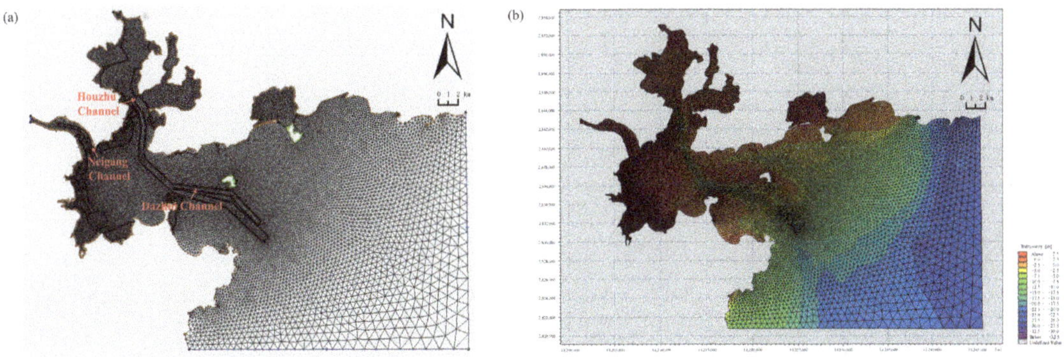

Figure 2. Mesh generation of the model. (**a**) Grid division of study area; (**b**) Bathymetric topography of the study area in 1954.

The shoals of Quanzhou Bay are mainly distributed from Xiutu to the west of Shihu, outside the Jinjiang River estuary, and on both sides of Luoyang River estuary. The tidal area is big, and the tidal channel is tortuous. The beaches are mainly distributed along the coasts on both sides of the bay mouth east of the Xiutu–Shihu line, with a gradient of 3°–5°. The main sandbanks are located on the west side of Dashui Island and outside the Jinjiang estuary. The underwater deep troughs are mainly distributed in the water channel from Dazhui Island to the Xiutu section [18].

In ArcGIS (Geographic Information System Information editing software developed by the Environmental System Research Institute) software, the marine map data and channel data were unified into Mercator Center projection coordinates, and the datum level was unified to the global sea level (Figure 2).

2.1.2. Hydrological Data

In order to ensure the accuracy of the model simulation, the engineering data were taken as the reference for the calibration of the model. The tidal level data were obtained from Chongwu and Jinyu tide stations in the Quanzhou Bay from 10:00 a.m. on 20 January 2016 to 12:00 a.m. on 9 February 2016. Tidal current velocity and flow direction data collected from five hydrological and sediment stations in the inner bay, which were observed from 10:00 a.m. on 26 January 2016 to 12:00 a.m. on 27 January 2016 (Spring tide) and from 10:00 a.m. on 3 February 2016 to 12:00 a.m. on 4 February 2016 (Neap tide). The interval of the data acquisition was 1 h. The geographical coordinates and location distribution of the tide station and hydrologic and sediment station are shown in Figure 3.

The measured tidal current data were collected by Acoustic Doppler Current Profiler (ADCP), which need to be converted into vertical mean velocity by weighted average processing. The data processing method is as follows [19]:

$$v = 0.1v_1 + 0.2v_2 + 0.2v_3 + 0.2v_4 + 0.2v_5 + 0.1v_6 \quad (1)$$

where v is the average velocity of vertical line (m/s) and v_i is the velocity of the different horizon stations (m/s).

2.2. Hydrodynamic Model of Quanzhou Bay

2.2.1. Model Meshing and Parameter Setting

Because the coastline of Quanzhou Bay was complex and changeable, the 1959 coastline was used as the base boundary to divide the grids in order to ensure the consistency of the grids in the calculation. The areas of the Houzhu, Neigang, and Dazhui channels was grid-encrypted (Figure 2). The coordinate system of the model adopted the Central Mercator projection. The extracted bathymetric data were interpolated into the grid in

Surface Water Modeling System (SMS), while uniformly converted into MIKE's software grid file (Figure 2).

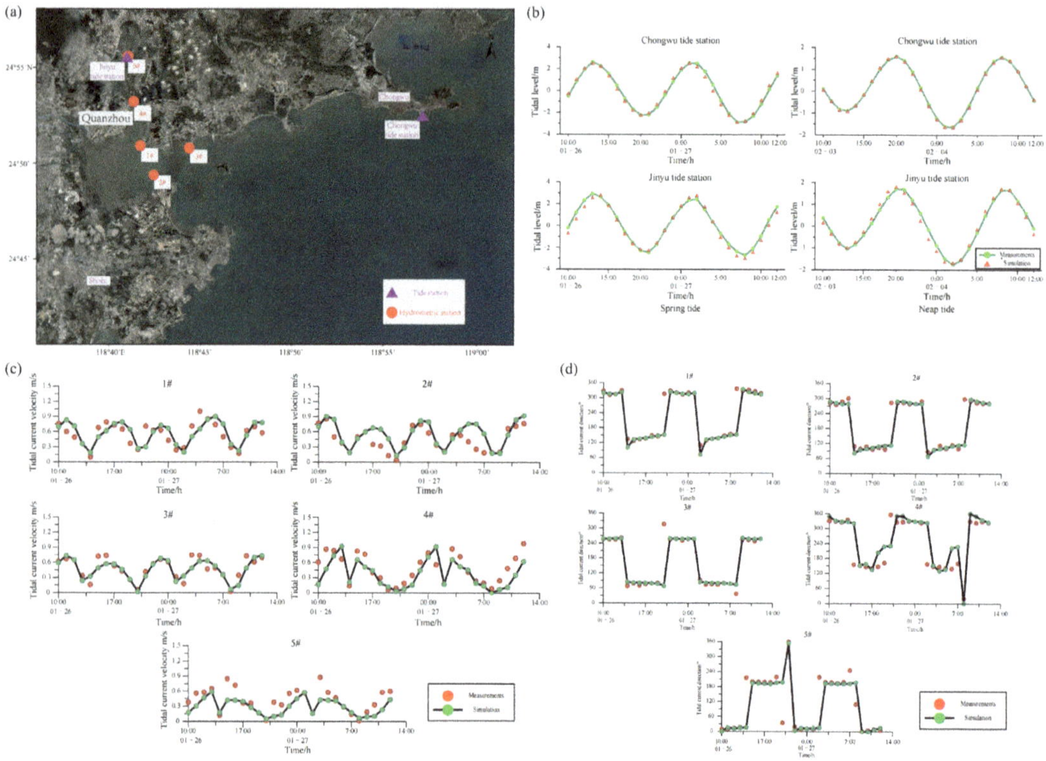

Figure 3. Calibration of the model. (**a**) Location of the tide stations and hydrological stations. (**b**) Comparison of the observed and model-simulated tide levels. (**c**) Comparison of the observed and model-simulated tidal current. (**d**) Comparison of the observed and model-simulated tidal current directions.

The model was simulated in a "Cold start" mode [20], with the initial water level and velocity set to 0. In this model, the Courant Friedrichs Lewy (CFL) number of the shallow water equation was set to 0.8, the main time step was 60 s, and the minimum time interval was 0.01 s. Flooding depth was set to 0.05 m, wetting depth was set to 0.1 m, and drying depth boundary was set to 0.005 m. The horizontal eddy viscosity coefficient was set to 0.28.

2.2.2. Water Exchange Capacity of the Bay

The tidal prism of the bay represents the total volume of the bay that can hold the tidal water, and can reflect the self-purification capacity of a bay [9]. The tidal prism is calculated by integrating the cross-sectional area and the product of the mean velocity of the cross-sectional area [21].

$$Q = \int_0^{t_c+\frac{T}{2}} \overline{v_i} S dt \tag{2}$$

where Q is tidal prism, $\overline{v_i}$ is cross-sectional average velocity (m/s), and S is the cross-sectional area (m^2).

2.3. Changes in Land-Use Structure

The landscape index comparison method was adopted to analyze the change in land-use structure [22]. Landscape indices (information entropy of land use structure) refers to the diversity of landscape elements or ecosystems regarding their structure, function, and change with time, which reflects the richness and complexity of green space landscape types.

$$H = -\sum_{i=1}^{n} (Pi \times lnPi) \tag{3}$$

where H is the landscape index, P is the proportion of landscape type i to the total land area, and n is the number of landscape types. In theory, when the landscape index is 0, the landscape is composed of a single element and the landscape is homogeneous; when the landscape index is the highest, the landscape is composed of more than two elements, and the proportion of each landscape type is equal. When the landscape index decreases, the difference in the proportion of each landscape type to the total land area increases.

3. Results

A large number of human activities have changed the coastline and underwater topography of Quanzhou Bay, which significantly affected the sedimentary dynamic environment.

3.1. Validation of Model Accuracy

The error was calculated (Table 3) according to the JTS-T231-2021 code for the simulation experiment of water transport engineering [23]. The model experimental deviation of tide level, velocity, and direction must meet the following requirements:

Table 3. Model error calculation.

Calculation of Tidal Level Error						
Time	Chongwu Tide Station	Jinyu Tide Station	Time		Chongwu Tide Station	Jinyu Tide Station
2016/1/26 13:00	0.08	0.29	2016/2/3 13:00		0.01	0.08
2016/1/26 19:00	0.09	0.02	2016/2/3 19:00		0.01	0.11
2016/1/27 1:00	0.09	0.17	2016/2/4 2:00		0.01	0.09
2016/1/27 8:00	0.18	0.11	2016/2/4 8:00		0	0.05
Calculation of Tidal Current Error						
	Time	**1#**	**2#**	**3#**	**4#**	**5#**
Spring tide Mean tidal current velocity	Flood	0.09	0.01	0.02	0.17	0.16
	Ebb	0.09	0.10	0.03	0.04	0.01
	Flood	0.06	0.01	0.08	0.05	0.19
	Ebb	0.19	0.22	0.03	0.08	0.12
	Flood	0.00	0.01	0.03	0.12	0.14
	Ebb	0.13	0.08	0.03	0.09	0.00
	Flood	0.03	0.04	0.01	0.11	0.23
	Ebb	0.17	0.18	0.07	0.10	0.10
	Flood	0.02	0.03	0.03	0.27	0.16
Calculation of Tidal Current Direction Error						
	Time	**1#**	**2#**	**3#**	**4#**	**5#**
Spring tide	2016/1/26 10:00	7.7	5.5	0.9	16.6	8.6
	2016/1/26 11:00	4.7	6.9	3.4	2.1	4.9
	2016/1/26 12:00	0.1	10.3	3.4	2.6	1.5
	2016/1/26 13:00	7.0	22.6	2.4	8.2	1.6
	2016/1/26 14:00	34.3	27.3	16.0	165.2	200.3
	2016/1/26 15:00	5.6	5.2	3.0	0.9	7.1

Table 3. *Cont.*

2016/1/26 16:00	2.0	9.4	10.2	10.8	6.8
2016/1/26 17:00	1.6	6.6	4.0	4.7	7.2
2016/1/26 18:00	5.9	4.5	0.1	56.5	7.4
2016/1/26 19:00	2.6	15.1	2.6	66.7	23.1
2016/1/26 20:00	165.2	172.6	248.8	125.2	163.0
2016/1/26 21:00	5.0	5.8	3.9	24.1	5.3
2016/1/26 22:00	1.3	5.1	1.1	23.0	16.0
2016/1/26 23:00	2.7	4.4	7.1	2.1	5.4
2016/1/27 0:00	6.4	0.1	0.8	0.1	0.2
2016/1/27 1:00	0.4	10.9	1.0	4.6	3.5
2016/1/27 2:00	38.2	20.3	8.6	165.4	204.3
2016/1/27 3:00	3.6	0.9	8.0	4.1	5.7
2016/1/27 4:00	0.2	13.5	4.9	17.3	5.6
2016/1/27 5:00	0.3	3.6	5.8	5.5	2.0
2016/1/27 6:00	4.5	5.5	1.0	83.7	7.2
2016/1/27 7:00	5.5	13.9	0.0	68.1	53.2
2016/1/27 8:00	184.8	184.2	36.4	20.0	87.8
2016/1/27 9:00	5.4	6.0	6.2	31.0	2.1
2016/1/27 10:00	9.3	7.5	4.6	25.7	5.4
2016/1/27 11:00	8.3	4.2	8.9	6.7	1.9
2016/1/27 12:00	4.8	1.3	0.2	1.7	8.0

Tide level: the deviation of high and low tide time and phase must be within ±0.5 h; the maximum and minimum tide level deviation must be less than ±0.1 m.

Velocity and direction of tide: the deviation of the mean velocity must be within ±10%; the shape of the velocity process line must be basically consistent; the deviation from the main tidal current direction must be within ±10°.

The simulation results (Table 3) show that the calculated data of the Chongwu and Jinyu Tide Stations were basically consistent with the measured data. The individual error exceeding ±0.1 m was caused by the small water depth of the Jinyu station located in the inner bay. The error between the measured value and the simulated value at other time is less than 0.1 m, which indicates that the model has a good fitting degree. The error between the measured and simulated tidal levels at high and low tides was less than 0.1 m, indicating that the model was relatively accurate for tidal level simulation. Station 5# is located in the northern side of the inner bay with a shallow water depth, and its depth varies greatly in the ebb and flood tide. This results in a discrepancy between the observed and simulated values at the site in the simulation of the speed and direction of tidal flow. The velocity process line basically coincides with the simulated value; the velocity and direction curves of the other 4 stations were basically consistent with the measured ones. The error in tidal current direction was kept within ±10°, and part of the larger error was mainly caused by the delay in the instrument measurement when the tidal currents was reversed. According to the test results, the simulation results of the model were accurate.

3.2. *Variation of Exchange Intensity of Bay Water*

Based on the hydrodynamic model, the tidal prism of Quanzhou Bay under different topographic conditions was calculated. The calculation area included the line from Xiangzhi to Qian'an to the Luoyang River and Jinjiang River in the bay (Figure 4). The calculated results show that the tidal prism decreased year by year from 1954 to 2017, and the total tidal prism decreases were 2.902×10^8 m^3 and 2.592×10^8 m^3 during the spring and neap tides, respectively. The change in tidal prism from 1968 to 1988 was the largest, and during this period, the tidal capacity decreased by 1.554×10^8 m^3 and 1.401×10^8 m^3, accounting for 53.5% and 54.1% of the total variation, respectively.

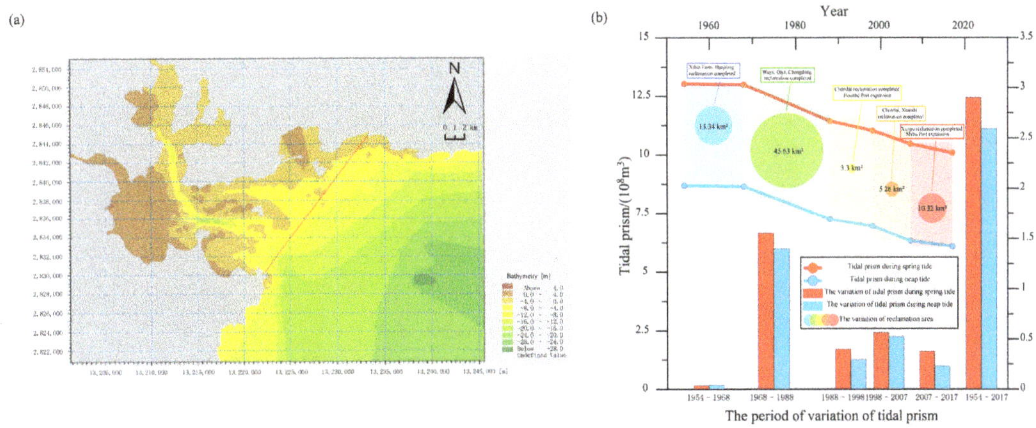

Figure 4. Tidal prism in different periods. (**a**) The calculation range of the tidal prism. (**b**) The tidal prism of Quanzhou Bay in different periods and reclamation at the same time.

3.3. Variation of Tidal Current Field

3.3.1. Change of Flow Field in Different Period

A model was used to simulate the characteristics of the tidal current field in Quanzhou Bay at different time periods (Figure 5). At flood tide time, the tidal current entered the inner bay from the outer bay along the North and South channels around Dazhui Island, and converged on the south side of Xiutu and then reached the Luoyang River and the Neigang channel along the Houzhu channel. In 1954, the maximum tidal current velocity appeared in the west of Xiutu and the north of Shihu. At that time, the tidal current velocities along the Xibin were all below 0.15 m/s, and the tidal current velocity in the tidal flat area of the lower Jinjiang River was between 0.15 and 0.75 m/s. Until 1968, the maximum tidal current velocity during flood tide still appeared in the Houzhu–Dazhui channel area, but the tidal current velocity decreased. The current velocity in the tidal flat region of the lower Jinjiang River was about 0.9–1.2 m/s. From 1968 to 1998, the current velocity in the channel area decreased, but the direction of the coastal tidal current changed from SSW to WNW. By 2007, the Neigang channel was opened for navigation, at which time the maximum tidal current velocity of the flood tide increased from 1.5 to 1.8 m/s. In 2017, the maximum tidal current velocity during flood tide still occurred in the northern side of Shihu. The same flow field changes during the ebb tide are shown in Figure 5.

3.3.2. Variation of Current Velocity in Channel Area

According to the variation in tidal current velocity and the position during flood tide, the area with a greater hydrodynamic difference was mainly from the Houzhu to Dazhui channels. The hydrodynamic changes in the outer bay area were small and stable. In order to estimate the hydrodynamic change in the channel area, the average current velocity in across channel's characteristic points during the flood and ebb tide was calculated by using the data of the point velocity in the channel area simulated by the model for different periods (Figure 6).

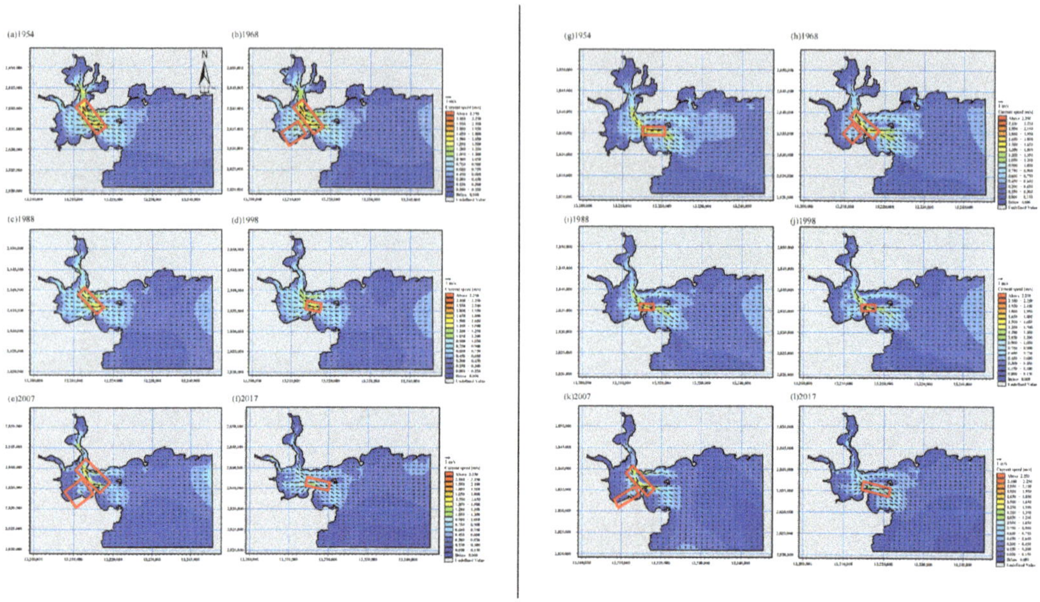

Figure 5. Changes in the tidal current field of the rising and falling tides in Quanzhou Bay during different periods. (**a**–**f**) The distribution of the hydrodynamic forces during the flood tide in different years. (**g**–**l**) The distribution of hydrodynamic forces during the ebb tide in different years.

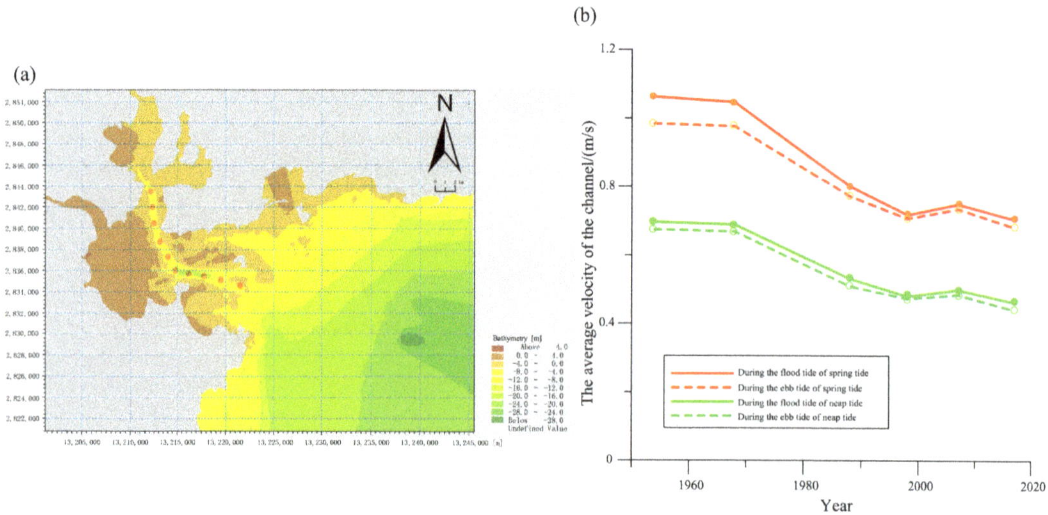

Figure 6. Changes in tidal current field of the flood and ebb tides in Quanzhou Bay during different periods. (**a**) Tidal current velocity calculation point distribution location; (**b**) The average velocity of the channel area in different years.

The results show that the average tidal current velocity in the channel area decreased continuously from 1954 to 1998 (Figure 6). During the spring tide from 1968 to 1988, the large-scale reclamation projects resulted in the hydrodynamic weakening of the inner bay. At this time, the average velocity during the flood and ebb tide in the channel decreases from 1.047 m/s and 0.977 m/s to 0.801 m/s and 0.772 m/s, decreasing by 23% and 21%, respectively. From 1998 to 2007, the navigable channel led to an increase in water depth in the channel area, and the flow converged in the channel area. During the spring tide, the average current velocity of the channel during flood tide and ebb tide increased by 0.031 m/s and 0.028 m/s, respectively. In 2017, the current velocity decreased to 0.705 m/s and 0.682 m/s, respectively.

4. Discussion

4.1. Effects of Human Activities on the Geomorphology and Hydrodynamic Evolution of the Bay

4.1.1. Influence of Sea Area Change on Exchange Capacity of Bay Waters

According to the calculation results of the model, it can be found that the main reason for the decrease in the tidal prism was the reclamation project. Reclamation reduced the area of the bay, which reduced the flow of tidal currents into the bay (Figure 4). From 1954 to 1968, the height of the shoals around the reclamation project was higher than the mean sea level. Although the reclamation area was large (Table 4), the impact of the reclamation on the variation in tidal capacity of the bay was minimal. During 1968 to1988, large-scale reclamation projects, such as "Wuyi reclamation" and "Qiyi reclamation", were carried out at this time, which directly blocked the tide from entering north of Baiqi village. As a result, the tidal prism was greatly reduced at this time. The reduction in tidal prism accounted for 52.7% and 53.8% of the total reduction. From 2007 to 2017, dredging in the channels and returning farmland to sea were carried out, which slowed down the trend of tidal prism reduction during this period.

Large-scale reclamation often results in a decrease in the area of the intertidal zone of the bay, reducing the tidal prism, which, in turn, will weaken the self-purification capacity of the water body [24]. Nevertheless, the development of industrialization has promoted the development of printing and dyeing, textile, electroplating, and other industries around the urban bay. Part of the industrial waste water and the sewage of the cities around the bay were discharged through sewage pipes or open channels, leading to a continuing increase in organic matter and heavy metal elements (such as Zn, Pb, etc.) in the bay [25,26]. In addition, the decrease in tidal prism prolonged the retention time of heavy metals and led to the chronic accumulation of pollutants, which had worsened the water quality of the bay [27]. With the continuous deposition of the pollutants in the bay, the retention of some nutrients and other artificial compounds may cause frequent red tide outbreaks [28]. This will affect the survival and reproduction of the original marine organisms, and also have an impact on coastal aquaculture [29,30], which makes the city's bay water quality problems increasingly significant.

Table 4. Reclamation area and landscape indices of Quanzhou Bay at different stages.

Year	Cities and Industries (m^2)	Agriculture (m^2)	Port (m^2)	Aquaculture (m^2)	Other Types of Land (m^2)	Total Area for Reclamation (m^2)	Landscape Indices
1972	5.64	6.71	0	1.01	0.23	13.588	0.976
1988	12.6	39.1	0.80	1.39	2.74	56.628	0.888
1995	17.95	33.75	1.29	4.20	3.00	60.188	1.103
2006	27.21	25.82	2.22	7.06	4.07	66.378	1.256
2019	41.02	22.61	3.69	5.56	6.09	78.968	1.226

4.1.2. Variation of Current Velocity in Channel Area

According to the hydrodynamic simulation results of Quanzhou Bay across different periods, the hydrodynamic forces in the bay as a whole appear to have weakened. Since the coastline and topography of Quanzhou Bay in different periods are different in the marine map information, the hydrodynamic changes may be related to both the coastline and the topography. In order to explore the influence of coastline change on the hydrodynamics of Quanzhou Bay, a hydrodynamics simulation was carried out by changing the coastline condition and keeping the water depth unchanged.

By comparing the simulated results of the changing coastline with the original hydrodynamic distribution (Figure 7), it was found that the coastline change has a significant impact on the hydrodynamics. In 1968, with the completion of the expansion of the military reclamation farms in the lower Jinjiang River and Xibin Farm, the average tidal current along Xibin decreased (Figures 5 and 7). There was a decrease in the tidal current velocity in the Dazhui channel area in northern Shihu. In 1988, after the "Wuyi reclamation" and "Qiyi reclamation" projects were completed, the tidal current of the channel from Houzhu to Dazhui decreased significantly. The tidal current velocity of Gufu and Xibin also decreased. Due to the completion of "Qiyi reclamation", the current velocity in the coastal area of Zhangban decreased. From 1998 to 2017, the excavation of the Neigang channel caused part of the water flow in Jinjiang River to converge with the water flow in the Houzhu channel on the western side of Xiutu, and the tidal currents in the northwest and western Shihu channels also decreased. Similarly, the contrast between the coastline changes and hydrodynamic distributions during neap tides is shown in Figures 5 and 7.

The variation in coastline has a greater impact on the hydrodynamics. As the width of the coast decreased, coastal currents increase and the seabed was continuously scoured, which will further deepen the passage between the coasts, such as the bay of Visant in France [31]. In some bay mouths, the reclamations have resulted in changes in the direction of the local flow fields, resulting in erosion of coastlines and seabed in the region under local wind waves and long-term natural storm surges [32]. At the same time, reasonable reconstruction of the bay line combined with the specific hydrodynamic conditions in the region can effectively prevent the erosion of the coastline [33]. Well-structured hydraulic structures (such as spur dikes and protective levees) can also ameliorate the problem of coastal deposition to some extent [34].

4.2. The Stage Development of Urban Bay Hydrodynamics and Geomorphology Evolution

According to previous research, the development of industrialization can be divided into the early stage, middle stage, late stage, and modern society [18]. According to the research results, we can find that the geomorphology and hydrodynamics of the bay show different characteristics at different development stages of the urban bay.

Before and during the early stage of industrialization, the geomorphologic changes of the bay were mainly controlled by natural factors. Marine hydrodynamic erosion and coastal sediment deposition were the main causes of coastline changes [35,36]. Before 1968, the tidal current velocity of Points 1#–3# in the river channel decreased due to the natural siltation of the river channel in the Houzhu Port area (Figure 8). In the Shihu Port area, there was almost no reclamation around the area, and the geomorphological changes were mainly influenced by natural factors. The regional tidal current velocity in the northern Shihu Port was high, and the channel presented a state of erosion. The tidal current velocity in Point 1# and 3# changed very little, and the tidal current velocity in Point 2# decreased due to the near-shore silting area. At this time, the reclamation land use was similar to that of most cities, which was mainly used for agricultural land (Table 4). As a result, the landscape index was low, and the reclamation land use was unitary.

From 1968 to 1998, the Quanzhou Bay was in the early to mid-stage of its industrialization. Large-scale reclamation activities such as "Wuyi reclamation" and "Qiyi reclamation" were carried out (Figure 4). The reclamation dam built on the west side of Baiqi village during the "Wuyi reclamation" project prevented water from entering the bay, resulting in a decrease in the influx of tidal water, which further weakened the originally enclosed hydrodynamic environment of Quanzhou Bay (Figure 8). Meanwhile, the construction of reclamation dams for reservoirs and sluices upstream of the Jinjiang and Luoyang Rivers has reduced the runoff and weakened the impact of runoff on tidal movements. Under the action of tidal current, some sediment particles ascend to the area of the Houzhu channel with tidal current due to re-suspension. Due to the reduced hydrodynamic forces in the bays, tidal forces were not sufficient to transport sediment particles out of the Houzhu channel, resulting in deposition in the channel area [37]. The development of the tertiary sector of the economy had increased the shipping demand and prompted the construction of new berths at the Shihu Port. The expansion of the berth extended the shoreline of the port and changed the hydrodynamic direction (Figure 8). Deposition of the channel reduced the tidal hydrodynamic forces and the narrow pipe effect of the tidal channel between the inner and outer bays [38,39], which attenuates ocean erosion around the Xiesha shoal. Due to the weakening of the hydrodynamic forces in the inner bay, the bay mouth was mainly eroded by ocean hydrodynamic forces, resulting in the landward movement of the 20 m isobath (Figure 8). At this time, the reclamation land increased, and the purpose of land use became more obvious with the land mainly used for agriculture, and the landscape index dropped (Table 4).

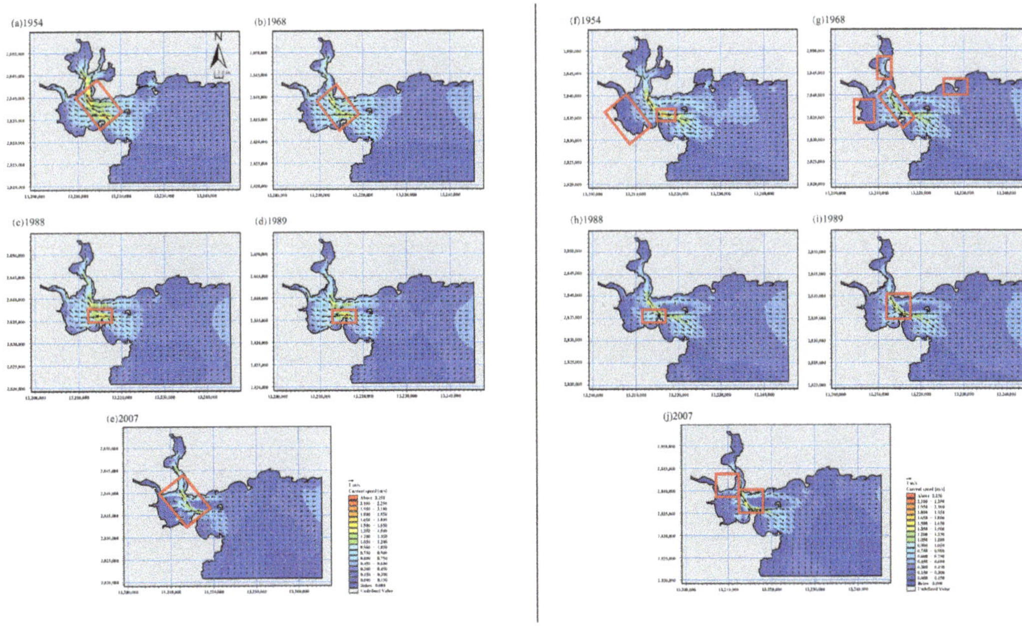

Figure 7. Hydrodynamics changes of Quanzhou Bay caused by coastline change over different periods. (**a–e**) The distribution of hydrodynamic forces during the flood tide in different years after the changes in the coastline. (**f–j**) The distribution of hydrodynamic forces during the flood tide over different years after the changes in the coastline.

From 1998 to 2007, the Quanzhou Bay was in the late stage of industrialization. The tidal current velocity decreased slightly in Points 1# and 2# of the Houzhu port, whereas that in Point 3# remained unchanged. The channel deposition thickness continued to rise. The excavation of the Neigang channel increased the influence of runoff on the tidal current, which increased the tidal current velocity. With the economic development and industrial upgrading, the area of industrial land, port land, and aquaculture land in the reclaimed land increased, which made the landscape index rise, with the types of reclaimed land showing the characteristics of diversification (Table 4).

During the period of 2007–2017, several channel dredging projects resulted in the re-deepening of the water depth, which caused the flow velocity in Houzhu channel to change again. This weak hydrodynamic environment increased the risk of deposition in the channel and affected the movement of ships in the channel. The Shihu ports continued to expand eastward. Under the action of the tides, a slow flow zone was formed in the bay. The sediment that was originally deposited was suspended under the convolution of the coastal waves [40], which migrated with the waves to the slow flow zone of the shoal. With the decrease in the whole hydrodynamic force in the bay area, a new deposit area was formed in the area around the Xiesha shoal, which made the Xiesha shoal area expand continuously. During the period from late industrialization to modern society, the reclamation land around the bay was mainly used for ecological restoration and port expansion. At this time, the reclamation land was mainly used for ecological restoration and urban development; the single use of land made the landscape index decrease.

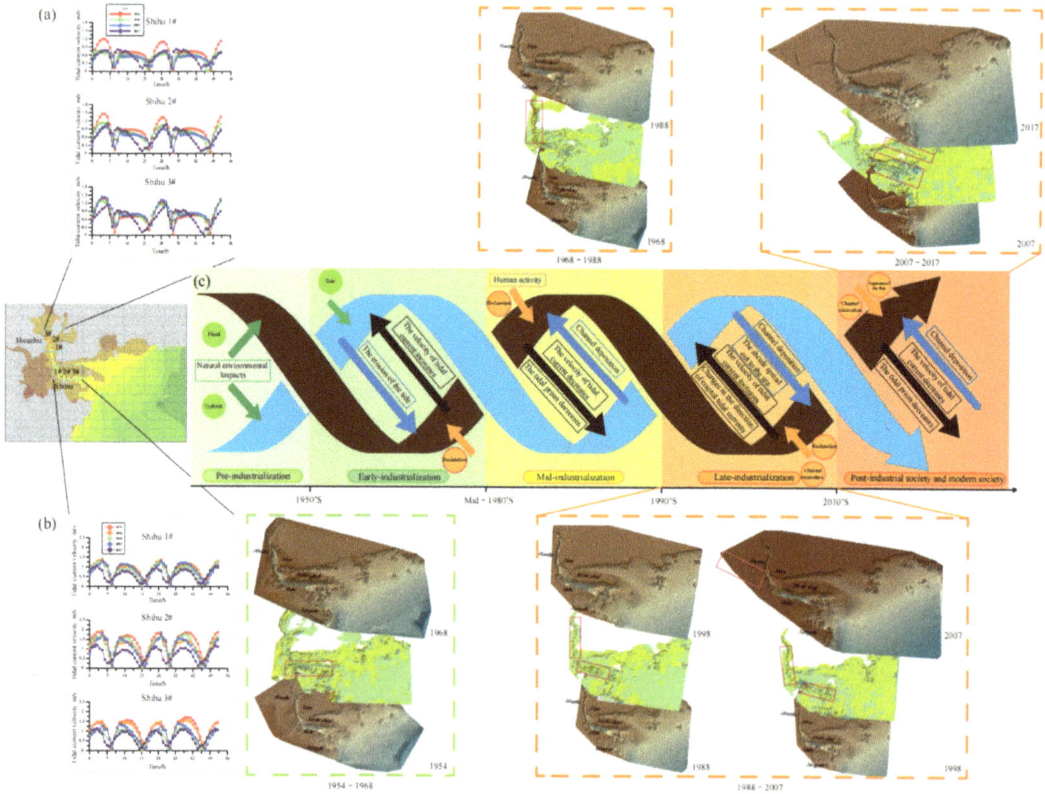

Figure 8. Coupled hydrodynamics and geomorphologic evolution in Quanzhou Bay. (**a**) Tidal currents in the Houzhu channel at different periods. (**b**) Tidal currents in the Shihu Port at different periods. (**c**) Coupling of geomorphologic changes and hydrodynamics in the Quanzhou Bay. The blue arrow shows the hydrodynamic development and the brown arrow shows the development and change of the bay geomorphology.

The industrialization of the bay has changed the shape of the coastline and sea area as well as changed the original conditions of the water and sediment transport, which caused changes in the hydrodynamic and sedimentary environment. The coastal construction of urban bays transforms most natural coastlines into artificial coastlines and changes the original hydrodynamic intensity and direction, which affects the natural development of topography in the bay. The coupling relationship between the sedimentary environment and hydrodynamics is a key to the future evolution of the bay geomorphology; indeed, the coupling relationship between the human activities and geomorphology has existed in other bays in the past, such as tidal flat deposition in the Pearl River estuary [41,42].

4.3. Future Urban Bay Development and Planning

With the continuous intervention by human activities, most urban bays have experienced different degrees of channel deposition and environmental pollution in bays. Currently, in China, the number of channels in most urban bays that do not meet the design standards is rising [43]. From the hydrodynamic model simulation results of Quanzhou Bay and the research status of other bays, it can be seen that repeated dredging projects fundamentally cannot solve the problem of channel deposition [44,45]. Therefore, more measures need to be taken to repair and transform the ecological environment of the urban bay to fundamentally solve the problem of the continuous deterioration of the bay environment.

The pollutants discharged into the bay are readily adsorbed by particles and then deposited in the bay [46,47]. Therefore, reducing the discharge of pollutants is the important way to reduce the pollution in the bay. For economically developed urban cities, large heavy industrial enterprises should upgrade and innovate industrial technologies and eliminate some backward industries, so as to reduce the input of heavy metals and artificial compounds into the bay [48]. This is of great significance to coastal agriculture and the surrounding ecological environment. For urban cities that are still in the process of industrialization, the carrying capacity of the bay environment should be considered in advance when expanding the scale of the enterprises around the bay [49].

In the area of ecological restoration, it is necessary to develop a regional development strategy to coordinate the development of the bay and regional ecological environment [1]. The coastal vegetation can absorb and transport heavy metal particles to a certain extent, so the establishment of nature reserves to restore the coastal vegetation has good ecological benefits [50]. The economically developed bay area should repair the damaged ecological environment in time. The economically underdeveloped bay area should consider adopting a balance of nature development options. The ecological development of the bay is combined with energy-saving and emission-reduction projects; examples include China's "Carbon neutrality", the "Blue Bay" restoration project and its "Peak carbon dioxide emissions" strategy, and offshore wind projects under development in many countries [51].

For the development policy, the government needs to plan development projects and activities and formulate the relevant legal policies to protect the urban bay ecosystem. Most of the economically developed bays are now working to reduce the adverse impacts of reclamation on the bays by setting up specialized bays regulatory agencies and establishing the relevant legal provisions, measures such as the establishment of a central government-led regulatory commission and the establishment of a marine coordination regulatory agency in the Netherlands, the enactment of the Public Surface Burial Act in Japan, and the Coastal Zone Management Act in the United States, to manage reclamation land [52]. In addition, the re-planning of completed reclamation is also an important tool for the restoration of the bay, as is the planning for the reconstruction of parts of old industrial zones in Spain [53]. However, it is also necessary to consider the differences between the socio-economic development levels and select the appropriate indicators to evaluate the development of the bay.

5. Conclusions

Based on the topographic and hydrological data of the bay, a numerical model was used to simulate the hydrodynamics of the bay under the influence of human activities. Finally, the paper discusses the coupling relationship between human activities and the evolution of geomorphology and hydrodynamics in the urban bay at different stages of industrialization. The main conclusions are as follows:

First, reclamation during the industrialization period are the main reasons for the decrease in tidal prism in the bay. During the whole industrialization period, with the increase in reclamation area, the water exchange capacity of Quanzhou Bay decreased. The tidal prism in Quanzhou Bay decreased 22.2% and 29.8% in the spring and neap tide, respectively.

Then, the main location of the hydrodynamic changes was in the channel of the bay. The mean tidal current velocity in the channel shows a trend of first decreasing, then increasing, and then decreasing.

Finally, there is an obvious coupling relationship between the geomorphologic evolution and the hydrodynamic forces in the urban bay, and the whole process of industrialization presents a special stage evolution process. There is an obvious coupling relationship between the geomorphologic evolution and the hydrodynamic forces in the urban bay, and the geomorphology shows a special stage evolution process in the whole process of industrialization. Before and during the early stage of industrialization, the geomorphologic changes of the bay were mainly influenced by natural factors. In the middle of industrialization, the coastal reclamation project increased the sediment content and weakened the hydrodynamic force in the bay, resulting in a large deposition in the channel area. The sedimentation of the seabed made the current velocity decrease, and formed a weak hydrodynamic environment in the channel area, which made the channel deposition more serious. In the late period of industrialization, the excavation and navigation of the Neigang channel made the flow of the channel converge, and the hydrodynamic force of the channel increased. In 2014, the further expansion of the reclamation area led to a decrease in hydrodynamic force. A reduced water exchange capacity allowed heavy metals and synthetic materials to deposit in the bay, causing changes in sediment composition, while human activities such as sand mining and dredging projects directly altered the underwater seabed. In future bay development, it is necessary to combine the specific situation of the bay with certain model simulation and prediction before development.

This research clarifies the stage evolution process of an urban bay in the process of industrialization, which can provide suggestions for the future development of the bay and ensure the sustainable development of an urban bays.

Author Contributions: Conceptualization, Y.L. (Yunhai Li) and X.X.; validation, J.T. and F.L.; formal analysis, F.S.; investigation, J.H. and L.W.; resources, B.Z.; writing—original draft preparation, X.X.; writing—review and editing, X.Z., W.C. and Y.L. (Yuting Lin). All authors have read and agreed to the published version of the manuscript.

Funding: This research was funded by the National Science Foundation of China (41976050, 42176220), and the Scientific Research Foundation of the Third Institute of Oceanography, MNR (TIO2019028, TIO2015014).

Institutional Review Board Statement: Not applicable.

Informed Consent Statement: Not applicable.

Data Availability Statement: Data inquiries can be directed to the corresponding author.

Acknowledgments: Many thanks to the editors and reviewers for their suggestions and amendments to this manuscript.

Conflicts of Interest: The authors declare no conflict of interest.

References

1. Liu, D.; Xu, Y.Y.; Faghihinia, M.; Kay, P.; Chan, F.K.S.; Wu, N. Evolving framework of studies on global gulf ecosystems with Sustainable Development Goals. *Environ. Sci. Pollut. Res.* **2022**, *29*, 18385–18397. [CrossRef] [PubMed]
2. Day, J.W.; Gunn, J.D.; Burger, J.R. Diminishing opportunities for sustainability of coastal cities in the Anthropocene, a review. *Front. Environ. Sci.* **2021**, *9*, 663275. [CrossRef]
3. Li, P.H.; Zhao, C.; Liu, K.; Xiao, X.T.; Wang, Y.J.; Wang, Y.T.; He, D. Anthropogenic influences on dissolved organic matter in three coastal bays, North China. *Front. Earth. Sci.* **2021**, *9*, 575. [CrossRef]
4. Wu, Z.Y.; Milliman, J.D.; Zhao, D.N.; Zhou, J.Q.; Yao, C.H. Recent geomorphic change in LingDing Bay, China, in response to economic and urban growth on the Pearl River Delta, Southern China. *Glob. Planet. Chang.* **2014**, *123*, 1–12. [CrossRef]
5. Shadrin, N.V. Coupling of Shoreline Erosion and Biodiversity Loss, Examples from the Black Sea. *Int. J. Mar. Sci.* **2013**, *3*, 43. [CrossRef]
6. Li, J.L.; Pu, R.L.; Yuan, Q.X.; Liu, Y.C.; Feng, B.X.; Guo, Q.D. Spatiotemporal change patterns of coastlines in Xiangshan Harbor (Zhejiang, China) during the past 40 years. *J. Coast. Res.* **2018**, *34*, 1418–1428. [CrossRef]

7. Yang, C.; Li, Q.Q.; Hu, Z.W.; Chen, J.Y.; Shi, T.Z.; Ding, K.; Wu, G.F. Spatiotemporal evolution of urban agglomerations in four major bay areas of US, China and Japan from 1987 to 2017, evidence from remote sensing images. *Sci. Total Environ.* **2019**, *671*, 232–247. [CrossRef]
8. Escudero, M.; Silva, R.; Mendoza, E. Beach Erosion Driven by Natural and Human Activity at Isla del Carmen Barrier Island, Mexico. *J. Coast. Res.* **2014**, *71*, 62–74. [CrossRef]
9. Yin, Y.; Jiang, L.F.; Zhang, Z.X.; Yu, H.B. Hydrodynamic changes due to coastal reclamation activities in Daya Bay. *Aquat. Ecosyst. Health* **2019**, *22*, 215–227. [CrossRef]
10. Lin, S.C.; Liu, G.P.; Niu, J.W.; Wei, X.; Cai, S.Q. Responses of hydrodynamics to changes in shoreline and bathymetry in the Pearl River Estuary, China. *Cont. Shelf. Res.* **2021**, *229*, 104556. [CrossRef]
11. Gallerano, F.; Cannata, G.; Scarpone, S. Bottom changes in coastal areas with complex shorelines. *Eng. Appl. Comput. Fluid Mech.* **2017**, *11*, 396–416. [CrossRef]
12. Yan, X.L.; Hu, Y.M.; Chang, Y.; Zhang, D.H.; Liu, M.; Guo, J.T.; Ren, B.H. Monitoring wetland changes both outside and inside reclamation areas for coastal management of the Northern Liaodong Bay, China. *Wetlands* **2017**, *37*, 885–897. [CrossRef]
13. De Muro, S.; Porta, M.; Pusceddu, N.; Frongia, P.; Passarella, M.; Ruju, A.; Buosi, C.; Ibba, A. Geomorphological processes of a Mediterranean urbanized beach (Sardinia, Gulf of Cagliari). *J. Maps.* **2018**, *14*, 114–122. [CrossRef]
14. Leonardo, H.N.N.; Noel, C.; Yovani, M.L. Bedload Transport of Sediments and Morphodynamics in the Northern Gulf of California. *J. Coast. Res.* **2014**, *30*, 228–236. [CrossRef]
15. Jia, R.; Lei, H.Y.; Hino, T.; Arulrajah, A. Environmental changes in Ariake Sea of Japan and their relationships with Isahaya Bay reclamation. *Mar. Pollut. Bull.* **2018**, *135*, 832–844. [CrossRef]
16. Xiao, X.B.; Li, Y.H.; Shu, F.F.; Wang, L.; He, J.; Zou, X.C.; Chi, W.Q.; Lin, Y.T.; Zheng, B.X. Coupling relationship of human activity and geographical environment in stage-specific development of urban coastal zone: A case study of Quanzhou Bay, China (1954–2020). *Front. Mar. Sci.* **2022**, *8*, 781910. [CrossRef]
17. Liu, C.L.; Chang, J.; Chen, M.J.; Zhang, T. Dynamic monitoring and its influencing factors analysis of coastline in the Laizhou Bay since 1985. *J. Coast. Res.* **2020**, *105*, 18–22. [CrossRef]
18. Lin, Y.P. *The Material "Source and Sink" Process of Quanzhou Bay under the Impact of Typhoon Matmo (2014)*; Academic Thesis of Third Institute of Oceanography: Xiamen, China, 2020; (In Chinese with English Abstract). [CrossRef]
19. Liu, X. *Study on the Marine sed i Mentary Env Ironment Evolution of the Southern of Laizhou Bay under the Impact or Port Projects*; Ocean University of China: Qingdao, China, 2014. (In Chinese with English Abstract).
20. Arasteh, A.M.; Dastdaran, M.; Morovvati, H.; Azad, M.T.; Lari, K. Investigation of the two-dimensional model of water flow in the Gorgan Bay under the influence of various hydrodynamic factors using Mike 21 software. *Ukr. J. Ecol.* **2017**, *7*, 366–373. [CrossRef]
21. Ji, H.D.; Pan, W.R.; Zhang, G.R.; Luo, Z.B. The calculation of volume of tidal prism and water exchange time of Yundang Lake. *J. Xiamen Univ. (Nat. Sci.)* **2006**, *5*, 660–663, (In Chinese with English Abstract).
22. Ma, Q.X. *A Study on Relation between the Level of Urbanization and the Changes of Land Utilization in Jinjiang City*; Fujian Normal University: Fuzhou, China, 2013. (In Chinese with English Abstract).
23. Management of Waterway Transportation Information System. Available online: https://mwtis.mot.gov.cn/syportal/sybz (accessed on 10 May 2022).
24. Kai, X.; Li, H.; Song, D.; Chen, Y.; Huang, Y. Field measurements for investigating the dynamics of the tidal prism during a spring-neap tidal cycle in Jiaozhou bay, China. *J. Coast. Res.* **2018**, *35*, 335–347. [CrossRef]
25. Yu, R.L.; Zhang, W.F.; Hu, G.R.; Lin, C.Q.; Yang, Q.L. Heavy metal pollution and Pb isotopic tracing in the intertidal surface sediments of Quanzhou Bay, southeast coast of China. *Mar. Pollut. Bull.* **2016**, *105*, 416–421. [CrossRef] [PubMed]
26. Homira, A. Ecological risk assessment of heavy metals in sediment, fish, and human hair from Chabahar Bay, Makoran, Iran. *Mar. Pollut. Bull.* **2021**, *169*, 112345. [CrossRef]
27. Wang, C.; Sun, Q.; Jiang, S.; Wang, J.K. Evaluation of pollution source of the bays in Fujian Province. *Procedia Environ. Sci.* **2011**, *10*, 685–690. [CrossRef]
28. Chen, B.H.; Wang, K.; Dong, X.; Lin, H. Long-term changes in red tide outbreaks in Xiamen Bay in China from 1986 to 2017. *Estuar. Coast. Shelf Sci.* **2021**, *249*, 107095. [CrossRef]
29. Azab, I.; Cdb, C.; Jv, D.; Vme, F.; Gi, B.; Asp, G.; Ac, H. Benthic microbial diversity trends in response to heavy metals in an oxygen-deficient eutrophic bay of the Humboldt current system offshore the Atacama Desert. *Environ. Pollut.* **2021**, *286*, 117281. [CrossRef]
30. Dong, J.Y.; Zhao, L.L.; Sun, X.; Hu, C.Y.; Wang, Y.H.; Li, W.T.; Zhang, P.D.; Zhang, X.M. Response of macrobenthic communities to heavy metal pollution in Laoshan Bay, China, A trait-based method. *Mar. Pollut. Bull.* **2021**, *167*, 112292. [CrossRef]
31. Alexa, L.; Arnaud, H.; Amandine, N.; Nicolas, P. Influence of shoreface morphological changes since the 19th century on nearshore hydrodynamics and shoreline evolution in Wissant Bay (northern France). *Mar. Geol.* **2020**, *422*, 106095. [CrossRef]
32. Tefa, B.; Vca, B.; Slgc, D.; Rsa, B.; Vdse, F.; Jllg, H. Decadal shoreline erosion and recovery of beaches in modified and natural estuaries. *Geomorphology* **2021**, *390*, 107884. [CrossRef]
33. Kuang, C.; He, L.; Gu, J.; Pan, Y.; Zhang, Y.; Yang, Y.; Qiu, R.; Zhang, J.B. Effects of submerged breakwater on hydrodynamics and shoreline change of the East Beach of Beidaihe, Bohai Bay, China. *J. Coast. Res.* **2014**, *30*, 598–614. [CrossRef]

34. Masria, A.A.; Negm, A.M.; Iskander, M.M.; Saavedra, O.C. Hydrodynamic modeling of outlet stability case study Rosetta promontory in Nile delta. *Water Sci.* **2013**, *27*, 39–47. [CrossRef]
35. Zhou, Y.; Ye, Q.; Shi, W.Y.; Yang, B.; Song, Z.K.; Yan, D.L. Wave characteristics in the nearshore waters of Sanmen Bay. *Appl. Ocean Res.* **2020**, *101*, 102236. [CrossRef]
36. Zarzuelo, C.; Ruiz, A.L.; Sánchez, M.O. The role of waves and heat exchange in the hydrodynamics of multi-basin bays: The example of Cádiz Bay (Southern Spain). *J. Geophys. Res. Ocean* **2021**, *126*, e2020JC016346. [CrossRef]
37. Liu, H.; Kang, H.X.; Yin, B.S. Sediment transport in the Quanzhou Bay. *Adv. Mater. Res.* **2014**, *2914*, 2388–2391. [CrossRef]
38. Wang, A.J. Hydrodynamics and associated sediment transport over coastal wetlands in Quanzhou Bay, China. *China Ocean Eng.* **2011**, *25*, 59–72. [CrossRef]
39. Sickbert, T.; Peterson, E.W. The effects of surface water velocity on hyporheic interchange. *J. Water Resour. Prot.* **2014**, *6*, 327–336. [CrossRef]
40. He, Z.X.; Jia, L.W.; Jia, Y.H.; He, J.W. Effects of flood events on sediment transport and deposition in the waterways of Lingding Bay, Pearl River Delta, China. *Ocean Coast. Manag.* **2020**, *185*, 105062. [CrossRef]
41. Shen, Y.M.; Zhang, H.X.; Tang, J. Hydrodynamics and water quality impacts of large-scale reclamation projects in the Pearl River Estuary. *Ocean. Eng.* **2022**, *257*, 111432. [CrossRef]
42. Liu, C.j.; Zhong, Z.Y.; Xiao, Y.Q.; Chen, X.Q.; Yu, M.H. Impacts of Human Activities on Hydrodynamic Structures during the dry season in the Modaomen Estuary. *J. Mar. Sci. Eng.* **2021**, *9*, 1327. [CrossRef]
43. China Dredging Association. Available online: http://www.chida.org/ (accessed on 10 August 2022).
44. Yang, L.; Liu, F.; Gong, W.; Cai, H.; Yu, F.; Pan, H. Morphological response of Lingding Bay in the Pearl River Estuary to human intervention in recent decades. *Ocean Coast. Manag.* **2019**, *176*, 1–10. [CrossRef]
45. Ilinskas, G.; Januait, R.; Jarmalaviius, D.; Pupienis, D. The impact of Klaipėda Port entrance channel dredging on the dynamics of coastal zone, Lithuania. *Oceanologia* **2020**, *64*, 489–500. [CrossRef]
46. Cheng, J.Y.; Wang, X.M.; Kong, L.N.; Li, Y.X.; Han, J.B.; Zhang, X.X. Source apportionment and specific risk assessment of heavy metals in sediment, Application to the Typical Bay, China. *Reg. Stud. Mar. Sci.* **2020**, *54*, 102452. [CrossRef]
47. Zeng, L.; Jiang, S.; Jing, L.H.; Xue, Y. Source Apportionment of heavy metal contamination in urban-agricultural-aquacultural soils near the Bohai Bay Coast, using land-use classification and google satellite tracing. *Remote Sens.* **2022**, *14*, 2436. [CrossRef]
48. Rey-Martínez, N.; Guisasola, A.; Baeza, J.A. Assessment of the significance of heavy metals, pesticides and other contaminants in recovered products from water resource recovery facilities. *Resour. Conserv. Recycl.* **2022**, *182*, 106313. [CrossRef]
49. Yu, C.; Sui, W.; Liang, B.; Bao, C.; Ma, M. Temporal trends of heavy metals in the sediments of Bohai Bay in China. *Environ. Monit. Assess.* **2022**, *194*, 79. [CrossRef]
50. Yan, X.X.; An, J.; Yin, Y.C.; Gao, C.C.; Wang, B.Y.; Wei, S.H. Heavy metals uptake and translocation of typical wetland plants and their ecological effects on the coastal soil of a contaminated bay in Northeast China. *Sci. Total Environ.* **2022**, *803*, 149871. [CrossRef]
51. Wen, Y.; Kamranzad, B.; Lin, P. Joint exploitation potential of offshore wind and wave energy along the south and southeast coasts of China. *Energy* **2022**, *249*, 123710. [CrossRef]
52. Hu, S.L. *A Study on Land Reclamation and Its Management Systems*; Ocean University of China: Qingdao, China, 2011. (In Chinese with English Abstract).
53. Nogués, S.; Arroyo, N.L. Alternative approach to prioritization of Brownfield reclamation attending to urban development potentialities, Case study in a depressed industrial district in Northern Spain. *J. Urban. Plan. Dev.* **2015**, *142*, 1–18. [CrossRef]

Journal of
*Marine Science
and Engineering*

MDPI

Article

Effects of Clay Mineral Composition on the Dynamic Properties and Fabric of Artificial Marine Clay

Yi Shan [1,2,3], Xing Wang [1,2,*], Jie Cui [1,2], Haihong Mo [4] and Yadong Li [1,2]

1 School of Civil Engineering, Guangzhou University, Guangzhou 510006, China; yshan@gzhu.edu.cn (Y.S.); jcui@gzhu.edu.cn (J.C.); liyadong@gzhu.edu.cn (Y.L.)
2 Guangdong Engineering Research Center for Underground Infrastructural Protection in Coastal Clay Area, Guangzhou 510006, China
3 Department of Geosciences, University of Padova, 35122 Padova, Italy
4 School of Civil Engineering and Transportation, South China University of Technology, Guangzhou 510641, China; cvhhmo@scut.edu.cn
* Correspondence: wangxing@gzhu.edu.cn

Abstract: Marine clays are easily affected by different mineral composition in cyclic load-based geological hazards. Therefore, based on analyzing the mineral composition of natural marine clay, it is the key to predict the dynamic properties of natural materials under cyclic loading by using quantitated artificial marine clay. In this study, the marine clay found in the South China Sea deltas was investigated. Based on the results of geological conditions and mineral composition analyses, raw non-clay minerals (such as quartz, albite) and clay minerals (such as Na-montmorillonite and kaolinite) were used to produce artificial marine clay, the dynamic properties of which were studied from the impact of mineral composition. Dynamic triaxial laboratory testing for artificial marine clay comprising various clay minerals was performed under identical test conditions. The artificial marine clay with high montmorillonite content exhibited slower development of strain, more sluggish growth in pore water pressure, more rounded hysteresis curves, greater stiffness, and more prolonged viscous energy growth than the clay with low montmorillonite content. In addition, the flocculated fabric of the artificial marine clay with high montmorillonite content demonstrated sufficient pore space changes, more uniform pore distribution, and larger specific surface area than the dispersed fabric of the clay with low montmorillonite content. The factors arising from the influence of montmorillonite may lead to microstructural and fabric changes, hinder the development of pore water, and increase intergranular contact stiffness as well as delay the cyclic strain amplitude at the breakpoint of viscous energy dissipation. In general, the results presented in this study confirm that clay minerals, especially montmorillonite, have significant influence on the dynamic properties of large strain.

Keywords: artificial marine clay; mineral composition; clay minerals; dynamic properties; sediment fabric

Citation: Shan, Y.; Wang, X.; Cui, J.; Mo, H.; Li, Y. Effects of Clay Mineral Composition on the Dynamic Properties and Fabric of Artificial Marine Clay. *J. Mar. Sci. Eng.* **2021**, *9*, 1216. https://doi.org/10.3390/jmse9111216

Academic Editor: Gemma Aiello

Received: 10 October 2021
Accepted: 31 October 2021
Published: 3 November 2021

Publisher's Note: MDPI stays neutral with regard to jurisdictional claims in published maps and institutional affiliations.

Copyright: © 2021 by the authors. Licensee MDPI, Basel, Switzerland. This article is an open access article distributed under the terms and conditions of the Creative Commons Attribution (CC BY) license (https://creativecommons.org/licenses/by/4.0/).

1. Introduction

The dynamic properties of marine clay under cyclic loading have received extensive attention to date. This is primarily due to marine clay being significantly affected by earthquakes [1–5], traffic loads [6–8], and wave loads [9–11]. For example, in the Alaska earthquake of 1964 [12] and the Mexican earthquake of 1985 [13], marine clays along the respective coastal and delta areas of these regions were severely affected, leading to the clay-related failure of marine infrastructures and slopes, as well as a large number of casualties and property losses. The studies relating to these disasters so far were aimed at informing on and understanding the cyclic behavior of marine clays [14,15].

Generally, due to the genesis of bedrock [16–18], the climate [19–21], and the sedimentation cycle of transgression and regression [22–24], the mineral composition of Quaternary marine clays in different deltas or local deltas is obviously different. The difference of mineral composition is the main reason for the difference of the Atterberg limits water content of marine clay. Meanwhile, the dynamic characteristics of marine clay under cyclic loading are affected by many factors [25,26], such as plasticity index (PI), confining pressure, void ratio, over-consolidation, and initial static shear stresses, etc. The extension parameter of the Atterberg limit water content, PI, directly determines the classification and physical properties of soils [27]. Additionally, the PI value is mainly influenced by the mineral composition and sediment fabric, which was also found to affect the dynamic properties of marine clay in laboratory dynamic testing [28].

Clay minerals in cohesive soils have frequently been included in research on dynamic properties. Marcuson and Wahls [29] studied the effects of consolidation time on dynamic shear modulus in remolded cohesive soils samples with kaolinite and calcium-based montmorillonite (Ca-montmorillonite). Subsequently, Fahoum et al. [30] and Park and Kim [31] used various artificial clay minerals to predict the dynamic properties of cohesive soil in their studies, which indicated that clay soils with a higher plastic mineral content provided better dynamic properties such as cyclic strength. Recent researches [32–35] summarized and suggested that a) a critical level of clay mineral content does exist; b) that clay minerals exhibit different dynamic property trends before and after this critical point; yet c) this critical point disappears when the clay particles consist in highly plastic minerals. The reason for this phenomenon is explained by the different microscopic fabric between clay minerals and coarse particles in the soil [36]. The above research has demonstrated that clay minerals have an important effect on the dynamic properties of cohesive soils, and also proved that various artificial clay minerals could be used as substitutes for clay particles in cohesive soils to study the influence of clay mineral composition on dynamic properties.

Although many studies of the dynamic properties achieved by adding artificial clay minerals to clays were conducted, a number of specific problems pertaining to dynamic properties were identified in marine clay—as a type of clay—comprising artificial clay minerals. These issues required urgent solutions which were sought in the present study.

- Previous research on the effects of mineral composition on the dynamic properties of cohesive soils aimed to identify the influence of mineral composition in sand–clay cohesive soils on their dynamic properties. However, little data exist to quantify the effect of clay minerals in marine clay on the latter's dynamic properties under large strain. According to Boulanger and Idriss [37,38], large strain and excess pore water pressure rapidly developed in sand–clay mixture with clay minerals, which was contrary to the gradual development of marine clay. This different development between sand–clay cohesive soil and marine clay was due to the failure of sand–clay mixture from excess pore water pressure, while the failure of marine clay arose from the development of strain and eventual fabric collapse. Therefore, we should re-study the dynamic properties of marine clay under large strain by focusing on mineral composition, especially clay minerals, and the fabric formed by clay minerals.
- Natural marine clay is generally composed of a variety of non-clay minerals (such as quartz and albite) and clay minerals (such as montmorillonite and kaolinite). Unlike other cohesive soils, the non-clay mineral particle size of marine clay is generally small and similar to that of clay mineral particles [39]. However, few studies [40,41] to date have investigated the dynamic properties of non-clay minerals with similar particle size to clay particles in marine clay. In addition, the difference in particle size between coarse quartz particles and fine clay particles in marine clay was found significantly to affect the mechanical properties of marine clay, (e.g., cyclic strength [42] and peak strength [43]). The mineral composition of natural marine clay is furthermore complex, and its content in various minerals cannot be controlled artificially. It is therefore difficult to investigate the influence of mineral content on dynamic properties. Studying the effects of mineral content on dynamic properties would require overcoming difficulties in achieving sample uniformity within laboratory dynamic testing. These

challenging problems result from differences in the deposition rate from different minerals [44] and from the effect of consolidation time under high cementation conditions [45]. These research shortcomings and complexities of marine clays indicate that the mineral composition and particle size of natural marine clay in the relevant local area should imperatively be considered when studying the influence of mineral composition on the large strain dynamic properties of marine clay.

In addition, clay minerals significantly impact on the microstructure and fabric of different clay minerals specimens. Some previous studies [46,47] found that the influence of mineral composition on mechanical properties, especially dynamic properties, could not be ignored for cohesive soil. Sun et al. [48] proved that the microstructure and fabric of natural marine clay with the same initial and loading states correlated with their mechanical properties by performing test using a field emission scanning electron microscope. Meanwhile, Cuisinier et al. [49] demonstrated the alkaline solution weakened the shear strength behavior of compacted artificial bentonite–argillite mixture by combining scanning electron microscopy (SEM) and mercury intrusion porosimetry (MIP) testing. However, these studies still do not consider the mineral composition as a variable, especially the content of clay minerals, when systematically studying the influence of microstructure and fabric on the dynamic properties of artificial marine clays.

Therefore, to investigate the effects of clay mineral composition on the dynamic properties and sediment fabric of marine clay under large strain, laboratory dynamic triaxial testing was conducted with clay mineral content as the primary research parameter, and the Atterberg limits and PI as auxiliary parameters. Based on the mineral composition of natural marine clays in two of the largest estuarine deltas (Pearl River Delta and Han River Delta) in Guangdong province, South China, four types of raw mineral powder (quartz, albite, Na-montmorillonite, and kaolinite) with similar particle sizes were used to produce the artificial marine clays. The research focused on the stiffness degradation and energy dissipation of different clay minerals. Subsequently, the sediment fabric of microstructure and microporosity in artificial marine clay specimens were examined using low-vacuum environmental scanning electron microscopy (LVESEM) testing and mercury intrusion porosimetry (MIP) testing. It was anticipated that the test results would help establish a research method to quantify the role of clay minerals in the dynamic properties and fabric of marine clay. The conclusions were proposed to help improve the understanding and application of mineralogy and engineering geology in studying the mechanical properties of marine clay.

2. Materials and Methods

2.1. Mineral Composition of Natural and Artificial Marine Clay

2.1.1. Geological Conditions and Mineral Composition of Natural Marine Clay

Since the Pliocene in the Neogene, the second Himalayan movement caused several northeast to southwest orientated fault zones under the two estuary (Pearl River and Han River) Deltas, which resulted in obstructing both the Pearl and Han Rivers in Guangdong Province, South China [50]. The existence of fault zones in close proximity of the South China Sea resulted in the more significant formation of Quaternary marine sediments along the delta front to the outside of the most extended shell sand bar, than continental sedimentation formation. Both deltas were formed during the Mindel-Riss and Riss-Wurm interglacial periods. Marine clay formed as a result of transgression during the Riss-Wurm interglacial, Atlantic, and Sub-Atlantic periods [51]. During these periods, three marine clay layers were formed from three sedimentary cycle of the Quaternary deposits in Pearl River Delta. However, due to only weak regression during the Sub-boreal period, which may have been negligible, continuous transgression occurred during the Atlantic and Sub-Atlantic periods, eventually leading to the formation of two sedimentary cycle of Quaternary deposits in Han River Delta [52].

In order to analyze the mineral composition of both estuarine deltas, three marine clay specimens were sampled from them and mineral composition test results were measured by X-ray diffraction. The basic physical properties of these three marine clays are listed in Table 1. The corresponding mineral compositions are shown in Table 2. These results were also compared to those of earlier studies of the mineral composition of marine clay found in Pearl River Delta [53] and in Han River Delta [54], reported in Table 3. The above tables show the mineral composition and geological age of the marine clays taken by drilling and sampling. The locations of the above sampling operations are shown in Figure 1.

Since all sampling locations were in the front delta area outside the most extensive shell sand bar, borehole analysis revealed these marine clays to be sediments formed by the transgression of the Quaternary Holocene period. It is well documented that the main sedimentary minerals in marine clay are non-clay minerals of quartz and albite, and clay minerals of montmorillonite, kaolinite, illite, and chlorite, along with small numbers of other minerals. For non-clay minerals, the ratio of quartz to albite ranged between 3:1 and 7:1, while for clay minerals, the ratios between the clay minerals varied among the diverse samples from the two deltas. To study the effects of clay mineral composition on the dynamic properties of artificial marine clay, the mineral composition of the natural marine clay in two estuarine deltas in the South China Sea was investigated. The mineral composition of natural marine clay in these two estuarine deltas was hence taken as the basis of and background for the study of dynamic properties.

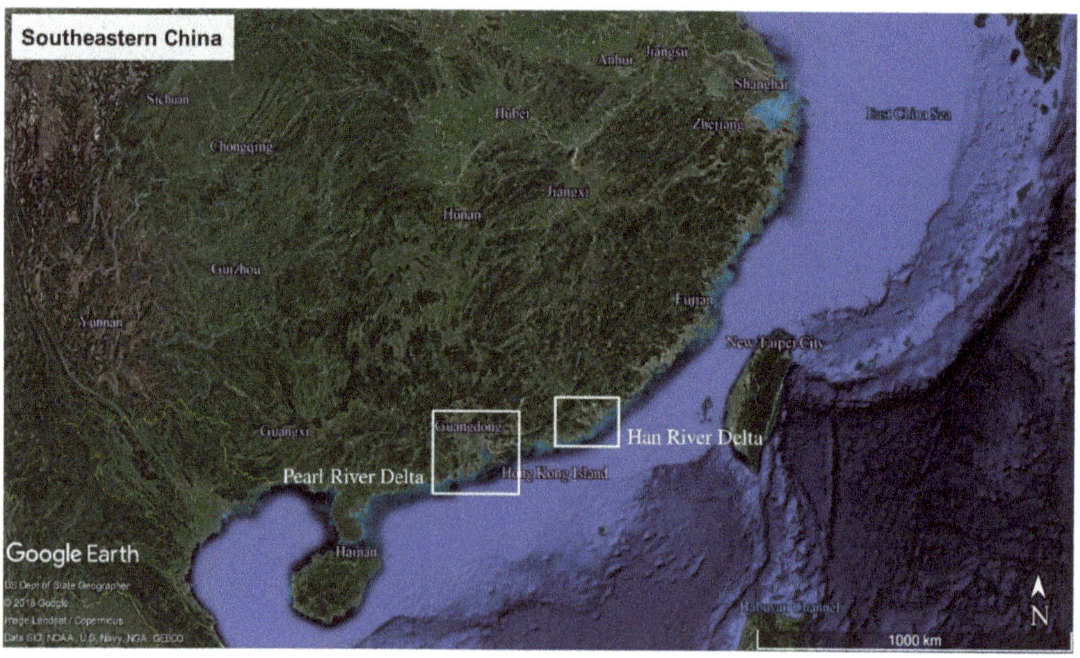

(a)

Figure 1. *Cont.*

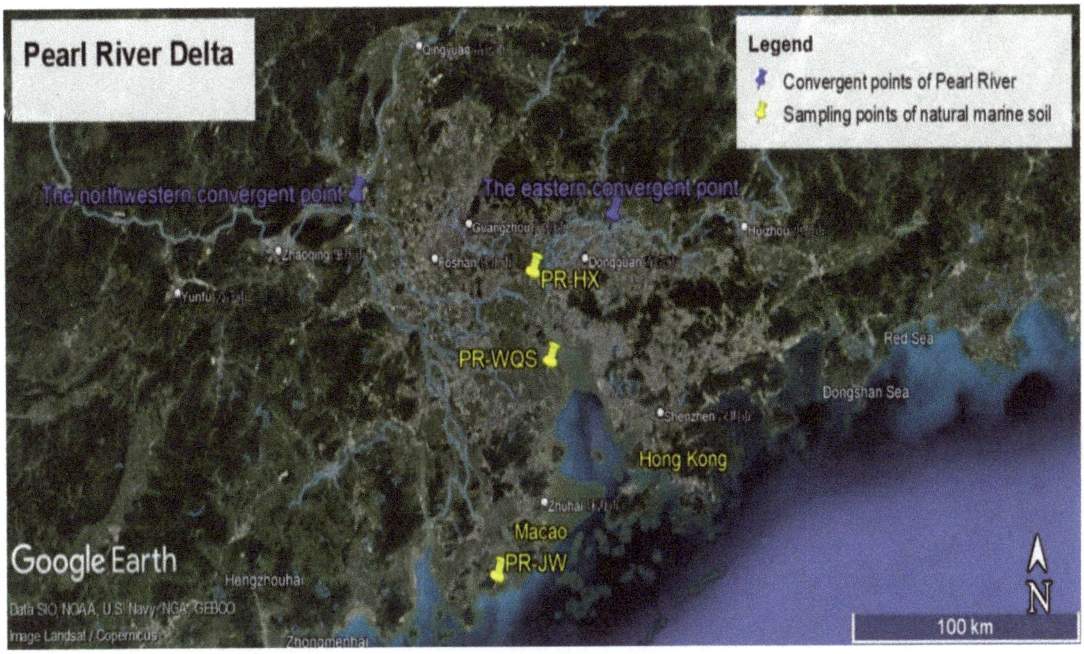

(**b**)

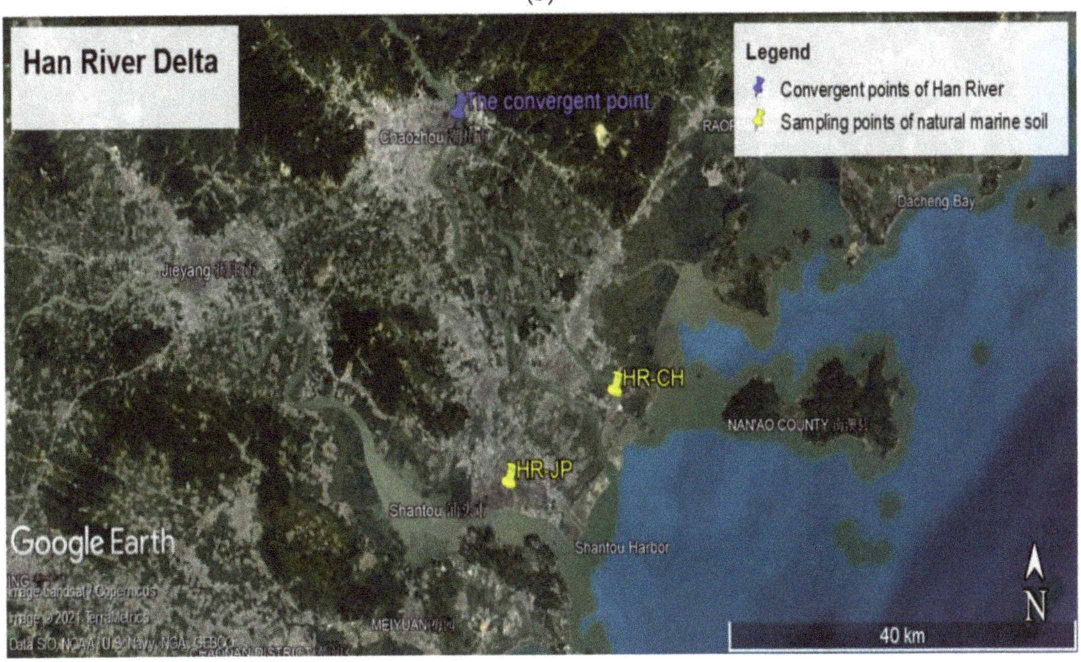

(**c**)

Figure 1. Geomorphologic maps and sampling locations of two typical estuarine deltas in the South China Sea. (**a**) Geomorphologic map of the southeastern coast of China; (**b**) Geomorphologic map and sampling locations of the Pearl River Delta; and (**c**) Geomorphologic map and sampling locations of the Han River Delta.

Table 1. The basic physical properties in marine clay of two deltas in the South China Sea.

Sample Name [1]	Specific Gravity G_s	Natural Void Ratio e	Plastic Limit PL (%)	Liquid Limit LL (%)	Plasticity Index PI (%)	Indexes of Unconsolidated Direct Shear	
						c (kPa)	φ (°)
HR-JP	2.67	1.771	21.0	54.3	33.3	11.1	5.3
PR-HX	2.73	1.584	29.6	54.8	25.2	12.4	2.3
PR-WQS	2.77	1.612	24.4	45.4	21.0	4.5	3.8

[1] HR and PR represent the Han River Delta and the Pearl River Delta, respectively. JP, HX, and WQS represent the names of sampling locations.

Table 2. The mass ratio of sedimentary mineral composition in marine clay of two deltas in the South China Sea.

Sample Name	Depth (m)	Quartz (%)	Albite (%)	Montmorillonite (%)	Kaolinite (%)	Illite (%)	Chlorite (%)	Calcite (%)	Pyrite (%)	Geological Age
HR-JP	8.2~8.7	20.1	11.5	-	20.8	23.9	23.7	-	-	Sub-Atlantic period (Q_4^3)
PR-HX	22.1~22.5	24.8	-	-	23.3	23.4	26.3	-	2.2	Atlantic period (Q_4^{2-1})
PR-WQS	7.8~8.3	34.1	-	-	19.3	17.5	24.1	5.0	-	Sub-Atlantic period (Q_4^3)

Table 3. The mass ratio of sedimentary minerals composition in marine clay of the Pearl River Delta in the Jinwan (JW) District of Zhuhai [53] and the Han River Delta in the Chenghai (CH) District of Shantou [54].

Sampling Location	Sample Name	Depth (m)	Quartz (%)	Albite (%)	Montmorillonite (%)	Kaolinite (%)	Illite (%)	Halite (%)	Calcite (%)	Pyrite (%)	Other Minerals (%)	Geological Age
Pearl River Delta	PR-JW (1-1)	3.0	41.2	7.4	7.4	22.2	19.8	-	-	-	2.0	Sub-Atlantic period (Q_4^3)
	PR-JW (1-8)	19.0	41.8	5.9	8.5	18.8	17.6	-	-	2.6	4.8	Atlantic period (Q_4^{2-1})
	PR-JW (2-1)	4.5	33.6	6.6	14.0	18.9	18.8	2.5	3.3	1.4	0.9	Sub-Atlantic period (Q_4^3)
	PR-JW (2-8)	18.5	40.4	5.1	11.8	16.0	17.7	2.4	4.2	2.3	0.1	Atlantic period (Q_4^{2-1})
Han River Delta	HR-CH (39-1)	4.0	25.6	7.9	19.8	15.5	13.4	17.6	-	-	0.2	Sub-Atlantic period (Q_4^3)
	HR-CH (39-4)	19.5	29.5	9.8	-	20.6	24.1	15.9	-	-	0.1	Atlantic period (Q_4^{2-1})

2.1.2. Mineral Composition of Artificial Marine Clay

Based on results of the above-mentioned mineral component analysis of marine clay, samples were obtained by mixing corresponding proportions of raw minerals for further laboratory dynamic testing and sediment fabric measurement. Four raw minerals were selected as the raw materials to create the artificial marine clay. Firstly, the non-clay minerals were selected from the main components: quartz and albite. Secondly, the clay minerals were selected from Na-montmorillonite and kaolinite, which are the main clay mineral components in natural marine clays [39]. In his study of clay mineralogy, Grim [55] found that these two clay minerals displayed extreme characteristics in terms of isomorphous substitution, specific surface area, and plasticity. According to earlier mineral analysis studies in the field [56], these four minerals were identified as the most common ones in natural marine clay. The raw minerals extracted from mining and their main parameters and mineral composition are shown in Table 4.

Table 4. The basic physical properties of raw minerals.

Raw Mineral	Specific Gravity G_s	Plastic Limit PL (%)	Liquid limit LL (%)	Plasticity Index [1] PI (%)	Specific Surface Area [2] S_s (m^2/g)	Mineral Composition [3] (%)
Quartz	2.656	9.1	15.7	6.6	2.66	Quartz 97.0, Mica 3.0
Albite	2.659	6.8	12.6	5.8	5.47	Albite 91.4, Anorthite 8.5
Na-montmorillonite	2.888	42.0	173.0	131.0	336.94	Na-montmorillonite 90.0, Quartz 8.3, Calcite 1.7
Kaolinite	2.500	36.8	64.6	27.8	11.81	Kaolinite 98.2, Plagioclase 1.8

[1] The Atterberg limit of non-clay minerals including the plastic limit and the liquid limit is due to the long-range interparticle attractive force, i.e., van der Waals forces [39]. The plasticity index is equal to the liquid limit minus the plastic limit. [2] The specific surface area of raw minerals is determined by using ethylene glycol monomethyl ether [57]. [3] The mineral composition of raw materials is determined by the X-ray diffraction based on the Normalized Reference Intensity Ratio Method [58].

Using a laser-scattering particle-size distribution analyzer, the particle size distribution of four types of raw minerals and three types of natural marine clays presented in Table 1 is shown in Figure 2. By screening out particles larger than 75 μm, the D50 (mean particle size) of raw minerals was explicitly controlled from 5 μm to 15 μm, including 10.31 μm for quartz, 7.67 μm for albite, 14.81 μm for Na-montmorillonite, and 5.47 μm for kaolinite. The mean particle sizes of the four raw minerals were consistent with the particle size of natural marine clays (HR-JP, PR-HX, and PR-WQS). The samples of the four minerals with similar mean particle sizes presented three advantages: (a) When all particle sizes were less than 75 μm, the artificial marine clay was classified as fine-grained soil; (b) If the size of particles approached 75 μm, the fine-grained soil would exhibit a rapid shaking reaction, thereby causing water slowly to appear on the surface of the specimen in liquefaction-like form during sample preparation and testing [59]. The mean particle sizes of typical natural marine clay in both deltas were furthermore primarily measured around 5 μm as shown in Figure 2. Therefore, the mean particle size of the four minerals was controlled to remain close to 5 μm rather than 75 μm. (c) If the particle size of each raw mineral presented different orders of magnitude, then particle size did become a relevant factor influencing the dynamic properties of fine-grained soil as mentioned in the Introduction of this paper. For these three reasons, the particle size of the four raw minerals was limited to a consistent narrow range.

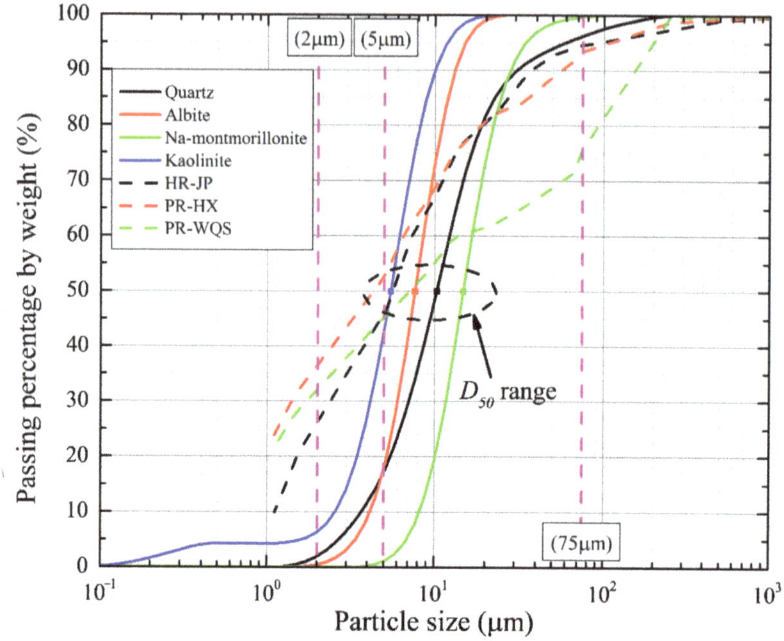

Figure 2. Particle size distribution curve of four types of raw minerals and three types of natural marine clay.

2.2. *Laboratory Dynamic Triaxial Test*

2.2.1. Specimen Preparation

On the basis of the mineral composition of natural marine clays, all artificial marine clays were made up of four raw minerals. The artificial marine clay was produced by mixing components according to specific mass ratios. The mass ratio of quartz to albite was fixed at 8:2 in non-clay minerals, while the mass ratio of Na-montmorillonite to kaolinite was adjusted to between 9:1 and 2:8 in clay minerals. The abbreviations for mass ratio of raw minerals (quartz: albite: Na-montmorillonite: kaolinite) were defined as the name of the corresponding artificial marine clay specimen. For example, the ratio of quartz, albite, Na-montmorillonite, and kaolinite was 8:2:9:1, hence the name of the associated specimen in this paper was 8291. The specimens were remolded into eight layers using a moist tamping method. This sample preparation method was used in preference to the slurry deposition method commonly used for marine clay to avoid the delamination caused by different mineral deposition rates. All samples intended for laboratory dynamic triaxial testing were 39.1 mm (diameter) $\times$ 80 mm $\pm$ 0.5 mm (height). Lastly, six test specimens were produced, the names and physical parameters of which are shown in Table 5. It should be noted that since the void ratio significantly influences the mechanical properties and dynamic characteristics of marine clay [60], all saturated specimens comprised the same void ratio (1.537), whilst retaining different raw mineral volume fractions in accordance with their variable mass ratios.

Table 5. The parameters of laboratory dynamic triaxial testing of artificial marine clays.

Test Speci-men	Mass Ratio of Raw Minerals (Quartz: Albite: Na-Montmorillonite: Kaolinite)	Specific Gravity G_s	Saturated Water Content $w_{initial}$ (%)	Plastic Limit PL (%)	Liquid Limit LL (%)	Plasticity Index PI (%)	Liquidity Index [1] LI	Cyclic Stress Ratio CSR	Consolidation Pressure: Axial Pressure σ_a (kPa)	Consolidation Pressure: Radial Pressure σ_r (kPa)	Consolidation Pressure: Back Pres-sure σ_B (kPa)	Quality of Specimen: Volume Change ε_{vol} (ml)	Quality of Specimen: Relative Void Ratio Change $\Delta e/e_o$	Terminated Cycle N_f	Maximum Cyclic Strain Amplitude [2] $\varepsilon_{cyc,max}$ (%)	$VEDR_{breakpoint}$ Cycle [3] $N_{breakpoint}$
DT-1	8:2:9:1 (8291) [4]	2.75	55.8	23.2	124.1	100.9	0.32	0.284	448.8	448.4	250.3	26.2	0.45	1000	0.91	-
DT-2	8:2:8:2 (8282)	2.73	56.2	22.8	119.0	96.2	0.35	0.282	449.2	449.0	249.6	19.2	0.33	1000	3.53	815
DT-3	8:2:6.7:3.3 (8273)	2.71	56.7	22.3	108.3	86.0	0.40	0.268	448.2	449.4	245.8	16.8	0.29	414	3.52	362
DT-4	8:2:5:5 (8255)	2.68	57.4	21.3	95.5	74.2	0.49	0.282	452.0	451.4	250.7	19.2	0.33	274	3.77	229
DT-5	8:2:3.3:6.7 (8237)	2.64	58.1	19.9	77.8	57.9	0.66	0.269	452.3	451.6	245.0	17.7	0.30	133	4.12	105
DT-6	8:2:2:8 (8228)	2.62	58.7	19.6	67.1	47.6	0.82	0.281	448.9	449.6	247.4	20.5	0.35	107	4.26	94

[1] The liquid limit index of the remolded sample is calculated at the void ratio of 1.537. [2] Maximum cyclic strain amplitude ($\varepsilon_{cyc} = (\varepsilon_{max} - \varepsilon_{min})/2$) is at the end of the test. [3] $N_{VEDRbreakpoint}$ is the number of cycles at $VEDR_{breakpoint}$. [4] The abbreviation for specimen is in parentheses.

In addition, a higher content of Na-montmorillonite was found to slightly increase the plastic limit, significantly raise the liquid limit, and was accompanied by a surge in the plasticity index, as shown in Table 5. The Montmorillonite activity [61] was introduced to reflect the linear effect of Na-montmorillonite content (Mc) on plasticity, as shown in Figure 3. When the Na-Montmorillonite content was nil, the mass ratio (quartz: albite: Na-montmorillonite: kaolinite) of mineral composition was 8:2:0:10. The artificial marine clay was composed of only non-clay minerals and kaolinite, and PI = 27.65. In order to express the natural marine clay and artificial marine clay from a geotechnical point of view, a plasticity chart [62] was produced and is shown in Figure 4, which demonstrates that HR-JP of natural marine clays and all artificial marine clays belonged to the category of fat clay, abbreviated CH. PR-HX is on the edge of "A line" and classified as elastic silt, abbreviated MH. PR-WQS is classified as lean clay with sand, abbreviated CL, due to the existence of about 25% coarse-grained particle from particle size distribution in Figure 2.

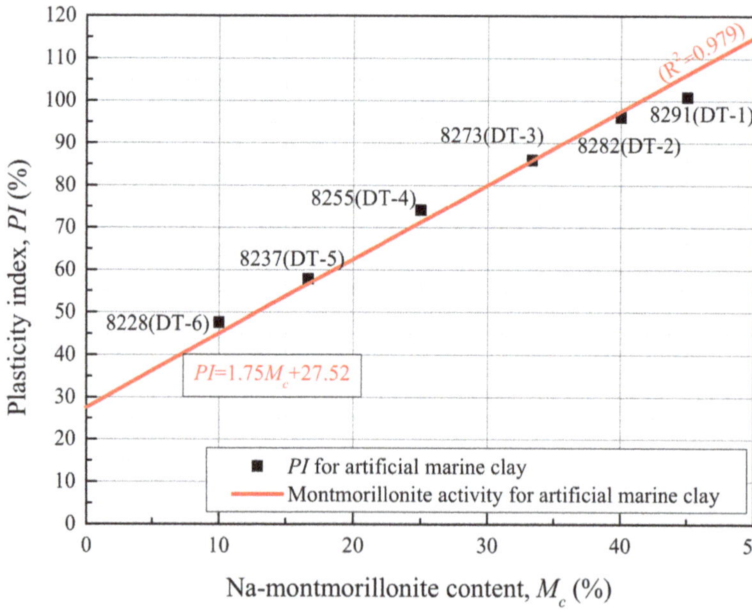

Figure 3. Relationship between Na-montmorillonite content and plasticity index (*PI*) in artificial marine clays.

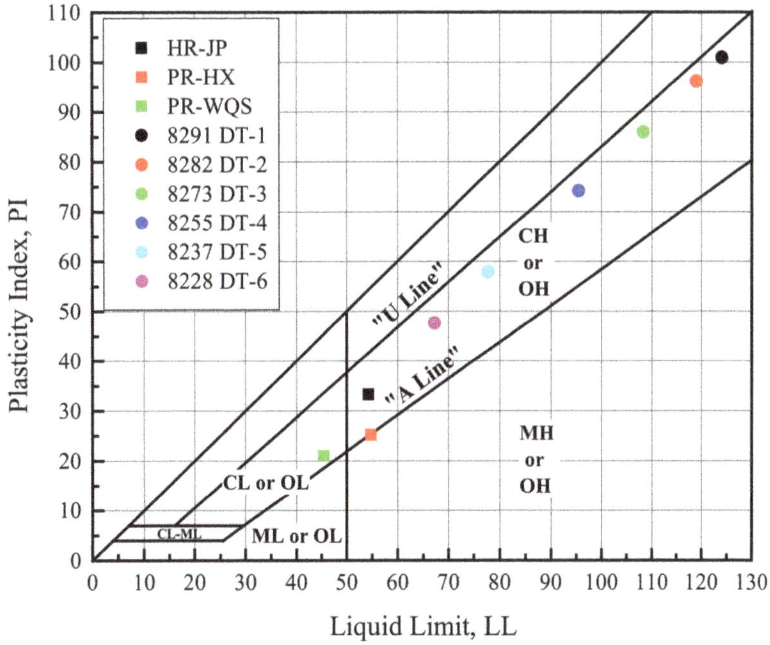

Figure 4. Plasticity chart of natural and artificial marine clays.

2.2.2. Experimental Procedure

In order to analyze the influence of clay mineral composition on the dynamic properties of marine clay under large strain, a series of consolidated undrained cyclic triaxial tests were conducted on soil specimens with different mineral compositions, using the stress-controlled DDS-70 microprocessor-controlled electromagnetic cyclic triaxial testing system (Binta Instrument Technology Co., Ltd., Beijing, China). The experimental apparatus and typical specimen diagram are shown in Figure 5. Existing studies showed that during hydration under high cementation conditions, the dynamic characteristics of marine clay were affected by the consolidation time [28,45]. Meanwhile, the Quaternary marine clay generally showed slight over-consolidation [1,63]. For these reasons, all specimens with saturation coefficient B greater than 0.97 after subsequent vacuum saturation and back pressure saturation were subjected to strict control of the consolidation time during the corresponding process. After 24 hours' consolidation under a back pressure of 200 kPa (the mean principal effective stress was 250 kPa), the primary consolidation was completed. The back pressure was then raised to 250 kPa for a further 24 h (the mean principal effective stress was 200 kPa). The resulting specimens showed the same consolidation time with an over-consolidation ratio of 1.25. As indicated in Table 5, cyclic triaxial tests were conducted on specimens DT-1 to DT-6 under a virtually identical cyclic stress ratio (CSR, namely the ratio of cyclic shear stress amplitude to the mean effective principal stress, τ_{cyc}/p'_0), from 0.268 to 0.284. This CSR range ensured that all specimens underwent at least 100 cycles by the end of the test. During the experiment, the cyclic loading waveform was sinusoidal, and the loading frequency was 0.1 Hz. The test was terminated when a single amplitude strain reached 5% or the number of cycles reached 1000. Figure 6 below shows the DT-6 results of dynamic triaxial testing of artificial marine clay.

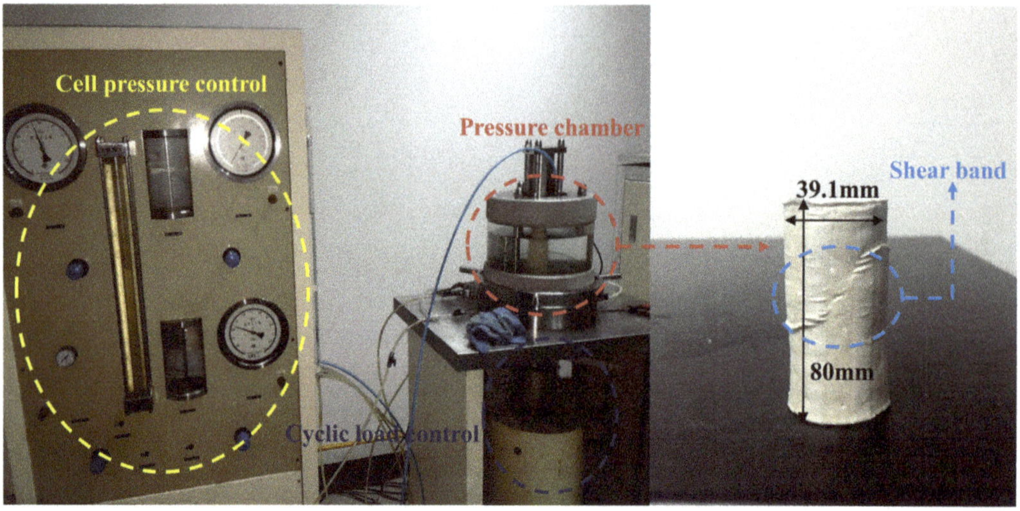

Figure 5. Experimental apparatus (DDS-70) and typical specimen diagram.

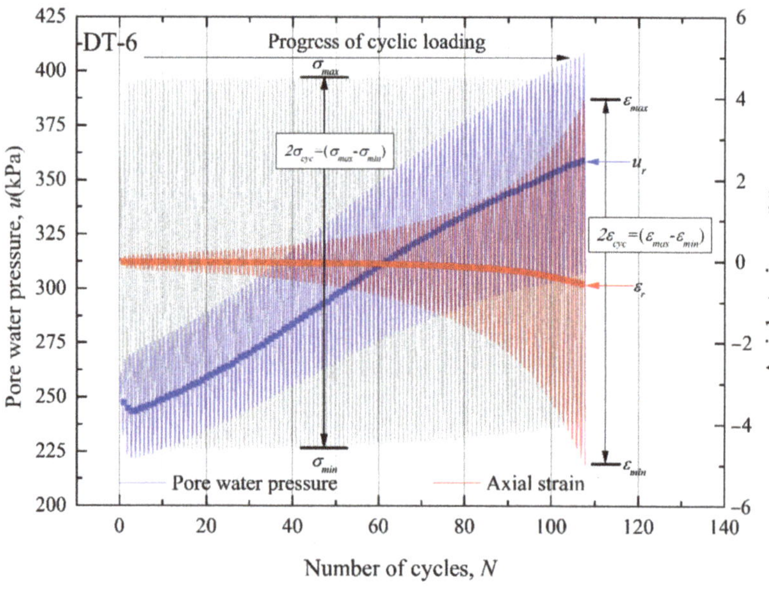

Figure 6. Results of dynamic triaxial test on artificial marine clay (DT-6).

To verify the repeatability of cyclic triaxial testing of artificial marine clays, two series of original and parallel trials were performed on the 8255 series ($CSR \approx 0.215$, test 1 and 2 are not listed in Table 5 as a test series) and 8228 series ($CSR \approx 0.281$) under identical test conditions, as shown in Figure 7. The figures reveal the results of two test components: (a) strain development and (b) hysteresis curve. Strain development included cyclic axial strain amplitude (the maximum and minimum axial strain amplitude at each cycle, $\varepsilon_{cyc} = (\varepsilon_{\max} - \varepsilon_{\min})/2$) and residual axial strain (cumulative strain when deviator stress is nil at each cycle, ε_r). The hysteresis curves represented the results of the 8255 series at the 500th cycle and the 8228 series at its failure cycle, respectively. As the figures illustrate, the differences in strain development and hysteresis curve of representative cycles were minimal in both series. The above results further demonstrated the reliability of cyclic triaxial testing of artificial marine clays.

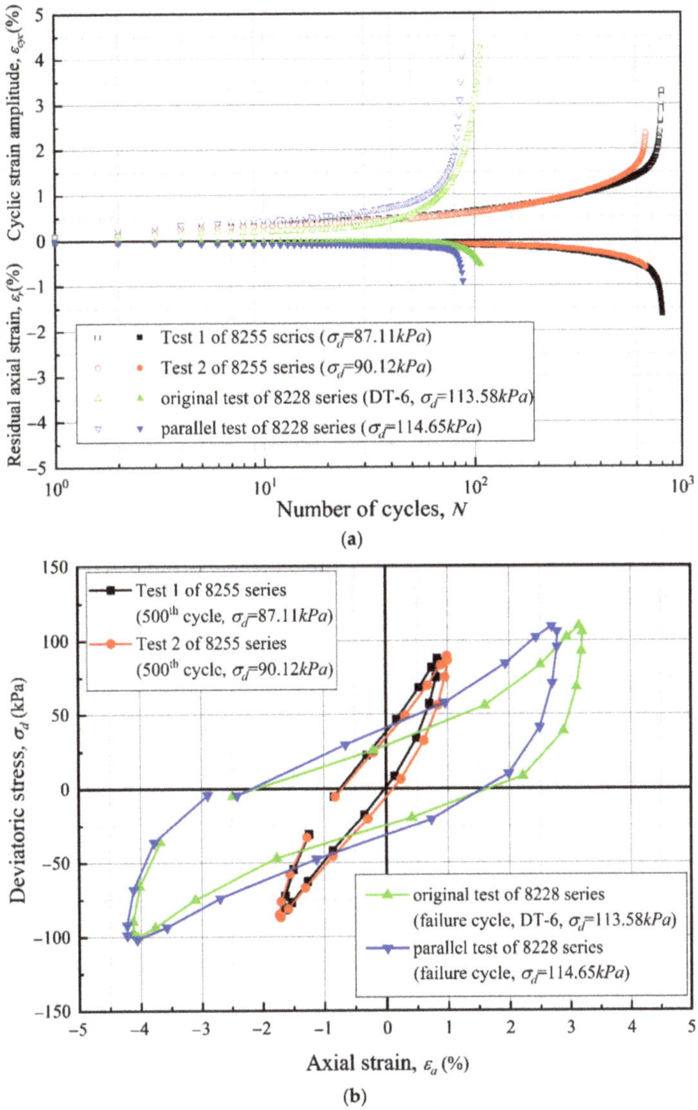

Figure 7. Dynamic characteristics of artificial marine clays in two series (test 1 and 2 of 8255 and original and parallel tests of 8228) under cyclic loading: (**a**) Relationship between number of cycles versus cyclic axial strain amplitude and residual axial strain (deviatoric stress is zero at each cycle); (**b**) Hysteresis curves of specific cycles.

3. Results

3.1. Cyclic Triaxial Test Results

The results of laboratory large strain dynamic testing of artificial marine clays were generally presented under three categories: (a) strain development; (b) pore water pressure change; and (c) hysteresis curve shape.

3.1.1. Strain

The strain development test results were presented in terms of cyclic axial strain amplitude and residual axial strain, as shown in Figure 8. The cyclic axial strain amplitude was used as an indicator of strain development, which increased with the loading cycle until completion of the test. The terminated cycle and maximum cyclic amplitude strain are presented in Table 5. As the proportion of Na-montmorillonite in the specimens decreased (from DT-3 to DT-6), the maximum cyclic amplitude strain at the failure cycle increased from 3.52% to 4.26% under identical test termination conditions (i.e., with the single amplitude strain of 5%). At the same time, the cyclic amplitude strain curve of the specimens with less Na-montmorillonite content changed from a gradual to an abrupt rise. The mineral composition of Na-montmorillonite was therefore found to hinder the rapid development of strain under unchanged remaining test conditions; increased Na-montmorillonite content was furthermore shown to further limit the development of strain significantly.

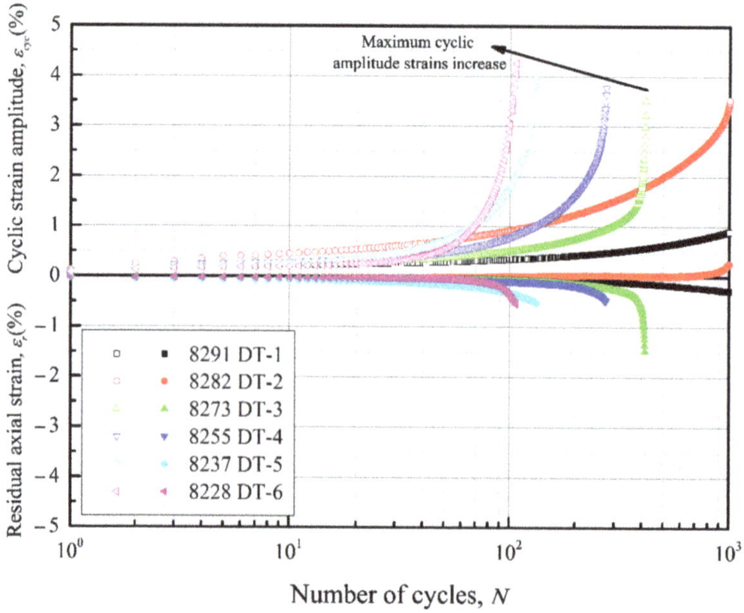

Figure 8. Relationship between number of cycles versus cyclic axial strain amplitude and residual axial strain for artificial marine clays.

3.1.2. Pore Water Pressure

The residual pore water pressure (PWP) and time-history of PWP are recorded in Figure 9. The residual PWP was equal to the PWP at zero deviatoric stress within the cycle. Due to the same back pressure and over-consolidation ratio, PWP of all specimens started from 250 kPa, initially developing in a negative direction and then increasing in a positive direction until the tests were terminated. The time-history PWP curves indicated that the development of pore water pressure was synchronous with the sinusoidal cyclic loading. The samples with high montmorillonite content demonstrated a slower cumulative increase rate in pore water pressure and lower residual pore water pressure at the end of the test. Meanwhile, the specimens with extremely high montmorillonite content, especially DT-1, presented a narrower pore water pressure amplitude (the width of time-history PWP) than low montmorillonite content specimens. Therefore, montmorillonite in mineral composition was also found to play a role in inhibiting the rise of PWP while limiting strain increase. The impeditive development of PWP affected both the cumulative results (residual PWP) and cyclic results (PWP amplitude) as the Na-montmorillonite content increased.

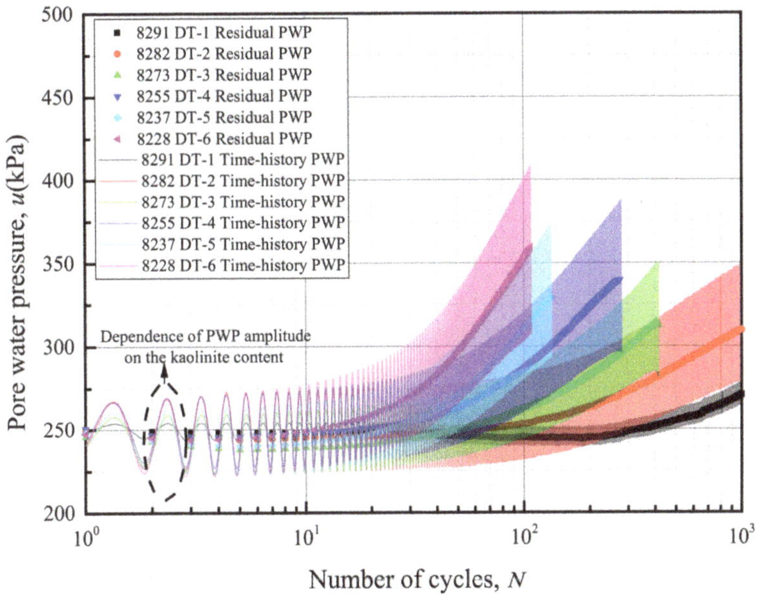

Figure 9. Relationship between number of cycles versus residual pore water pressure (deviatoric stress is zero at each cycle) and time-history pore water pressure for the mineral composition of artificial marine clays.

3.1.3. Hysteresis Curve Shape

Although the CSRs of all specimens were similar; their strains exhibited significant differences due to their various mineral compositions. In order to eliminate the differences in stress and strain, the double-normalized hysteresis curve at the failure cycle (or terminated cycle) was used to compare the shapes of the hysteresis curves. In Figure 10a, the evolution of the double-normalized hysteresis curve process is presented and explained by moving the hysteresis curve back to the origin of the X-axis and double-normalizing. The normalized deviatoric stress in the Figure is equal to the deviatoric stress divided by the cyclic axial stress amplitude ($\sigma_{cyc} = (\sigma_{\max} - \sigma_{\min})/2$). The normalized axial strain is equal to the axial strain after the hysteresis curve has been moved back to the origin ($Movement = \varepsilon_a - (\varepsilon_{\max} + \varepsilon_{\min})/2$), divided by the cyclic axial strain amplitude. Subsequently, the double-normalized hysteresis curves of artificial marine clays at the failure cycle are presented in Figure 10b. Here, some specimens with high montmorillonite content (DT-1 and DT-2) reveal narrow hysteresis curves resembling a "banana shape", because they only meet the requirements for 1000 terminated cycles, but not those for large strain failure. In other specimens (DT-3 to DT-6), the samples with high montmorillonite content such as DT-3 and DT-4, comprise a more rounded "olive shape" than the specimens with low montmorillonite content such as DT-5 and DT-6. A corresponding relationship can furthermore be seen in Figure 10b between the shape of the double-normalized hysteresis curve and viscous energy dissipation, which will be discussed in the following Section 4.2.

3.2. Stiffness and Energy Dissipation

Previous research [3,64,65] indicated that the dynamic properties of stiffness degradation and energy performance in artificial marine clays could be further calculated and explored on the basis of test results of strains and hysteresis curves under cyclic loading.

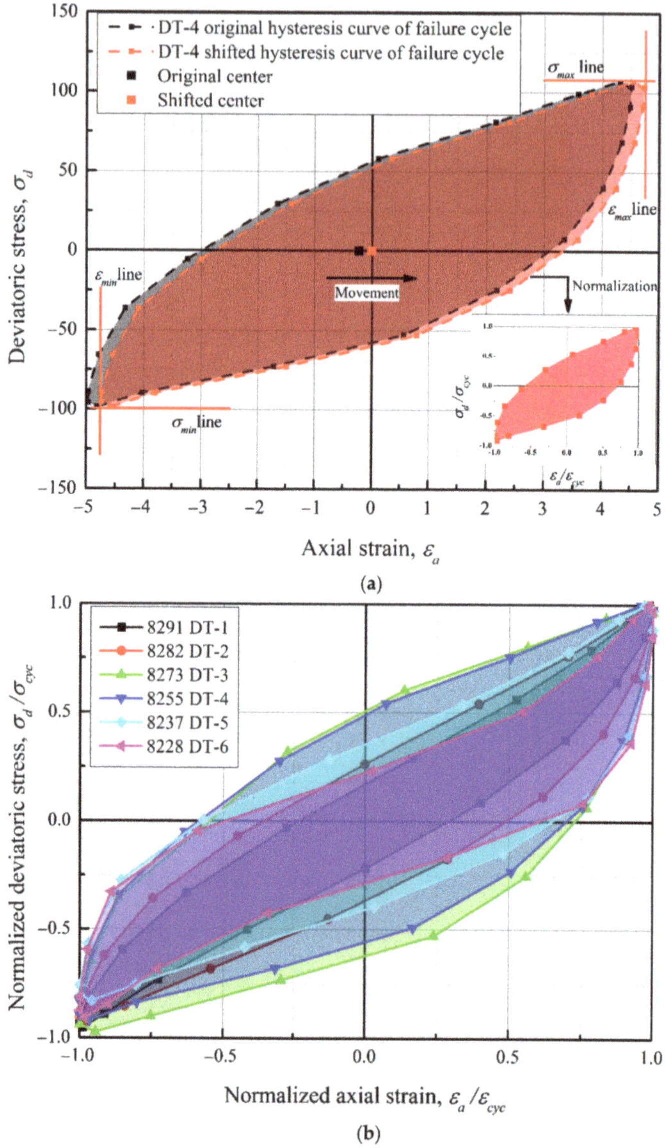

Figure 10. Double-normalized hysteresis curve of failure (or terminate) cycle: (**a**) Schematic diagram of normalization; (**b**) Double-normalized hysteresis curve of failure (or terminate) cycle for the mineral composition of artificial marine clays.

3.2.1. Stiffness Degradation

The cyclic axial strain amplitude (ε_{cyc}) was transformed into cyclic shear strain amplitude (γ_{cyc}) by Poisson's ratio (μ) as shown in Equation (1) below. The shear modulus under the cyclic triaxial test was obtained from the slope of the stress–strain top-line in Equation (2) below and Poisson's ratio conversion in Equation (3) below. Based on the assumption of undrained conditions, Poisson's ratio was assumed to be 0.5 in this study.

$$\gamma_{cyc} = (1 + \mu)\varepsilon_{cyc} \tag{1}$$

$$E_d = \frac{\sigma_{cyc}}{\varepsilon_{cyc}} = \frac{(\sigma_{\max} - \sigma_{\min})/2}{(\varepsilon_{\max} - \varepsilon_{\min})/2} \tag{2}$$

$$G_d = \frac{E_d}{2(1+\mu)} \tag{3}$$

In order furthermore to eliminate the influence of the maximum dynamic shear modulus of various artificial marine clays, the normalized dynamic shear modulus was presented. The maximum dynamic shear moduli of artificial marine clay were determined from the results of resonance column tests [41] and shown in Table 6. The relationship between cyclic shear strain amplitude in large-scale strain and normalized shear modulus was obtained, as shown in Figure 11.

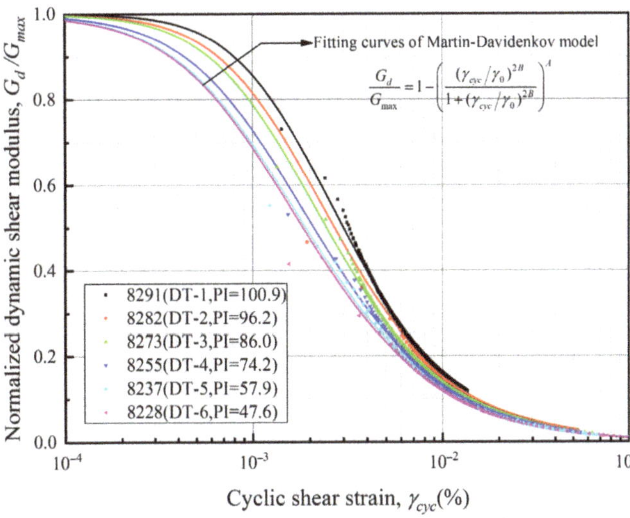

Figure 11. Relationship between cyclic shear strain and normalized dynamic shear modulus curves for the mineral composition of artificial marine clays.

Table 6. The parameters result of the Martin–Davidenkov model of artificial marine clays.

Test Specimen	Plasticity Index I_P	Maximum Dynamic Shear Modulus [1] G_{max} (MPa)	Martin–Davidenkov Model Parameters			Correlation Coefficient R^2
			γ_0	A	B	
DT-1	100.9	8.56	1.587×10^{-3}	1.889	0.636	0.997
DT-2	96.2	8.08	1.394×10^{-3}	1.859	0.599	0.993
DT-3	86.0	8.03	1.318×10^{-3}	1.787	0.598	0.997
DT-4	74.2	6.29	1.234×10^{-3}	1.594	0.566	0.995
DT-5	57.9	5.71	1.140×10^{-3}	1.556	0.554	0.991
DT-6	47.6	4.94	1.104×10^{-3}	1.552	0.553	0.999

[1] The maximum dynamic shear modulus of artificial marine clays refers to Shan et al. [41] by resonance column test.

With the increased cyclic shear strain, the normalized shear modulus exhibited a tendency towards degradation. The cyclic shear strain ranged from 1.24×10^{-3} to 7.17×10^{-2}, a relatively large strain. Lastly, under a cyclic shear strain of 10^{-1}, the shear modulus of all specimens degraded from around 0.6 to nearly zero. The samples with high montmorillonite content displayed greater normalized shear moduli under the same cyclic shear strain. Since stiffness resulted from the transmission of shear waves between particles and intergranular contact, the specimens with high montmorillonite content (high plasticity) displayed greater contact, even once the shear strain had occurred. This conclusion was deemed consistent with the small strain stiffness test results of resonant column testing in artificial marine clays [41]. Meanwhile, combined with the plasticity index of artificial marine clays, the above conclusion, was deemed consistent with the relationship between stiffness degradation and soil plasticity for natural normal and over-consolidated soils studied by Vucetic and Dobry [66].

In addition, the fitting curves of the nonlinear Martin–Davidenkov model [67] was adopted to describe the stiffness degradation, as shown in Figure 11. The Martin–Davidenkov model was expressed as follow:

$$\frac{G_d}{G_{\max}} = 1 - \left(\frac{(\gamma_{cyc}/\gamma_0)^{2B}}{1 + (\gamma_{cyc}/\gamma_0)^{2B}} \right)^A \tag{4}$$

where γ_0, A, and B were fitting parameters related to stiffness degradation, and the corresponding parameter results are presented in Table 6.

The plasticity index (*PI*) rose with the increase in montmorillonite content within the mineral composition. At that point, all three parameters in the Martin–Davidenkov model showed transformation of power functions (Equations (5)–(7) below) trend increases with the rise of *PI*, as shown in Figure 12, which was deemed consistent with the above stiffness degradation.

$$\begin{gathered} \gamma_0 = \gamma_{0,\min} + a_\gamma \cdot (PI)^{b_\gamma} \\ (\gamma_{0,\min} = 1.109 \times 10^{-3},\ a_\gamma = 7.237 \times 10^{-12},\ \text{and } b_\gamma = 4.882) \end{gathered} \tag{5}$$

$$\begin{gathered} A = A_{\min} + a_A \cdot (PI)^{b_A} \\ (A_{\min} = 1.502,\ a_A = 5.706 \times 10^{-8},\ \text{and } b_A = 3.420) \end{gathered} \tag{6}$$

$$\begin{gathered} B = B_{\min} + a_B \cdot (PI)^{b_B} \\ (B_{\min} = 0.549,\ a_B = 6.532 \times 10^{-11} \text{ and } b_B = 4.533) \end{gathered} \tag{7}$$

where $\gamma_{0,\min}$, a_γ, b_γ, $A_{\min}$, a_A, b_A, $B_{\min}$, a_B, and b_B are fitting parameters. When the mass ratio (quartz: albite: Na-montmorillonite: kaolinite) of mineral composition was 8:2:0:10, the lower limit reached by the domain of *PI* in the three equations was 27.52, as can be seen in Figure 3. At the same time, γ_0, *A*, and *B* were 1.110×10^{-3}, 1.507, and 0.550, respectively, which were all minimum values.

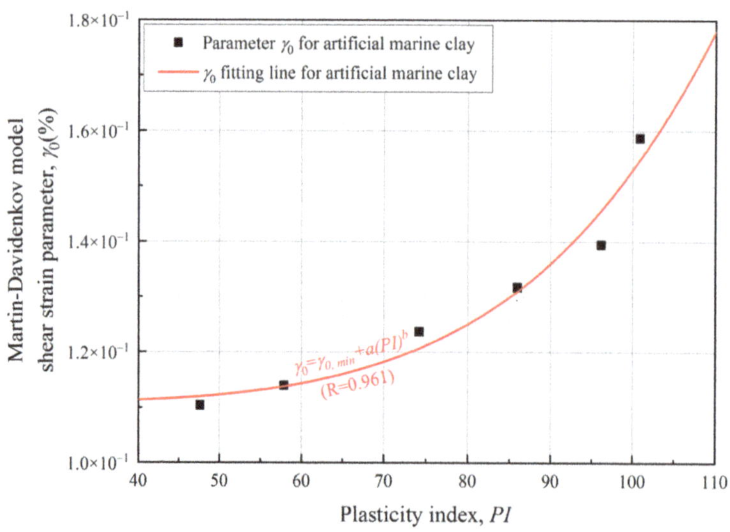

(a)

Figure 12. *Cont.*

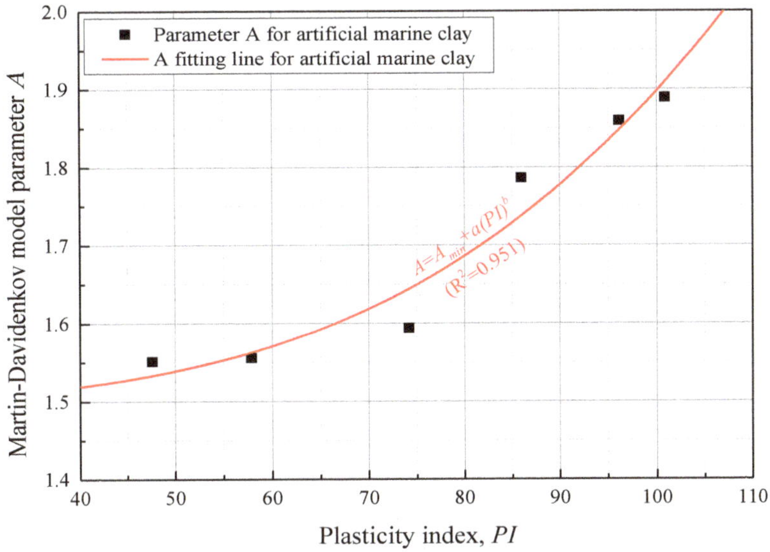

(b)

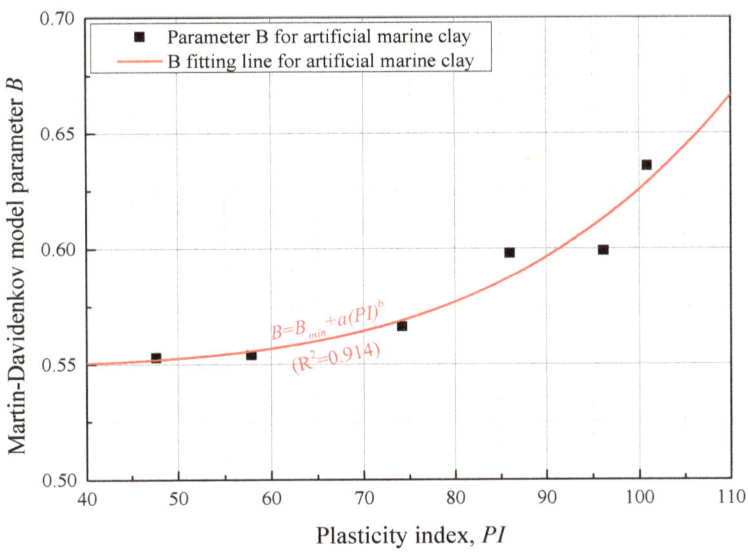

(c)

Figure 12. Relationship between plasticity index and Martin–Davidenkov Model parameter curves for mineral composition of artificial marine clays: (**a**) shear strain parameter, γ_0; (**b**) parameter A; (**c**) parameter B.

3.2.2. Viscous Energy Dissipation

The area contained within the hysteresis curve is taken as viscous energy dissipation of each cycle has been widely recognized by researchers [68–70]. The calculation method of viscous energy dissipation for each cycle is presented in Equation (8). Furthermore, the viscous energy dissipation ratio (VEDR), a non-dimensional energy index to eliminate stress and strain [71–73], was introduced to investigate the viscous energy dissipation characteristics of artificial marine clay. VEDR is dimensionless in terms of viscous energy dissipation in each cycle. It is similar to calculating the area ratio between hysteretic curve and circumscribed square in double-normalized hysteresis curves. The expression and schematic diagram of VEDR are shown in Equation (9) and Figure 13 below, respectively.

$$\Delta W_i = A_{Hysteresis} = \sum_{i=1}^{n-1} \frac{(\sigma_{d,i+1} + \sigma_{d,i})}{2}(\varepsilon_{a,i+1} - \varepsilon_{a,i}) \tag{8}$$

$$\text{VEDR} = \frac{\Delta W_i}{S_{\text{rectangle}}} = \frac{\Delta W_i}{(\sigma_{\max,i} - \sigma_{\min,i})(\varepsilon_{\max,i} - \varepsilon_{\min,i})} \tag{9}$$

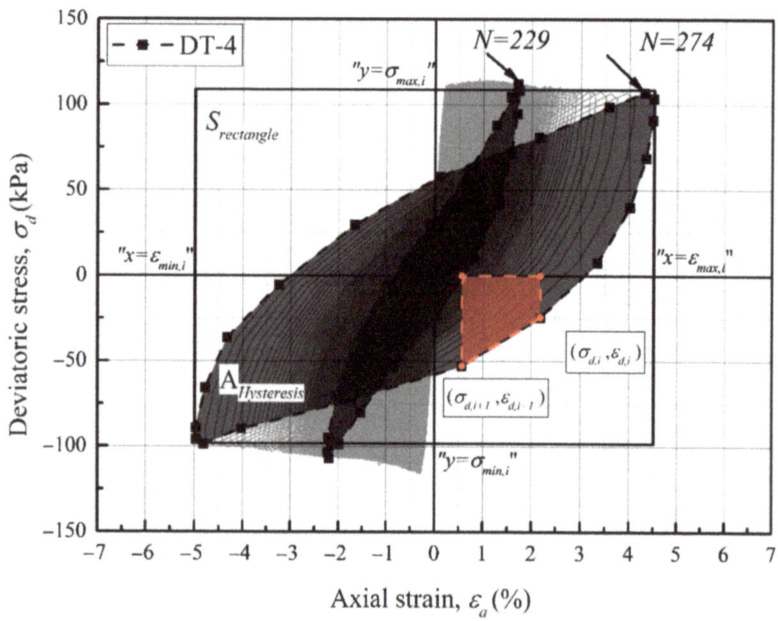

Figure 13. Typical hysteresis curve for calculation of viscous energy dissipation ratio (VEDR).

Figure 14 shows the relationship between number of cycles and VEDR for artificial marine clays. The VEDR curves follow a slight increase–slow decrease–slow increase pattern before undergoing a sudden, rapid increase at the obvious breakpoint, except in the case of DT-1. DT-1 is where failure does not occur due to the required single amplitude strain, hence the final stage of rapid increase is not reached. In general, the VEDR curves showed a rising trend, indicating that viscous energy dissipation in artificial marine clay increased overall with cyclic loading. The wave-like formation and rising trend of VEDR curves described above were confirmed to be fully consistent with the research results of Shan et al. [61].

At the same time, the artificial marine clay specimens with various clay mineral contents displayed different VEDRs in the early and failure cycles. In the early cycles, the development of VEDRs was limited by the content in montmorillonite. VEDRs with high montmorillonite content were lower than those with high kaolinite content in the same cycles. In the failure cycles, VEDRs (for DT-3 to DT-6) increased with the rise in montmorillonite content. This indicated that a high montmorillonite content reflected more prominent viscous energy dissipation capability under the same failure strain. However, the VEDRs in the terminal cycles of samples with extremely high montmorillonite content (DT-1 and DT-2) were lower. The differences in shape of the double-normalized hysteretic curves also reflected the energy dissipation status of artificial marine clays. The rounder, olive-shaped double-normalized hysteretic loops reflected greater viscous energy dissipation. In contrast, the double-normalized hysteresis curves of nonfailure specimens with extremely high montmorillonite content were banana-shaped.

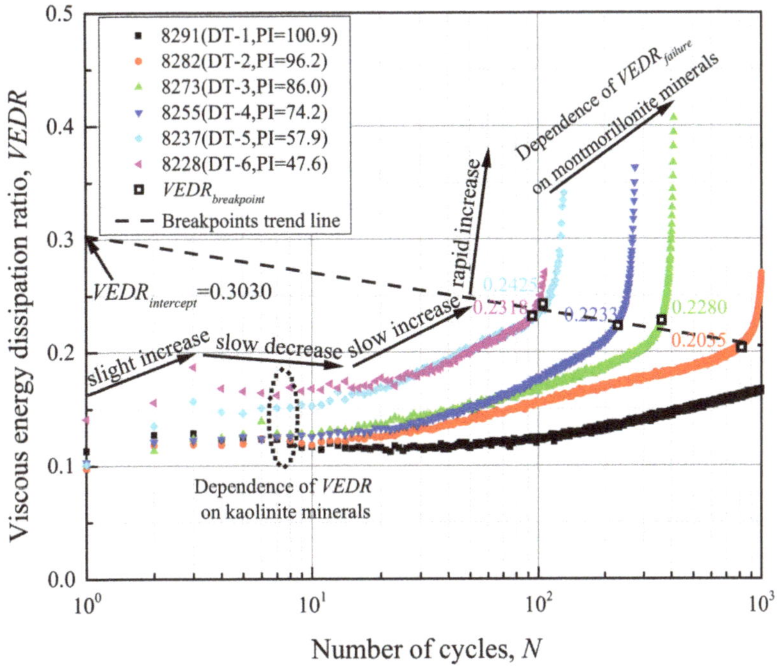

Figure 14. Relationship between number of cycles and VEDR for artificial marine clays.

Furthermore, the $VEDR_{breakpoint}$ of artificial marine clays was affected by the montmorillonite content of the clay. Shan et al. [61] indicated that $VEDR_{breakpoint}$ was the point at which the gradient of the VEDR curve became very steep and marked the energy turning point at a late stage of cyclic loading. The $VEDR_{breakpoint}$ of artificial marine clay is illustrated in Figure 14. With reduced montmorillonite content and increased kaolinite content, the $VEDR_{breakpoint}$ rose slowly. Eventually, the intercept of $VEDR_{breakpoint}$ trend line in artificial marine clay was 0.3030. It should be noted that this intercept of VEDR represents the energy ratio at the point when soil specimens failed after only one cycle and should only correlated with the composition and proportion of non-clay minerals. In order to further illustrate the dynamic characteristics of the $VEDR_{breakpoint}$, the linear relationship between cyclic strain amplitude at breakpoint ($\varepsilon_{breakpoint}$) and plasticity index (*PI*) for artificial marine clays is shown in Figure 15. This provided further evidence that the dissipated viscous energy of high montmorillonite content specimens was inhibited, and the cyclic strain of energy breakpoint was delayed.

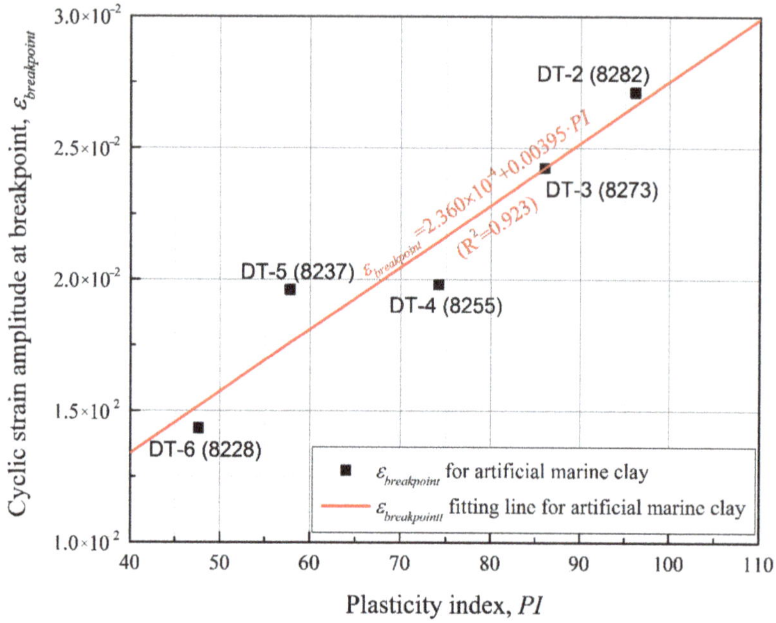

Figure 15. Relationship between plasticity index and cyclic strain amplitude at breakpoint for artificial marine clays.

4. Microstructure and Fabric

The low-vacuum environmental scanning electron microscopy testing (LVESEM, Quanta 200, Center for Electron Microscopy and Analysis (CEMAS), Columbus, OH, USA) and MIP testing (AutoPore 9510, Micromeritics Instrument Corporation., Norcross, GA, USA) were conducted to investigate artificial marine clay fabric microstructure and microporosity in artificial marine clay specimens. Four out of the six consolidated artificial marine clays (8291, 8282, 8255, and 8228 specimens) were selected to undergo the sediment fabric tests. Freeze-drying pretreatment [74] was carried out on the artificial marine clay specimens after the consolidated treatment before the sediment fabric tests. The LVESEM and MIP test specimens were immersed for about 1 min in a container filled with isopentane as a transition. Then, the specimens were transferred to a container filled with liquid nitrogen (−196 °C) and soaked for about 1 min. After complete freezing, the pore water was extracted using a vacuum pump that provided an absolute pressure of 6 Pa. The dried specimens were then kept in a vacuum dryer pending the sediment fabric tests.

4.1. LVESEM Analysis

The horizontal section of the dried specimens was broken off by hand, and the flat observation area was selected as the test area for the LVESEM experiment. Figure 16 shows the LVESEM test results for artificial marine clay. Based on the SEM petrology atlas [75], the atlas comparison method was used to distinguish the mineral particles and analyze artificial marine clay fabric in the LVESEM images. The LVESEM images of the four artificial marine clay specimens, the consolidated specimens with high montmorillonite content, especially specimen 8291, demonstrated pronounced flocculated and aggregated structures with edge-to-face and edge-to-edge contacts at the same initial void ratio. At the same time, the consolidated specimens with low montmorillonite content, especially specimen 8228, displayed dispersed and deflocculated structures with face-to-face association. As suggested by Gumaste et al. [44], the reason proposed for this sediment fabric phenomenon was that montmorillonite comprise a larger specific surface area and greater capacity for isomorphous substitution than kaolinite. This was attributed to the fact that montmorillonite particles show greater Coulomb attraction and London–van der Waals forces to overcome diffuse double-layer models of repulsive forces, and eventually form a flocculated sediment fabric.

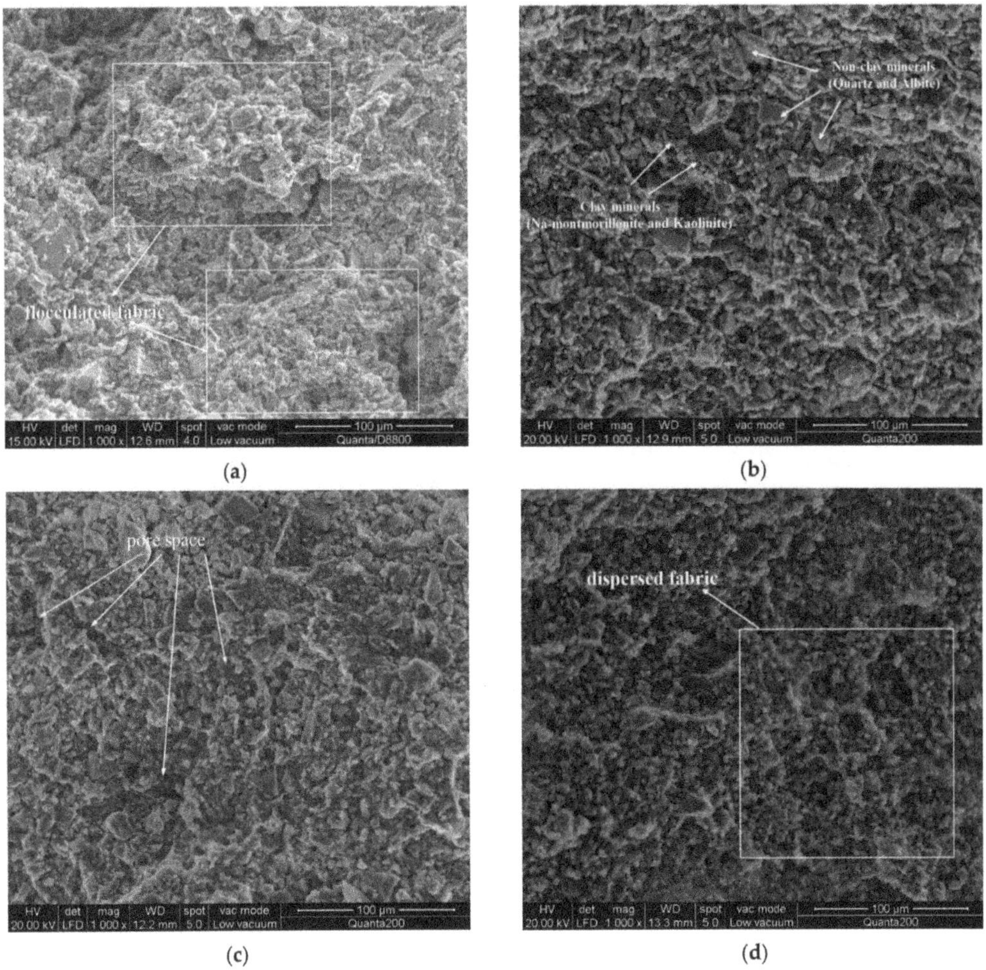

Figure 16. Quanta200 environmental scanning electron microscope (LVESEM) photo of artificial marine clays with a magnification of 1000: (**a**) 8291; (**b**) 8282; (**c**) 8255; (**d**) 8228.

Chen et al. [76] proposed that the unloaded flocculated structures in the sediment fabric tended to show greater random particle orientation and deformability and were thus more sensitive to load. As a result, pore variations of the specimens with higher montmorillonite content were more significant than those with lower montmorillonite content after consolidation under the same effective stress. For example, although samples 8291, 8282, 8255, and 8228 shared the same initial void ratio prior to consolidation, their respective pore area percentages after consolidation were 55.8%, 56.7%, 57.8%, and 61.4% according to LVESEM images from the Particles (Pores) and Cracks Analysis System (PCAS, Nanjing University, Nanjing, Jiangsu, China) [77] which can be processed into binarized images and help calculate the proportion of pore spaces. Compared with kaolinite, montmorillonite comprised a thicker adsorbed water film (stern layer) and narrower free water passage (diffusion layer) between weakly adsorbed water layers. The narrow path and low porosity of montmorillonite hindered the development of pore water pressure under cyclic loading. Research by Pillai et al. [78] furthermore indicated that dispersed microfabric was found to offer reduced resistance under cyclic loading.

As marine clay with lower void ratio was shown to have a higher dynamic shear modulus [79], the low porosity in specimens with high montmorillonite content offered a reasonable explanation for the high stiffness under the same strain conditions from Figure 11. From the perspective of energy dissipation, viscous energy dissipation in cohesionless particles with liquefaction potential is mainly provided by the sliding and rolling occurring between particles [80]. Similarly, in artificial marine clays, the advantage of montmorillonite compared to kaolinite in terms of specific surface area and capacity for isomorphous substitution may result in a significant increase in intergranular contact between particle–particle, particle–aggregate, and aggregate–aggregate. Therefore, this increased intergranular contact increases the difficult of the sliding and rolling between montmorillonite particles in the consolidated flocculated low-porosity specimens under the same cyclic loading. Lastly, the specimens with high montmorillonite content were shown to experience a more prolonged rise in viscous energy before failure on the basis criterion of single amplitude strain, resulting in greater VEDR at the failure cycle ($VEDR_{failure}$). At the same time, the VEDR of high-montmorillonite content specimens remained low, and the cyclic strain amplitudes at breakpoint were delayed.

4.2. MIP Analysis

Dried specimens weighing between 0.8 g and 1 g were tested in the automatic mercury porosimeter. Since mercury is a non-wetting liquid when in contact with mineral particles, the relationship between mercury intrusion pressure and pore diameter could be derived from the Washburn equation [81]. On the basis of Diamond's [82] experimental study of mercury intrusion into clay with kaolinite and illite, the contact angle between mercury and clay minerals was determined as 147 degrees, and the surface tension coefficient as 0.485 $N \cdot m^{-1}$. At the same time, with the increased mercury intrusion pressure, the pressure-associated pore diameter decreased continuously. The results of MIP tests were therefore plotted in terms of the variation of cumulative ($V_{cumulative}$) and incremental ($V_{incremental}$) intrusion volumes of mercury in the specimen with respect to pore diameter (D_i) as shown in Figures 17 and 18, respectively. The cumulative intrusion volume of mercury ($V_{cumulative} = \sum_i \Delta V_i$) reflected the cumulative mercury content of the intrusion curve under different pore diameters. The incremental intrusion volume of mercury ($V_{incremental} = -\Delta V_i / \Delta logD_i$) expressed the characteristics of pore distribution. The MIP test results are listed in Table 7.

Table 7. The result of the fabric in artificial marine clays by LVESEM test and MIP test.

Test Specimen	Plasticity Index I_P	Percentage of Pore Area [1] P_{area} (%)	Final Cumulative Intrusion Volume of Mercury $V_{cumulative}$ (mL/g)	Intra-Aggregate Pore Diameter $D_{intra\text{-}aggregate}$ (nm)	Incremental Intrusion Volume of Mercury at Intra-Aggregate Pore $V_{incremental}$ (mL/g)
8291	100.9	55.8	0.196	456	0.263
8282	96.2	56.7	0.237	564	0.360
8255	57.9	57.8	0.307	566	0.629
8228	47.6	61.4	0.369	718	0.623

[1] The percentage of pore area of artificial marine clays is calculated by Particles (Pores) and Cracks Analysis System (PCAS) [77].

Figure 17 shows how the cumulative intrusion curves of the specimens with high montmorillonite content always remain below those with low montmorillonite content. This study result proved that the total pore volume of high montmorillonite specimens such as samples 8291 and 8282, is lower than that of low montmorillonite specimens, as samples 8255 and 8228. This conclusion was demonstrably consistent with the pore areas percentage findings in LVESEM. The relationships between *PI* and final cumulative intrusion volume porosity in MIP test results and percentage of pore areas in LVESEM test results are shown in Figure 19. With the increase in plasticity, the pore volumes and pore areas of artificial marine clay decreased gradually. However, the fitting line of pore volume is steeper than that of pore area in Figure 19. It should be remembered that cumulative intrusion volume porosity is a typical three-dimensional pore space, while percentage of pore area is a two-dimensional pore space on the observed section. Therefore, the pore volume is more sensitive than pore area to the influence of plasticity index.

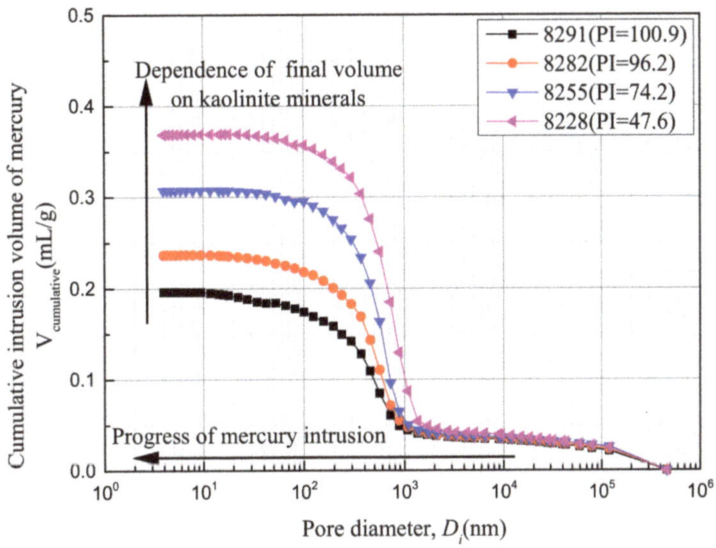

Figure 17. Pore-size distribution with respect to cumulative intrusion volume of mercury.

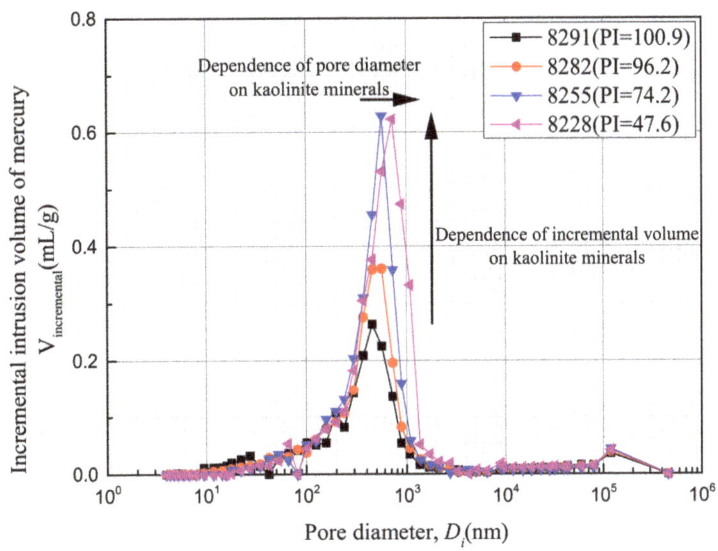

Figure 18. Pore-size distribution with respect to incremental intrusion volume of mercury.

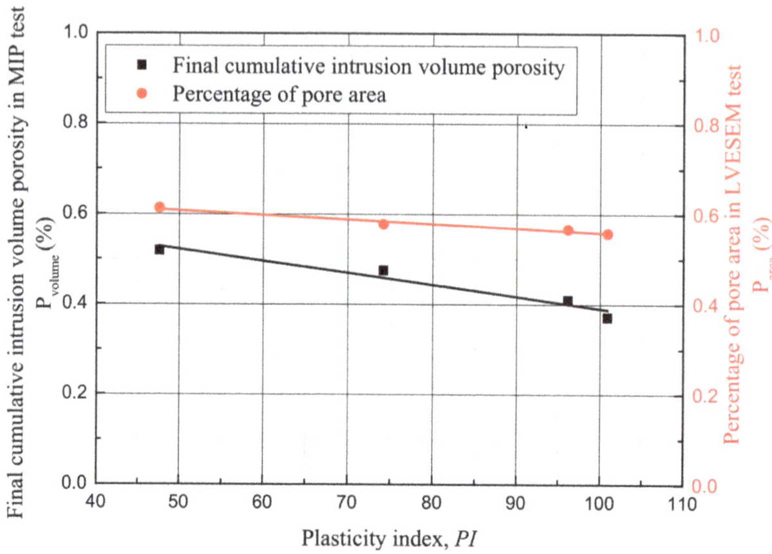

Figure 19. Relationship between plasticity index and final cumulative intrusion volume porosity in MIP test and percentage of pore areas in LVESEM test for artificial marine clays.

From Figure 18, it can be seen that the incremental mercury intrusion volumes in all artificial marine clays showed as a unimodal curve. Garcia-Bengochea et al. [83] proposed the existence of bimodality between inter-aggregate pore and intra-aggregate pore in natural fine-grained soil. However, Delage and Lefebvre [84] reported the results of the Champlain experiment involving scanning reconstituted clay after consolidation by electron microscopy (SEM) and performing an MIP test. They found that the consolidation behavior resulted in the collapse of the larger inter-aggregate pore while the finer intra-aggregate pore remained intact. The inter-aggregate pore was unmistakably seen to have disappeared, while the size of intra-aggregate pore had remained between 466 and 718 nm in the consolidated artificial marine clay. As shown in Table 7, the diameters of intra-aggregate pore and incremental intrusion volumes of the specimens with high montmorillonite content (samples 8291 and 8282) were furthermore found to be lower than those of the specimens with low montmorillonite content (samples 8255 and 8228). These findings indicated that montmorillonite not only constituted the principal determining factor of (intra-aggregate) pore size of the finer artificial marine clay, but also contributed to a more uniform pore distribution within the specimens.

The MIP test results and LVESEM test results in microstructure and microporosity provided mutual confirmation. These tests result fully reflected the sediment fabric of artificial marine clay observed from two experimental perspectives. Meanwhile, the sediment fabric of artificial marine clay comprising various clay minerals significantly influenced their dynamic properties. This further proved that the sediment fabric that resulted from clay mineral composition was primarily responsible for the variations in dynamic properties, and that the plasticity index could only be used as a reference parameter.

5. Conclusions

According to the geological conditions and mineral composition analysis results of marine clay from two typical estuarine deltas in Guangdong Province, South China, artificial marine clays composed of different clay mineral contents were remolded. Under identical laboratory dynamic triaxial testing conditions, the dynamic properties of artificial marine clays at large strain were investigated with particular consideration of the effects of changes in montmorillonite and kaolinite content in clay minerals. Meanwhile, low-vacuum environmental scanning electron microscopy (LVESEM) test and mercury intrusion porosimetry (MIP) tests were carried out on artificial marine clays to explore the sediment fabric of microstructure and microporosity.

Based on observations from the laboratory dynamic triaxial testing results on artificial marine clays, the dynamic properties of the specimens with high montmorillonite content were: (a) a smoother strain increase rate and smaller maximum cyclic strain amplitude; (b) an inhibited increase in pore water pressure including a slower cumulative increase rate in pore water pressure, reduced residual pore water pressure, and narrower pore water pressure amplitude; (c) a more rounded double-normalized hysteresis curve at the failure cycle. Artificial marine clays with high montmorillonite content were found to present very high Atterberg limits and plasticity indices, which was attributed to the fact that strong and loose-adsorbed water (stern layer) was more easily adsorbed on clay surfaces by montmorillonite in the plastic state. It was therefore deemed unconducive to the development of strain caused by aggregate particles sliding and rolling and the rise of free water-induced pore water pressure.

Accordingly, both the stiffness degradation curve and the viscous energy dissipation ratio curve were affected by clay mineral composition. The specimens with high montmorillonite content displayed greater normalized shear moduli under the same cyclic shear strain and experienced a longer period of increased viscous energy before failure. The breakpoint of viscous energy dissipation with high montmorillonite content occurred at greater cyclic strain amplitude. Moreover, due to the influence of specific surface area and isomorphous substitution, the flocculated fabric of the high montmorillonite content specimens and the dispersed fabric of the low montmorillonite content specimens were presented under a low-vacuum environmental scanning electron microscopy test and mercury intrusion porosimetry test. After the same initial void ratio and effective stress consolidation, the flocculated low porosity of the high montmorillonite specimens ensured the opportunity for contact among aggregates under an identical cyclic strain, which provided favorable conditions to improve stiffness. Furthermore, montmorillonite minerals were found to impede sliding and rolling between particles. This reaction is the reason for the slow increase of viscous energy before failure and the hysteretic strain at the breakpoint of viscous energy dissipation in high montmorillonite specimens.

It should be noted that previous research [28,66,85] primarily focused on the influence of the plasticity index on the dynamic properties in marine clay when discussing mineral composition. In the present authors' view, the plasticity index should be considered as an extension parameter based on Atterberg's limit water content according to differing mineral compositions. Although soil plasticity variations are reflected by strong and loose absorbed water and are important indicators in practical engineering, they are unable to reflect the actual component of marine clays. Recent studies [35,40] have shown that even if cohesive soils share a similar plasticity index, their dynamic characteristics are quite different. The results of the present study have shown that the composition of clay mineral does indeed influence the dynamic properties of high plasticity marine clay under large strain due to differences in microstructure and fabric. Thus, this research on clay mineralogy may provide guidance for future engineering geology studies.

Author Contributions: Conceptualization, Y.S. and X.W.; methodology, Y.S.; software, Y.S.; validation, J.C., H.M., and Y.L.; formal analysis, Y.S.; investigation, Y.S.; resources, X.W.; data curation, Y.S.; writing—original draft preparation, Y.S.; writing—review and editing, Y.S. and X.W.; visualization, Y.S.; supervision, H.M.; project administration, X.W.; funding acquisition, Y.S., J.C., and Y.L. All authors have read and agreed to the published version of the manuscript.

Funding: This research was funded by the National Natural Science Foundation of China, grants number 51878192 and 52008121), International (Regional) Cooperation and Exchange Program of National Natural Science Foundation of China, grant number 52011530394, Major International (Regional) Joint Research Project, grant number 52020105002, Chinese Postdoctoral Science Foundation (Grants No. 2020M682652), and the International Training Program Foundation for Young Outstanding Scientific Research Talents in Guangdong Province, China.

Institutional Review Board Statement: Not applicable.

Informed Consent Statement: Not applicable.

Data Availability Statement: Data can be obtained from the corresponding author.

Acknowledgments: Mineral composition testing via X-ray diffraction support from Guangzhou Institute of Geochemistry, Chinese Academy of Sciences (CAS) is highly appreciated by the present study authors.

Conflicts of Interest: The authors declare no conflict of interest.

References

1. Hanna, A.M.; Javed, K. Design of foundations on sensitive Champlain clay subjected to cyclic loading. *J. Geotech. Geoenviron.* **2008**, *134*, 929–937. [CrossRef]
2. Mortezaie, A.R.; Vucetic, M. Effect of Frequency and Vertical Stress on Cyclic Degradation and Pore Water Pressure in Clay in the NGI Simple Shear Device. *J. Geotech. Geoenviron.* **2013**, *139*, 1727–1737. [CrossRef]
3. Zapata-Medina, D.G.; Finno, R.J.; Vega-Posada, C.A. Stress history and sampling disturbance effects on monotonic and cyclic responses of overconsolidated Bootlegger Cove clays. *Can. Geotech. J.* **2014**, *51*, 599–609. [CrossRef]
4. Mortezaie, A.R.; Vucetic, M. Threshold Shear Strains for Cyclic Degradation and Cyclic Pore Water Pressure Generation in Two Clays. *J. Geotech. Geoenviron.* **2016**, *142*, 04016007. [CrossRef]
5. Hu, X.Q.; Zhang, Y.; Guo, L.; Wang, J.; Cai, Y.Q.; Fu, H.T.; Cai, Y. Cyclic behavior of saturated soft clay under stress path with bidirectional shear stresses. *Soil Dyn. Earthq. Eng.* **2018**, *104*, 319–328. [CrossRef]
6. Tang, L.S.; Chen, H.K.; Sang, H.T.; Zhang, S.Y.; Zhang, J.Y. Determination of traffic-load-influenced depths in clayey subsoil based on the shakedown concept. *Soil Dyn. Earthq. Eng.* **2015**, *77*, 182–191. [CrossRef]
7. Cai, Y.; Guo, L.; Jardine, R.J.; Yang, Z.; Wang, J. Stress–strain response of soft clay to traffic loading. *Géotechnique.* **2017**, *67*, 446–451. [CrossRef]
8. Guo, L.; Cai, Y.Q.; Jardine, R.J.; Yang, Z.X.; Wang, J. Undrained behaviour of intact soft clay under cyclic paths that match vehicle loading conditions. *Can. Geotech. J.* **2018**, *55*, 90–106. [CrossRef]
9. Li, L.L.; Dan, H.B.; Wang, L.Z. Undrained behavior of natural marine clay under cyclic loading. *Ocean Eng.* **2011**, *38*, 1792–1805. [CrossRef]
10. Wang, Y.K.; Gao, Y.F.; Guo, L.; Cai, Y.Q.; Li, B.; Qiu, Y.; Mahfouz, A.H. Cyclic response of natural soft marine clay under principal stress rotation as induced by wave loads. *Ocean Eng.* **2017**, *129*, 191–202. [CrossRef]
11. Ren, X.W.; Xu, Q.; Teng, J.D.; Zhao, N.; Lv, L. A novel model for the cumulative plastic strain of soft marine clay under long-term low cyclic loads. *Ocean Eng.* **2018**, *149*, 194–204. [CrossRef]
12. Stark, T.D.; Contreras, I.A. Fourth Avenue landslide during 1964 Alaskan earthquake. *J. Geotech. Geoenviron.* **1998**, *124*, 99–109. [CrossRef]
13. Mendoza, M.; Auvinet, G. The Mexico earthquake of September 19, 1985—behavior of building foundations in Mexico City. *Earthq. Spectra* **1988**, *4*, 835–853. [CrossRef]
14. Ye, J.H.; Wang, G. Numerical simulation of the seismic liquefaction mechanism in an offshore loosely deposited seabed. *Bull. Eng. Geol. Environ.* **2016**, *75*, 1183–1197. [CrossRef]
15. Tavanaei, F.; Hassanpour, J.; Memarian, H. The behavior and properties of Tehran alluvial soils under cyclic loading of urban vibrations—a case study: Arash-Esfandiar tunnel. *Bull. Eng. Geol. Environ.* **2020**, *79*, 4245–4263. [CrossRef]
16. Bjørlykke, K.; Bue, B.; Elverhøt, A. Quaternary sediments in the northwestern part of the Barents Sea and their relation to the underlying Mesozoic bedrock. *Sedimentology* **1978**, *25*, 227–246. [CrossRef]
17. Jeong, G.Y.; Yoon, H.I. The origin of clay minerals in soils of King George Island, South Shetland Islands, West Antarctica, and its implications for the clay-mineral compositions of marine sediments. *J. Sediment. Res.* **2001**, *71*, 833–842. [CrossRef]
18. Jeong, G.Y.; Cheong, C.S. Recurrent events on a Quaternary fault recorded in the mineralogy and micromorphology of a weathering profile, Yangsan Fault System, Korea. *Quat. Res.* **2005**, *64*, 221–233. [CrossRef]
19. Clayton, T.; Pearce, R.B.; Peterson, L.C. Indirect climatic control of the clay mineral composition of Quaternary sediments from the Cariaco basin, northern Venezuela (ODP Site 1002). *Mar. Geol.* **1999**, *161*, 191–206. [CrossRef]
20. Pal, D.; Bhattacharyya, T.; Sinha, R.; Srivastava, P.; Dasgupta, A.; Chandran, P.; Ray, S.; Nimje, A. Clay minerals record from Late Quaternary drill cores of the Ganga Plains and their implications for provenance and climate change in the Himalayan Foreland. *Palaeogeogr. Palaeocl.* **2012**, *356*, 27–37. [CrossRef]
21. Thiry, M. Palaeoclimatic interpretation of clay minerals in marine deposits: An outlook from the continental origin. *Earth-Sci. Rev.* **2000**, *49*, 201–221. [CrossRef]
22. Jennings, A. The Quaternary history of Cumberland Sound, southeastern Baffin Island: The marine evidence. *Géogr. Phys. Quat.* **1993**, *47*, 21–42. [CrossRef]
23. Sivan, D.; Gvirtzman, G.; Sass, E. Quaternary stratigraphy and paleogeography of the Galilee coastal plain, Israel. *Quat. Res.* **1999**, *51*, 280–294. [CrossRef]
24. Sinsakul, S. Late quaternary geology of the Lower Central Plain, Thailand. *J. Asian Earth Sci.* **2000**, *18*, 415–426. [CrossRef]

25. Hardin, B.O.; Drnevich, V.P. Shear modulus and damping in soils: Measurement and parameter effects. *J. Geotech. Geoenviron.* **1972**, *98*, 667–692.
26. Dobry, R.; Vucetic, M. Dynamic properties and seismic response of soft clay deposits. In Proceedings of the International Symposium on Geotechnical Engineering of Soft Soils, Mexico City, Mexico, 13–14 August 1987; pp. 51–86.
27. *Geotechnical Investigation and Testing-Laboratory Testing of Soil–Part 12: Determination of Liquid and Plastic Limits*; European Committee for Standardization: Brussels, Belgium, EN ISO 17892-12:2018; CEE-CENELEC Management Centre: Brussels, Belgium, 2018.
28. Kokusho, T.; Yoshida, Y.; Esashi, Y. Dynamic properties of soft clay for wide strain range. *Soils Found.* **1982**, *22*, 1–18. [CrossRef]
29. Marcuson, W.F.; Wahls, H.E. Time effects on dynamic shear modulus of clays. *J. Soil Mech. Found. Div.* **1972**, *98*, 1359–1373. [CrossRef]
30. Fahoum, K.; Aggour, M.; Amini, F. Dynamic properties of cohesive soils treated with lime. *J. Geotech. Geoenviron.* **1996**, *122*, 382–389. [CrossRef]
31. Park, S.S.; Kim, Y.S. Liquefaction Resistance of Sands Containing Plastic Fines with Different Plasticity. *J. Geotech. Geoenviron.* **2013**, *139*, 825–830. [CrossRef]
32. Gratchev, I.B.; Sassa, K.; Osipov, V.I.; Sokolov, V.N. The liquefaction of clayey soils under cyclic loading. *Eng. Geol.* **2006**, *86*, 70–84. [CrossRef]
33. Beroya, M.; Aydin, A.; Katzenbach, R. Insight into the effects of clay mineralogy on the cyclic behavior of silt–clay mixtures. *Eng. Geol.* **2009**, *106*, 154–162. [CrossRef]
34. Wang, S.Y.; Luna, R.; Yang, J.S. Reexamination of effect of plasticity on liquefaction resistance of low-plasticity fine-grained soils and its potential application. *Acta Geotech.* **2016**, *11*, 1209–1216. [CrossRef]
35. Ajmera, B.; Brandon, T.; Tiwari, B. Characterization of the Reduction in Undrained Shear Strength in Fine-Grained Soils due to Cyclic Loading. *J. Geotech. Geoenviron.* **2019**, *145*, 04019017. [CrossRef]
36. Ajmera, B.; Brandon, T.; Tiwari, B. Influence of index properties on shape of cyclic strength curve for clay-silt mixtures. *Soil Dyn. Earthq. Eng.* **2017**, *102*, 46–55. [CrossRef]
37. Boulanger, R.W.; Idriss, I.M. Liquefaction susceptibility criteria for silts and clays. *J. Geotech. Geoenviron.* **2006**, *132*, 1413–1426. [CrossRef]
38. Boulanger, R.W.; Idriss, I.M. Evaluation of cyclic softening in silts and clays. *J. Geotech. Geoenviron.* **2007**, *133*, 641–652. [CrossRef]
39. Mitchell, J.K.; Soga, K. *Fundamentals of Soil Behavior*; John Wiley & Sons: New York, NY, USA, 2005; Volume 3.
40. Shan, Y. Mineral Composition Based Experimental Study of Dynamic Behaviors of Quaternary Marine Fine-Grained Soil in the Typical Estuary Deltas of Guangdong. Ph.D. Thesis, South China University of Technology, Guangzhou, China, 2018.
41. Shan, Y.; Chen, J.; Ke, X.; Mo, H. Resonant Column Test Study of the Effect of Clay Minerals on Maximum Dynamic Shear Modulus in Marine Clay. In Proceedings of the 7th International Conference on Earthquake Geotechnical Engineering, Rome, Italy, 17–20 June 2019; pp. 17–20.
42. Huang, Y.; Zhao, L.Y. The effects of small particles on soil seismic liquefaction resistance: Current findings and future challenges. *Nat. Hazards* **2018**, *92*, 567–579. [CrossRef]
43. Feng, D.L.; Fang, Y.G. Theoretical Analysis and Experimental Research on Multiscale Mechanical Properties of Soil. *Int. J. Geomech.* **2016**, *16*, 04015094. [CrossRef]
44. Gumaste, S.D.; Iyer, K.R.; Sharma, S.; Channabasavaraj, W.; Singh, D.N. Simulation of fabric in sedimented clays. *Appl. Clay Sci.* **2014**, *91–92*, 117–126. [CrossRef]
45. Anderson, D.; Stokoe, K. Shear modulus: A time-dependent soil property. In *Dynamic Geotechnical Testing*; ASTM International: West Conshohocken, PA, USA, 1978. [CrossRef]
46. Jiang, M.J.; Zhang, F.G.; Hu, H.J.; Cui, Y.J.; Peng, J.B. Structural characterization of natural loess and remolded loess under triaxial tests. *Eng. Geol.* **2014**, *181*, 249–260. [CrossRef]
47. Chen, R.F.; Cai, G.; Congress, S.S.C.; Dong, X.; Duan, W. Dynamic properties and environmental impact of waste red mud-treated loess under adverse conditions. *Bull. Eng. Geol. Environ.* **2021**, *80*, 93–113. [CrossRef]
48. Sun, H.; Hou, M.X.; Chen, C.; Ge, X.R. Microstructure investigation of soft clay subjected to triaxial loading. *Eng. Geol.* **2020**, *274*, 105735. [CrossRef]
49. Cuisinier, O.; Deneele, D.; Masrouri, F. Shear strength behaviour of compacted clayey soils percolated with an alkaline solution. *Eng. Geol.* **2009**, *108*, 177–188. [CrossRef]
50. Xiong, H.X.; Zong, Y.Q.; Qian, P.; Huang, G.Q.; Fu, S.Q. Holocene sea-level history of the northern coast of South China Sea. *Quat. Sci. Rev.* **2018**, *194*, 12–26. [CrossRef]
51. Huang, Z.; Li, P.; Zhang, Z.; Li, K.; Peng, Q. *Formation, Development and Evolution of the Pearl River Delta*; The Guangzhou Branch of Popular Science Press: Guangzhou, China, 1982.
52. Li, P.; Huang, Z.; Zong, Y. *The Han River Delta*; Chinese Ocean Press: Guangzhou, China, 1987.
53. Wang, J. Macroscopic and Microscopic Experimental Study on Consolidation Properties of Zhuhai Soft Soil with Mechanism Analysis. Ph.D. Thesis, South China University of Technology, Guangzhou, China, 2013.
54. Yu, S. Study of the Clay Mineral Effects on Small-Strain Dynamic Properties of Remolded Saturated Clay Soils. Master's Thesis, Shantou University, Shantou, China, 2015.
55. Grim, R.E. *Clay Mineralogy*; McGraw-Hill: New York, NY, USA, 1968.

56. Liu, Z.F.; Colin, C.; Huang, W.; Chen, Z.; Trentesaux, A.; Chen, J.F. Clay minerals in surface sediments of the Pearl River drainage basin and their contribution to the South China Sea. *Chin. Sci. Bull.* **2007**, *52*, 1101–1111. [CrossRef]
57. Carter, D.; Heilman, M.; Gonzales, C. Ethylene glycol monoethyl ether for determining surface area of silicate minerals. *Soil Sci.* **1965**, *100*, 356–360. [CrossRef]
58. Chung, F.H. Quantitative interpretation of X-ray diffraction patterns of mixtures. II. Adiabatic principle of X-ray diffraction analysis of mixtures. *J. Appl. Crystallogr.* **1974**, *7*, 519–525. [CrossRef]
59. Leonards, G.A. *Engineering Properties of Soils, Foundation Engineering*; McGraw-Hill Book Co.: New York, NY, USA, 1962; pp. 66–240.
60. Seed, H.B.; Woodward, R.J., Jr.; Lundgren, R. Clay mineralogical aspects of the Atterberg limits. *J. Soil Mech. Found. Div.* **1964**, *90*, 107–131. [CrossRef]
61. Shan, Y.; Meng, Q.L.; Yu, S.M.; Mo, H.H.; Li, Y.D. Energy based cyclic strength for the influence of mineral composition on artificial marine clay. *Eng. Geol.* **2020**, *274*, 105713. [CrossRef]
62. ASTM Standard D2487-17e1. *Standard Practice for Classification of Soils for Engineering Purposes (Unified Soil Clas-Sification System)*; ASTM International: West Conshohocken, PA, USA, 2017. [CrossRef]
63. Zhou, J.; Gong, X.N. Strain degradation of saturated clay under cyclic loading. *Can. Geotech. J.* **2001**, *38*, 208–212. [CrossRef]
64. Wichtmann, T.; Andersen, K.H.; Sjursen, M.A.; Berre, T. Cyclic tests on high-quality undisturbed block samples of soft marine Norwegian clay. *Can. Geotech. J.* **2013**, *50*, 400–412. [CrossRef]
65. Yang, Q.; Ren, Y.B.; Niu, J.L.; Cheng, K.; Hu, Y.X.; Wang, Y. Characteristics of soft marine clay under cyclic loading: A review. *Bull. Eng. Geol. Environ.* **2018**, *77*, 1027–1046. [CrossRef]
66. Vucetic, M.; Dobry, R. Effect of soil plasticity on cyclic response. *J. Geotech. Eng.* **1991**, *117*, 89–107. [CrossRef]
67. Martin, P.P.; Seed, H.B. One-dimensional dynamic ground response analyses. *J. Geotech. Eng.* **1982**, *108*, 935–952. [CrossRef]
68. Nemat-Nasser, S.; Shokooh, A. A unified approach to densification and liquefaction of cohesionless sand in cyclic shearing. *Can. Geotech. J.* **1979**, *16*, 659–678. [CrossRef]
69. Green, R.; Mitchell, J.; Polito, C. An energy-based excess pore pressure generation model for cohesionless soils. In Proceedings of the John Booker Memorial Symposium, Sidney, NSW, Australia, 16–17 November 2000; AA Balkema Publishers: Rotterdam, The Netherlands, 2000.
70. Javdanian, H. Evaluation of soil liquefaction potential using energy approach: Experimental and statistical investigation. *Bull. Eng. Geol. Environ.* **2019**, *78*, 1697–1708. [CrossRef]
71. Ke, X.Q.; Chen, J.S.; Shan, Y. A new failure criterion for determining the cyclic resistance of low-plasticity fine-grained tailings. *Eng. Geol.* **2019**, *261*, 105273. [CrossRef]
72. Ke, X.; Chen, J.; Pan, W.; Shan, Y. An energy-based process evaluation for low-plasticity fine-grained soil during cyclic loading. In Proceedings of the Geo-Congress 2020: Geotechnical Earthquake Engineering and Special Topics, Minneapolis, MN, USA, 25–28 February 2020; pp. 79–86.
73. Shan, Y.; Ke, X.Q. Reexamination of collapse failure of fine-grained soils and characteristics of related soil indexes. *Environ. Earth Sci.* **2021**, *80*, 402. [CrossRef]
74. Romero, E.; Gens, A.; Lloret, A. Water permeability, water retention and microstructure of unsaturated compacted Boom clay. *Eng. Geol.* **1999**, *54*, 117–127. [CrossRef]
75. Welton, J.E. *SEM Petrology Atlas*; The American Association of Petroleum Geologists: Tulsa, OK, USA, 2003.
76. Chen, H.; Jiang, Y.L.; Niu, C.C.; Leng, G.J.; Tian, G.L. Dynamic characteristics of saturated loess under different confining pressures: A microscopic analysis. *Bull. Eng. Geol. Environ.* **2019**, *78*, 931–944. [CrossRef]
77. Liu, C.; Shi, B.; Zhou, J.; Tang, C.S. Quantification and characterization of microporosity by image processing, geometric measurement and statistical methods: Application on SEM images of clay materials. *Appl. Clay Sci.* **2011**, *54*, 97–106. [CrossRef]
78. Pillai, R.J.; Robinson, R.G.; Boominathan, A. Effect of Microfabric on Undrained Static and Cyclic Behavior of Kaolin Clay. *J. Geotech. Geoenviron.* **2011**, *137*, 421–429. [CrossRef]
79. Hardin, B.O.; Black, W.L.; Division, F. Vibration modulus of normally consolidated clay. *J. Soil Mech. Found. Div.* **1968**, *94*, 353–369. [CrossRef]
80. Thevanayagam, S.; Kanagalingam, T.; Shenthan, T. Intergrain friction, contact density, and cyclic resistance of sands. In Proceedings of the Pacific Conference on Earthquake Engineering, Christchurch, New Zealand, 13–15 February 2003.
81. Washburn, E.W. The dynamics of capillary flow. *Phys. Rev.* **1921**, *17*, 273. [CrossRef]
82. Diamond, S.J.C. Pore size distributions in clays. *Clays Clay Miner.* **1970**, *18*, 7–23. [CrossRef]
83. Garcia-Bengochea, I.; Altschaeffl, A.G.; Lovell, C.W. Pore distribution and permeability of silty clays. *J. Soil Mech. Found. Div.* **1979**, *105*, 839–856.
84. Delage, P.; Lefebvre, G. Study of the structure of a sensitive Champlain clay and of its evolution during consolidation. *Can. Geotech. J.* **1984**, *21*, 21–35. [CrossRef]
85. Kagawa, T. Moduli and damping factors of soft marine clays. *J. Geotech. Eng.* **1992**, *118*, 1360–1375. [CrossRef]

Journal of
Marine Science and Engineering

MDPI

Article

Sedimentary Environmental Evolution of the Western Taiwan Shoal Area since the Late Pleistocene

Chipeng He [1], Longbin Sha [1,2], Dongbo Zhao [3], Lu Dai [1,2], Zheng Li [1], Jiabing Tang [1], Xianfu Li [1] and Dongling Li [1,2,*]

1 Department of Geography and Spatial Information Technology, Ningbo University, Ningbo 315211, China; 1911073004@nbu.edu.cn (C.H.); shalongbin@nbu.edu.cn (L.S.); dailu@nbu.edu.cn (L.D.); 1911073009@nbu.edu.cn (Z.L.); 51213904003@stu.ecnu.edu.cn (J.T.); xianfuli21@hotmail.com (X.L.)

2 Institute of East China Sea, Ningbo University, Ningbo 315211, China

3 Fujian Provincial Key Laboratory of Coast and Island Management Technology Study, Fujian Institute of Oceanography, Xiamen 361013, China; zdb@fjio.net

* Correspondence: lidongling@nbu.edu.cn; Tel.: +86-574-8760-5546; Fax: +86-574-8760-5546

Abstract: A new pollen analysis and major and trace element contents were conducted on a 40 m long gravity core recovered from the Taiwan Shoal (sand ridges), south of the Taiwan Strait, beginning in the Late Pleistocene. The changes in the pollen assemblage and concentration represent the climate change around the Taiwan Shoal and the strength of the Zhe-Min Coastal Current, whereas variations in major and trace element contents can imply the source of the sediments in the Taiwan Shoal, which are correlated with the rise or fall of the sea level with increased marine dinoflagellate cysts. The interval of 40–30 m was characterized by high pollen and spore concentrations, and evergreen *Quercus* was dominant taxon, which indicates a warm sedimentary environment, and the surrounding area of the Taiwan Shoal were covered by a tropical and subtropical broad-leaved forest. There were no pollen and spores from 30–24 m, which indicates a strong hydrodynamic sedimentary environment, and most of the Taiwan Shoal might have been experience subaerial exposure. The interval of 24–17 m was characterized by the reappearance of pollen and spores, as well as marine dinoflagellate cysts and foraminifera, suggesting the climate was warm and wet in the study area and an apparent marine sedimentary environment with relatively high sea level. Deciduous *Quercus* dominated the interval of 17–12 m, which indicated that the climate was relatively cool, corresponding to the end of Marine isotope stages3 (MIS3) to the Last Glacial Maximum accompanied by weathering and denudation. Above 12 m, the low pollen concentration with increased marine dinoflagellate cysts and foraminifera abundance suggested a marine sedimentary environment in the Taiwan Shoal. The high concentrations in *Pinus* corresponds to Holocene high sea level.

Keywords: Late Pleistocene; major and trace elements; pollen; principal component analysis; Taiwan Shoal

Citation: He, C.; Sha, L.; Zhao, D.; Dai, L.; Li, Z.; Tang, J.; Li, X.; Li, D. Sedimentary Environmental Evolution of the Western Taiwan Shoal Area since the Late Pleistocene. *J. Mar. Sci. Eng.* **2021**, *9*, 1150. https://doi.org/10.3390/jmse9101150

Academic Editor: Gemma Aiello

Received: 4 September 2021
Accepted: 14 October 2021
Published: 19 October 2021

Publisher's Note: MDPI stays neutral with regard to jurisdictional claims in published maps and institutional affiliations.

Copyright: © 2021 by the authors. Licensee MDPI, Basel, Switzerland. This article is an open access article distributed under the terms and conditions of the Creative Commons Attribution (CC BY) license (https://creativecommons.org/licenses/by/4.0/).

1. Introduction

The Taiwan Strait, a waterway between Mainland China and Taiwan, is a relatively shallow shelf sea (average 60 m in water depth) dominated by tidal currents [1]. It is considered an important channel for material and energy resource exchange (Figure 1). Interaction between the Kuroshio Current, South China Seawater and the Zhe-Min Coastal Current, and strong tidal currents prompts a large amount of sediment to accumulate at the narrows or shoals in the strait [2]. Wu [3] found that the distribution pattern of rare-earth elements in the Taiwan Shoal is similar to that of granite in Fujian and the East China Sea continental shelf. The foraminifera analysis of 16 cores in the western Taiwan Strait shows that the strata of the Taiwan Strait since the late Quaternary can be divided into five layers, and the Taiwan Strait was a neritic environment in the Last Glacial period [4]. Lan et al. [5,6] also reported that the coarse sand at ~20 ka BP in the Taiwan Shoal contained exotic substances, and there was no "land bridge" in the Taiwan Strait since the Late Pleistocene. At the same time, the sea level of the Taiwan Strait had been rising since the Holocene, reaching the modern level at 4–3 ka BP. In addition, according to Yang et al. [7], the magnetic susceptibility of surface sediments declined from the nearshore area to the seaward in the western Taiwan Strait, and the magnetic susceptibility was controlled by the source of sediments and sedimentation. Dong et al. [8] discussed the history of the sea-level change and coastline migration in southwest Taiwan during the Holocene, and the results show that southwest Taiwan experienced marine transgression three times in the Holocene (~6.2 ka, ~3.1 ka, and ~1.8 ka). Wang et al. [9] analyzed the stratigraphic sequence and the paleoenvironmental development of the Taiwan Shoal based on records of the grain size and the Accelerator Mass Spectrometry (AMS) ^{14}C dating results. However, all the above studies are based on one or two proxies with deficient chronological results. In order to clarify the sedimentary environmental evolution of the Taiwan Shoal since the Late Pleistocene, multiproxy research is considered indispensable.

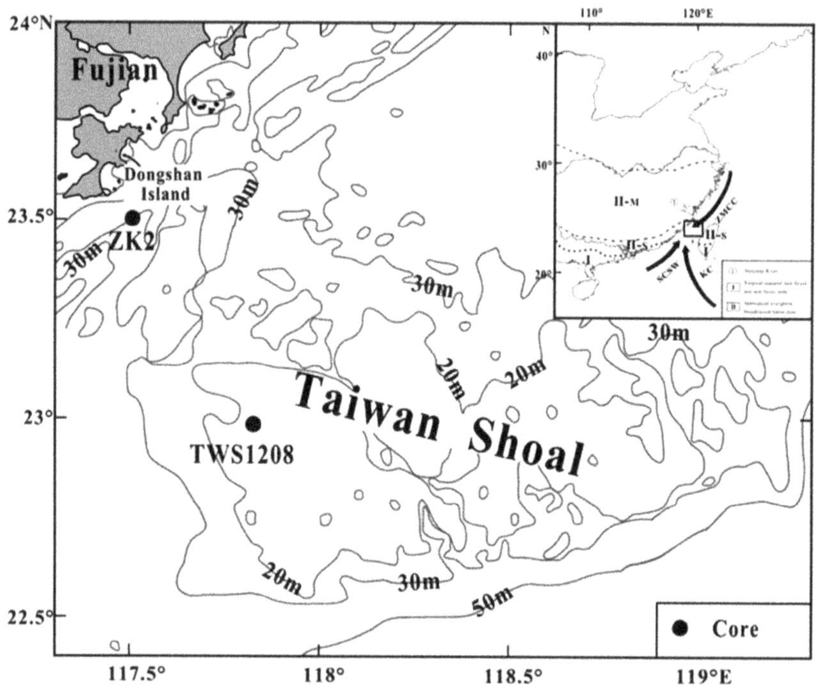

Figure 1. Map of the Taiwan Shoal and location of core ZK2 and TWS1208 [9]. ZMCC: Zhe-Min Coastal Current. SCSW: South China Sea water. KC: Kuroshio Current. Figure of vegetation zones in tropical and subtropical China is modified from Zheng [10] and Yue et al. [11].

Pollen in marine sediments comes from the surrounding land and coastal areas, so the surrounding vegetation is one of the basic factors that affect the distribution of pollen assemblages in marine sediments. In order to better clarify the process of the pollen transport from land to marine areas, many scholars have performed studies on pollen in marine areas to provide an adequate reference for the reconstruction of the paleoenvironment, such as in the Bohai Sea [12,13], the East China Sea [14], and the South China Sea [15–19]. However, in other areas, several studies have shown a good correlation between marine pollen signals and the nearby terrestrial vegetation [20–22]. However, in other areas, such as in the Gulf of Lions in southeastern France (the western Mediterranean Sea), the relationship between pollen signals and the inland vegetation is not evident [23]. Jing et al. [24] studied the distribution patterns of pollen and spores in the northwestern Taiwan Strait based on 338 surface sediment samples. The results show that the pollen abundance is mainly affected by water depth and sediment grain size (texture). Lan et al. [6] studied the pollen assemblages of gravity cores in the western Taiwan Strait since the Late Pleistocene. However, the above studies have mainly analyzed the transportation and deposition processes in the Taiwan Strait, but a few of them have combined the paleontological records and geochemical proxies to infer the environmental and climatic change.

Previous works have mainly been focused on the modern sedimentary process of the Taiwan Strait, which was the junction of the East China Sea and the South China Sea. There are few research reports on the long-term stratigraphic structure and sedimentary environment evolution of the Taiwan Strait. Despite that the Taiwan Shoal in the southern Taiwan Strait is one of the world's famous submarine shoals, the changes in the paleoenvironment of the Taiwan Shoal have been scarcely studied because of discontinuous formation and lack of precise and accurate age-depth model. In this study, we obtained a sedimentary core ZK2 from the Taiwan Shoal area and reconstruct the paleoenvironmental evolution of the Taiwan Shoal by palynological and geochemical analysis. The aim of research was to provide a multi-proxy reconstruction of the evolution of the sedimentary environment in the Taiwan Shoal since the Late Pleistocene and provide scientific references for future studies.

2. Geographical Setting

The Taiwan Shoal in the southern Taiwan Strait represents one of the sedimentary subregions [1,25], and it is the shallowest part of the Taiwan Strait. Meanwhile, as one of the world's famous submarine shoals, the Taiwan Shoal is the junction of the East China Sea and the South China Sea. Furthermore, the Taiwan Shoal is adjacent to the Zhangpu coastal area of the southern Fujian in the west and extends to the Penghu Islands in the middle of the Taiwan Strait in the east. It measures about 250 km long from east to west, 135 km wide from north to south, and covers an area of about 13,000 km^2. Moreover, the bottom of the Taiwan Shoal is mainly sandy coarse-grained sediments [26], and it is obviously different from the sedimentary layers of other sea areas in China. The Taiwan Shoal, with a water depth of only 10–25 m, is an uplifted area of the Taiwan Strait, and some geological records are missing due to the repeated transgressions and retreats [2].

The direction of the surface and bottom ocean currents shows apparent differences in the Taiwan Shoal due to the combined effect of the South China Seawater, the Kuroshio Current, and the Zhe-Min Coastal Current. The surface ocean current possesses the characteristics of the wind-driven current in summer and flows northeast, whereas it flows southward in winter under the control of the Zhe-Min Coastal Current. Meanwhile, the bottom ocean current flows northward throughout the year. In winter, it is affected by the extended South China Sea warm current and presents a different flow direction from the surface current [27]. The study area, with an intense tide action, experiences an irregular semi-diurnal tide, which is dominated by the M2 sub-tide, the average annual tidal level is 0.46 m, the average tidal range is 2.33 m, the average annual tidal current velocity is 0.46 m/s, and the maximum can be up to 0.8 m/s [28]. The Taiwan Shoal belongs to the southern tropical oceanic monsoon climate, with a mild and humid climate and plentiful rainfall. The annual average air temperature is estimated as 21.2 °C, and the average air temperatures of the hottest months (July–August) and the coldest months (January–February) are estimated as 27 °C and 13 °C, respectively. The annual average precipitation is estimated as 1071 mm, and the precipitation in spring and summer accounts for 61% of the total precipitation throughout the year. The prevailing wind is mostly in the northeaster wind direction and it is strong throughout the year. The multi-year average of windy days is estimated as 122 days, with an annual average wind speed of 7.1 m/s, and typhoons and storm surge activities in the study area are frequent [2].

Vegetation in the southeastern monsoon region of China belongs to tropical and subtropical forests. The natural forest components from south to north in southern China are mainly rainforests, seasonal rainforests, evergreen seasonal rainforests, and evergreen broad-leaved forests. The vegetation partition is shown in Figure 1 [10,11]. As a marine sedimentary area, the palynology of the Taiwan Shoal comes from I (tropical seasonal rain forest and rain forest), II-M (evergreen broadleaved forest), II-S (monsoon subtropical evergreen broadleaved forest) [10,11,29].

3. Materials and Methods

A 40 m long gravity core, core ZK2 (23°30′38.44″ N, 117°30′55.94″ E), was collected from the coast of the Dongshan Island, Fujian, in June 2015. The core is located in the northwest of the Taiwan Shoal, with a water depth of 37.5 m and a column length of 40 m. In this analysis, samples were collected at an interval of 2 m, and a total of 20 samples were collected.

All pollen samples were dried and weighed. Then, an exotic *Lycopodium*-spore tablet as a maker containing 27,637 ± 593 grains was added to each sample in order to permit the estimation of the pollen concentration before preparation, followed by an alternative treatment with HF (40%) and HCl (10%). For sandy samples, they are first elutriated, and suspended solids are separated, precipitated, and then chemically treated. Subsequently, the residue of the sample is cleaned and placed in an ultrasonic oscillator. The remaining impurities are removed by using a 7 μm mesh sieve, and finally, pollen grains are deposited. The pollen identification process was performed under a 400× Leica optical microscope, and each sample counted to at least 200 grains. Fern spores, the freshwater phytoplanktonic remains and dinoflagellate cysts were also identified on the pollen slides.

ELAN 9000 ICP-MS (PE Company, Groton, CT, USA) was used for the micro-element determination. The operating steps were as follows: 0.0400 g of the sample was accurately weighed and placed in a Teflon cup, 1.5 mL of HF and 0.5 mL of HNO_3 were added, and the mixture was sealed and digested to at 150 °C for 12 h and then cooled. The Teflon cup was taken out, the product was weighed and diluted to 40 g (dilution factor was about 1000), and then applied to the micro-element determination by using ICP-MS. The sensitivity of the ICP-MS instrument was adjusted to 1 ng mL^{-1} 115 In 30,000 cps. The testing results of GBW07315 (sediment), GBW07316 (sediment), and BHVO-2 (basalt) were found to be consistent with the recommended values. The relative errors of most element determination results were within 5–10%.

The trace element determination was performed by using the IRIS Intrepid II XSP ICP-OES (Thermo Electron, Waltham, MA, USA), and the steps were the same as those of the trace element determination. The testing results of Al_2O_3, CaO, MnO, TiO_2, GBW07315 (sediment), GBW07316 (sediment), and BHVO-2 were identical to those of the recommended values. The relative errors of the testing results were smaller than 2%.

Five accelerator mass spectrometry (AMS) ^{14}C data were measured by using the core ZK2. The age-dating materials were benthic foraminifera and charcoal (surface). Samples were sent to Beta Analytic, Miami, FL, USA, for completing the test. All ^{14}C ages were corrected into the calendar years by using the OxCal 4.4 software (available online: https://c14.arch.ox.ac.uk/oxcal/OxCal.html (accessed on 15 October 2021)) with 95.4% probability (High Probability Density Range Method (HPD): MARINE 20) (Table 1). The global oceanic carbon reservoir deviation (ΔR) was 113 ± 37 an according to the data on the northwest coast of Taiwan [30]. The ^{14}C age at layer 36.5 m was >43,500 a BP, out of the AMS^{14}C dating range, so the age-dating result at layer 36.5 m was excluded.

In this study, Past, a free software [31], is used to conduct Principal component analysis (PCA) on the contents of major and trace elements in core ZK2, in order to determine the ecological environmental implication of the principal component axis and provide a quantitative/semi-quantitative study of environmental indicators.

Table 1. AMS ^{14}C ages of core ZK2. IRMS: Isotope Ratio Mass Spectrometry.

Core	Depth (m)	Laboratory Code	Dated Materials	^{14}C Age (a BP)	IRMS δ13C (‰)	Calibrated Age (cal a BP)
ZK2	3.5	Beta-496141	Benthic foraminifera	3780 ± 30	−0.6	3701–3388
ZK2	7.5	Beta-496142	Benthic foraminifera	4540 ± 30	0.1	4599–4186
ZK2	17.5	Beta-496143	Benthic foraminifera	35,790 ± 280	0.3	40,410–39,304
ZK2	20.5	Beta-491937	Benthic foraminifera	35,910 ± 280	−0.5	40,494–39,383
ZK2	36.5	Beta-491938	Charcoal	>43,500	−25.4	>43,500

4. Results

4.1. Palynomorph Composition and Concentration

All samples from the core ZK2 (excluding samples ZK2-25, ZK2-27 and ZK2-29) were found to contain pollen, and 56 pollen taxa were identified (Figure A1). Fern spores, the freshwater phytoplanktonic remains (mainly *Concentricystes*), and dinoflagellate cysts (mainly *Spiniferites*) were also found in the core ZK2. Among them, *Concentricystes* is an algal zygospore of the probable zygnematacean affinity [32]. Fossils of *Concentricystes* are widely distributed in the Cenozoic of eastern China [33] and are commonly found in the soil and black silt in the lake or river sections. *Concentricystes* is thus indicative of freshwater lakes or swamps on flood plains.

Arboreal pollen dominates the pollen assemblages, in which tropical and subtropical broad-leaved pollen taxa are the most abundant (47.34% in average), followed by the *Pinus* pollen (14.54% in average). However, changes in the contents of *Pinus* do not provide any climate information because *Pinus* communities are very common [34]. The average content of the temperate broad-leaved pollen is 10.38%. In addition, there was a low content of coniferous components at high altitudes and tropical and subtropical coniferous components, with an average content of 0.33% and 2.31%, respectively. Tropical and subtropical broad-leaved pollen taxa are dominated by evergreen *Quercus* with an average percentage of 35.19%. The most abundant temperate broadleaved taxa are deciduous *Quercus*, *Alnus*, *Ulmus*, and *Carpinus*. As compared to arboreal pollen, the contents of the herb pollen were relatively low (24.12% on average). The most abundant terrigenous herb taxa are Poaceae, Cyperaceae, Chenopodiaceae, and *Artemisia*, whereas the aquatic herb (mainly *Typha*) appears occasionally.

High concentrations of pollen were mainly observed in depth ranges of 22.5–12.5 m and 38.4–32.4 m; especially they were higher between the depths of 12.5 and 22.5 m. In addition, marine dinoflagellate cysts and the freshwater phytoplanktonic remains were mainly observed above the depth of ~22.5 m.

Palynomorph assemblages recovered from the Taiwan Shoal can be divided into the following three general groups (Figure 2):

4.1.1. Zone I: (40–30 m)

High pollen concentrations (9530 grains/g on average) between samples in this zone indicate a stable sedimentary environment in the Taiwan Shoal. In the pollen assemblage, evergreen *Quercus* was absolutely dominant, with an average content of 55.40%, and other tropical and subtropical broad-leaved forests were found sporadically. *Pinus* is the second dominant component, with an average content of 16.54%. However, high-altitude coniferous forest, tropical and subtropical coniferous forest, temperate components, herbs and fern spores were all found sporadically in this zone. In addition, no marine dinoflagellate cysts were found in this zone.

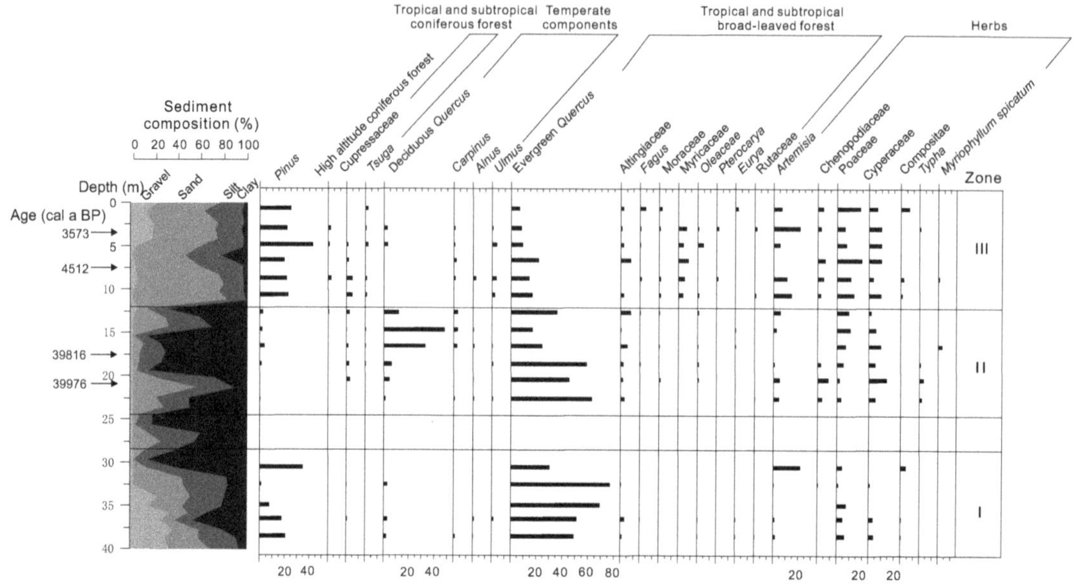

Figure 2. Pollen concentration diagram of core ZK2, and figure of sediment composition is modified from Zhao et al. [2].

4.1.2. Zone II (24.5–12.5 m)

The pollen concentration was also relatively high but lower than that in zone I. Evergreen *Quercus* also dominated the pollen assemblage, with the average content as high as that in zone I, reaching 52.34%. Another tropical and subtropical broad-leaved tree taxon Altingiaceae was obviously increased in this zone. The content of temperate broad-leaved pollen increased remarkably as compared to that in zone I, and the main component was deciduous *Quercus*, with an average content of 17.57%. The content of *Pinus* shows an obvious decrease (1.75% on average). The content of terrestrial herbs shows a clear increase relative to that in zone I (19.05% on average), where the main components were Cyperaceae (7.06%) and Poaceae (6.36%). The aquatic herbs in this zone were more abundant than those in zone I. The contents of high-altitude conifers, tropical and subtropical conifers, and fern spores were also found occasionally. However, the concentration of the freshwater phytoplanktonic remains was high in this zone, and marine dinoflagellate cysts were found sporadically.

4.1.3. Zone III (12.5–0 m)

The pollen concentrations in this zone were much lower than those in zones I and II. The content of the most abundant tropical and subtropical broad-leaved pollen taxa evergreen *Quercus* decreased remarkably, with an average content of 12.86%. However, the contents of the other tropical and subtropical broad-leaved pollen taxon increased as compared to those in zones I and II, especially the content of Myricaceae. The concentration of *Pinus* was much higher than that in zone II, with an average content of 25.66%. The content of temperate broad-leaved pollen was lower than those in zones I and II, with an average content of 4.45%. The contents of terrestrial herbs, mainly Poaceae (12.66%), Cyperaceae (8.32%), and *Artemisia* (9.76%), were higher than those in zones I and II, with an average content of 35.87%. The content of fern spores was 27.31%, which was also higher than those in zones I and II. The concentration of the freshwater phytoplanktonic remains was much lower than that in zone II, whereas the concentrations of marine dinoflagellate cysts were slightly lower than that in zone II.

4.2. Major and Trace Element Contents

The vertical variations of major element contents in the core ZK2 are shown in Figure 3. The average content of oxides of major elements was found in the following order: SiO_2 > Al_2O_3 >CaO > TiO_2 > MnO. Among them, SiO_2 was the most abundant composition, ranging from 61.04% to 89.11% (74.24% on average). Al_2O_3 ranged between 2.55% and 16.77% (10.08% on average). The content of CaO was observed between 0.1% and 3.02% (0.9% on average). The variation range of TiO_2 was between 0.09% and 0.96% (0.57% on average). The MnO content was between 0.02% and 0.19% (0.05% on average).

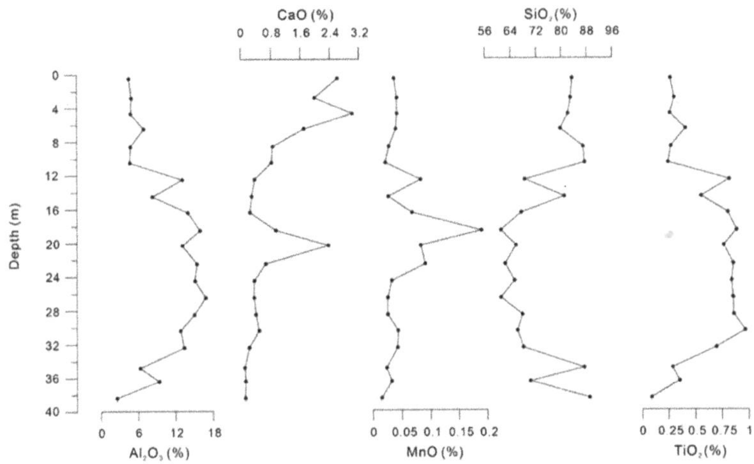

Figure 3. Changes in major element contents of core ZK2.

Figure 4 shows the variability of trace element contents in the core ZK2 since the Late Pleistocene. The contents of Ba were high, and most depths were above 200 mg/kg, with an average content of 342.9 mg/kg. The contents of Rb, Sr, Cu, Ni, and V were relatively low, with the average contents of 96.2, 85.8, 13.2, 23.1, and 62.8 mg/kg, respectively. The variation characteristics of trace elements were found generally consistent, indicating a gradually increasing trend at a depth of 40–20 m, followed by a reduction above the depth of ~20 m.

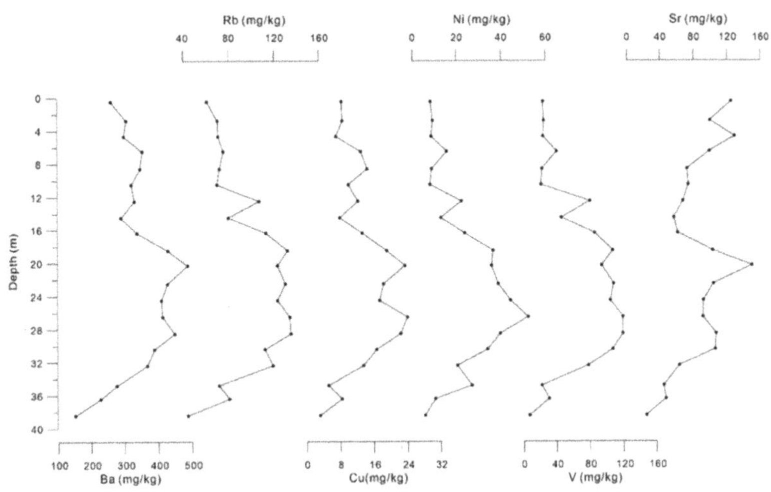

Figure 4. Changes in trace element contents of core ZK2.

4.3. Principal Componenet Analysis

Principal component analysis was carried out in order to clarify the environmental implications of major and trace elements in core ZK2, and the scores of each element on the PCA axes (axis 1 and axis 2) are shown in Figure 5. The principal component loading diagram (Figure 5) shows the relationship between different elements, where each vector axis represents each element, and the horizontal and vertical axes (axis 1 and 2) represent the most important principal components. The length of the vector axis is proportional to the content of the element it represents. The angle between each vector and the coordinate axis represents the correlation between elements and between elements and each principal component. When the included angle is 90°, it indicates that there is no obvious correlation between the two elements. Less than or greater than 90° indicates that the two elements are positively or negatively correlated, and the quantitative relationship between the elements and environmental variables is obtained by its projection on the principal component axis.

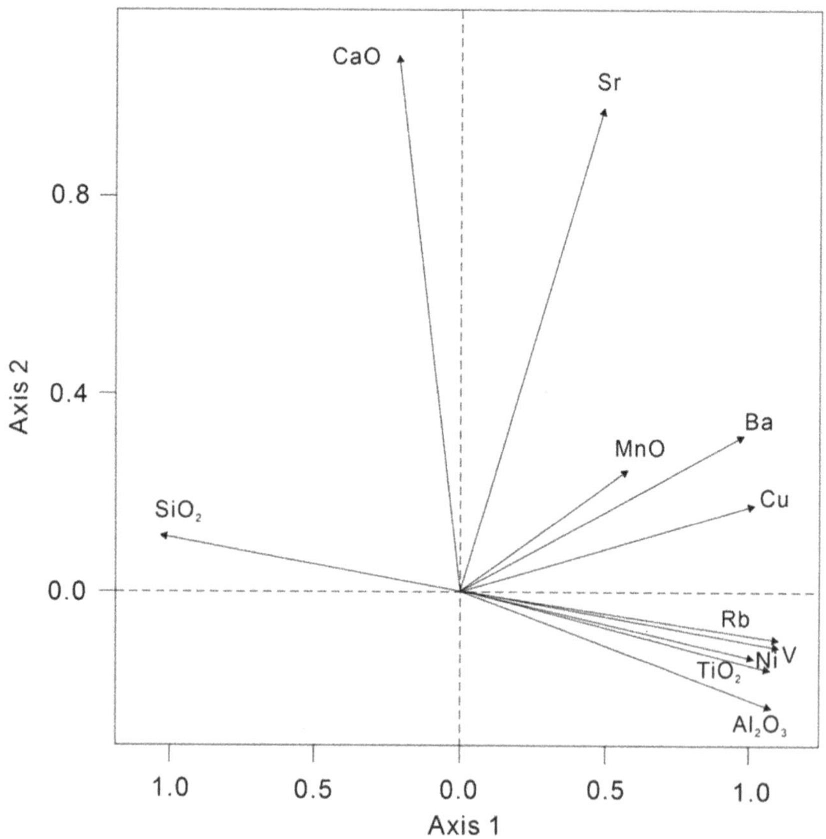

Figure 5. Principal component analysis (PCA) of major and trace elements scatter diagram.

The cumulative contribution rates of the first and second principal components to the total variance of all elements accounted for 87.36%, with 69.44% of axis 1 and 17.91% of axis 2, respectively. The contribution rates of other principal components are relatively low. This shows that the first and second principal components can explain most of the environment information contained in major and trace elements.

In the scatter diagram of axis 1 against axis 2, axis 1 is positively correlated with Al_2O_3, TiO_2, MnO, Cu, Rb, Ni, V, and Ba, but negatively correlated to SiO_2 (Figure 5). Changes in the contents of SiO_2 were negatively correlated with the other elements, and the content of SiO_2 was generally proportional to the sediments with large grain sizes, whereas other elements were proportional to the finer sediments [35]. MnO and Al_2O_3 are elements with relatively stable geochemical properties under the supergene condition, and they are mainly found in terrigenous debris and clay minerals. TiO_2 is believed to be derived from terrestrial debris [36–39]. Generally, the alkali earth element Ba in marine sediments is often associated with autogenous effects, such as biological productivity [39–41], and Cu is considered an essential element for phytoplankton [42,43]. Therefore, axis 1 may be associated with the input of terrigenous materials or primary productivity and authigenic deposition. Axis 2 is positively correlated to CaO and Sr (Figure 5), which are closely related to the action of marine organisms [42,44]. CaO is considered an important constituent of biological carbonate, and Sr transforms into carbonate shells and bones after it is absorbed by organisms. Therefore, it was concluded that axis 2 could be represented by marine biogenic matters.

5. Discussion

The pollen record of the core ZK2 indicated that arboreal pollen dominated the Late Pleistocene, which was possibly carried by the Zhe-Min Coastal Current and the surrounding rivers near the study area [6,24]. The core ZK2 can be divided into the following five deposition units according to the comprehensive analysis of pollen and major and trace element results (Figure 6).

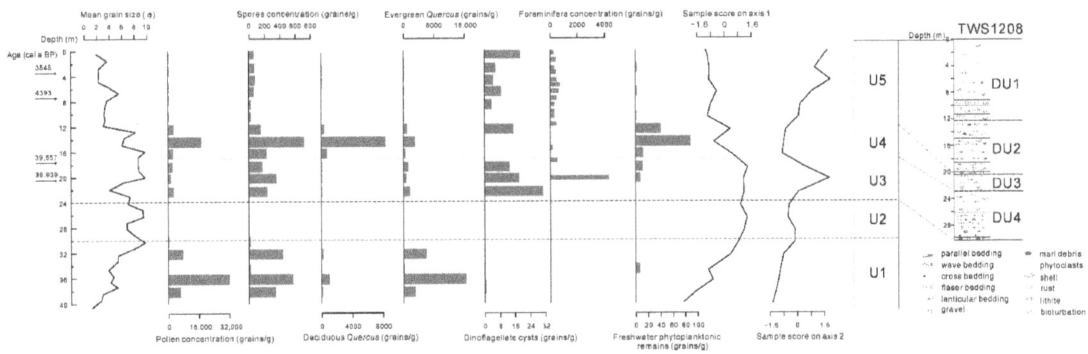

Figure 6. Mean grain size, major and trace elements and pollen and foraminifera concentrations of core ZK2 compared with the stratigraphic of core TWS1208 [9]. Figure of the mean grain size and TWS1208 is modified from Zhao et al. [2] and Wang et al. [9], respectively.

In phase U1 (40–30 m), pollen and spore concentrations were high, which indicates an almost unchanged terrestrial sedimentary environment. The high abundance of evergreen *Quercus* in this phase indicated that the surrounding area was covered by a tropical and subtropical broad-leaved forest (Figure 2). This is consistent with a pollen record in the Minjiang River estuary, where the pollen species of the Late Pleistocene strata are mainly *Pinus*, Cupressaceae, *Quercus* and Polypodiaceae [45]. Lan et al. [6] studied the pollen assemblages of gravity cores in the western Taiwan Strait since the Late Pleistocene, which reveals a *Quercus*, *Pinus*–Polypodiaceae assemblage during the MIS 3. The vegetation evolution study of Fuzhou Basin shows that the primary and mixed forests with deciduous and evergreen elements covered the mountainous areas with relatively warm and humid climate in MIS3 period, which is also consistent with the pollen record of this study [11]. In addition, no marine dinoflagellate cysts and foraminifera [2] were found in this phase (Figure 6), accompanied by a very low content of the representative marine sedimentary element CaO (Figure 3), which suggest no marine influence in the Taiwan Shoal. The low sample score on axis 2 also suggests a lack of the marine biogenic matter source. In this phase, the high chemical index of illite indicated that chemical weathering was strong and the contents of unstable minerals epidote and hornblende also shows the near-source characteristics [2]. Therefore, we can conclude that the Taiwan Shoal might have been a terrigenous sedimentary environment in this phase, corresponding to the stratigraphic D layer of the late Quaternary in the western Taiwan Strait [46].

There was a disappearance of pollen and spores, as well as marine dinoflagellate cysts and foraminifera [2] in phase U2 (30–24 m). It has been reported that once pollen and spores are deposited into water, they behave similarly to small sediment particles of clay and fine silt [13]. This observation suggests that the sedimentation of pollen and spores in marine waters is controlled by mechanisms similar to those that control the sedimentation of clay and fine silt [20]. Strong boundary currents or coastal currents have a considerable impact on pollen sediment [47]. Phase U2 with coarse particle size (Figure 6) without pollen is deposited because of the strong hydrodynamic suspension and screening for sediments and pollen. A slight increase in sample score on axis 1, accompanied by a slight decrease in sample score on axis 2, suggested that the source of sediments in the Taiwan Shoal area fluctuated slightly during this period. Most intense chemical weathering results in a preponderance of aluminous clay minerals, and physical weathering leads to silicate enrichment [48,49]. Therefore, a slight increase in Al_2O_3 and TiO_2 as compared to a decrease in SiO_2 in this phase would imply the enhanced chemical weathering under warm and humid weather [50]. In addition, according to Zou et al. [51] and Liang et al. [52], the elements Ni and V can be easily accumulated under a reducing environment and positively correlated to fine-grained sediments. The contents of CaO and Sr, as well as sample score on axis 2, showed little change in this phase (Figures 3, 4 and 6), which implies no marine influence in the Taiwan Shoal area.

The reappearance of pollen and spores, including evergreen *Quercus* in phase U3 (24–17 m) implied a warm and wet conditions, and the study area was covered by the tropical and subtropical broad-leaved forest, corresponding to the sedimentary unit DU4 (Coastal-intertidal deposits) of the core TWS1208 in the western Taiwan Shoal [9]. The pollen record of the Pingnan County, Fujian Province of southern China, also indicates that Fujian was covered by a mixed forest containing both subtropical evergreen and temperate deciduous broad-leaved trees prior to the Last Glacial Maximum [11], supporting the appearance of evergreen *Quercus* at this stage in the Taiwan Shoal area. Marine dinoflagellate cysts and foraminifera appeared with an abrupt increasing trend, suggesting an enhanced marine influence in the study area, corresponding to the stratigraphic C layer of the late Quaternary in western Taiwan Strait [46]. The contents of Al_2O_3, MnO, TiO_2 and Rb were low, indicating a slight reduction in the input of terrigenous debris at that time (Figures 3 and 4). In contrast, the contents of CaO and Sr reached their peak values, which indicates an enhanced marine influence in the 24–18 m interval. The sample score on axis 1 was stable, whereas the sample score on axis 2 increased abruptly and reached the maximum in this phase, which suggests a remarkable increase in the marine influence in the Taiwan Shoal. Furthermore, a pollen record along the Fujian seashore shows a subtropical evergreen broad-leaved forest during that period, which represents the warm and humid climate [46]. Since the age control of 20.5 m was 40 cal ka BP, the 24–18 m interval possibly corresponds to the MIS 3 [53], and the Taiwan Shoal appeared in a subtidal environment during that period. The mean grain size fluctuated greatly during the 24–18 m interval [2] (Figure 6), which suggests an unstable depositional environment in the Taiwan Shoal. Previous studies from the coastal areas of Fujian also recorded the last transgression of the Late Pleistocene at 44–22 cal ka BP [53–55]. Coincidentally, the Chao-shan Plain has also recorded sea level rise, indicated by marine microfossils such as diatoms, foraminifera, and mangroves representing the coastal environment [56]. Therefore, the Taiwan Shoal was a coastal environment in phase U3 and herbs deposited in this area were carried by nearby rivers [57].

In phase U4 (17–12 m), deciduous *Quercus* concentrations increased gradually, reaching the maximum at a depth of 15 m. This indicated a relatively cold and dry condition in the Taiwan Shoal, which would have allowed the temperate deciduous taxa to expand. The pollen record in Fujian also suggests that a deciduous forest prevailed during this period [11], and there was a *Pinus*–Polygonaceae–*Pteridium* assemblage in the western Taiwan Strait during the MIS 2 [6]. There were no marine dinoflagellate cysts and foraminifera found in this phase [2], whereas the freshwater phytoplanktonic remains were relatively abundant, which implies that the Taiwan Shoal was a fresh water environment between 15 and 12 m, possibly corresponding to the terrestrial deposit in the MIS 2 [2] and the sedimentary unit DU3 (Valley filling deposit) of the core TWS1208 in the western Taiwan Shoal [9]. In addition, an increase in the content of SiO_2 in this phase (Figure 3) indicated a high physical weathering degree, whereas a low content of Al_2O_3 indicated a weak chemical weathering degree (Figure 3). According to the analysis of clay minerals, the increase in contents of unstable minerals epidote and hornblende indicates that the distance of sediment transport was closer, suggesting a continental sedimentary environment at the end of MIS3 or the beginning of the LGM [2]. Furthermore, low contents of CaO and Sr (Figures 3 and 4), as well as obviously declined sample score on axis 2, suggest little marine influence in this phase. Although the age of the top of phase U4 is unable to be examined accurately, clay minerals records show that the top of phase U4 was characterized by relative higher contents of illite and lower content of kaolinite than that in the modern, suggesting a close distance of sediment transport and strong physical weathering [2]. The reflected seismic data of the nearby core TWS1208 show the unit DU3 was exposed and subjected to intense weathering erosion [9]. In addition, the LGM strata with strong weathering and denudation also occurred in Chao-shan Plain, southwest of the study area [56]. Consequently, a long time interval between the two ^{14}C dating points on the upper part of the core may be contributed to the strong weathering and denudation between U4 and U5 (Figure 2).

Pollen and spore concentrations declined dramatically in phase U5 until they were absent in the upper part of the core, whereas the pollen record shows a mixed South subtropical evergreen coniferous-broad-leaved forest along the Fujian seashore [46]. The grain size of ZK2 analysis shows that this phase mainly consists of medium-coarse sands and gravelly medium-coarse sands [2] (Figure 2). Previous studies showed that the sediment environment with larger grain size was not conducive to the deposition of palynology; pollen is scarce in coarse clastics and abundant in marine sediments composed primarily of fine silt and clay-size particles [20]. In addition, ZMCC as a strong boundary current of the area is also one of the main reasons for the decrease of palynology concentration because the strong hydrodynamic environment is not conducive to the deposition of palynology [13]. However, the content of *Pinus* was 19.6–42.7% in phase U5, which is much higher than that in U4 (2–3.7%) (Figure 2). Previous studies have shown that *Pinus* can be used as an indicator of sea level changes during the glacial-interglacial period, and the proportion of *Pinus* will increase as the increase of the distance from the shore because of its airbags, which facilitate the propagation of wind [18]. Therefore, the increase in *Pinus* at this phase is probably related to the sea level rise. The offshore distance of the study area increased, and the amount of terrestrial pollen brought by land runoff decreased greatly, while the proportion of airborne pollen (*Pinus*) increased.

Relatively high abundance of marine dinoflagellate cysts, foraminifera [2], as well as the contents of CaO and Sr (Figures 3 and 4) all exhibited an increasing trend in the 10.5–0 m interval, which suggests an increased marine influence on the study area. According to the dating control points, phase U5 corresponded to the Holocene. Increased sample score on axis 2 also indicates the enhanced marine influence or sea level rise (Figure 6). All the relative sea level reconstructions from Longhai Plain, Fuzhou Basin and Singapore show sea level rise at about 9 cal ka BP, with a minimum sea-level rise of 14.5 m between 9 and 7.3 cal ka BP in the Fuzhou Basin and about 17 m rise in sea level from 9.5 to 7 cal ka BP in Singapore [58–61], supporting the relatively high sea level in the Taiwan Shoal inferred by pollen and elements data in this study. Therefore, we deduced that the study area was below the sea level during phase U5, corresponding to the Holocene high sea-level stage.

This phase also corresponded to the sedimentary unit DU2 (transgressive sand body) and DU1 (modern tidal current sand body) of the core TWS1208 in the western Taiwan Shoal, which may be affected by strong tidal currents during the Holocene [9].

6. Conclusions

(1) The pollen record of the core ZK2 shows that arboreal pollen dominated since the Late Pleistocene, among them tropical and subtropical broadleaved pollen were the dominant taxa, accompanied with relatively low abundance of temperate broadleaved pollen. This implies that the Taiwan Shoal area has always had a subtropical climate.

(2) The principal component analysis (PCA) of the major and trace elements in the core ZK2 shows that changes in sample score on axis 1 are possibly associated with the input of terrigenous debris carried by the southward Zhe-Min Coastal Current and sample score on axis 2 could be indicated by marine biogenic matters in the Taiwan Shoal area.

(3) Five evolution phases of the sedimentary environment in the Taiwan Shoal area are divided on the basis of the analysis of pollen and major and trace element results with the combination of the AMS ^{14}C dating results. Phase U1 (40–30 m) possibly corresponded to a terrestrial sedimentary environment. Phase U2 (30–24 m) was a transition between terrestrial and marine sedimentary environments, and coarser grain size and strong hydrodynamics sedimentary environment are not conducive to pollen preservation. Phase U3 (24–17 m) was a transgression period with a coastal-intertidal deposits sedimentary environment. Phase U4 (17–12 m) was a terrestrial and marine alternative phase and the land was exposed in the Last Glacial Maximum, with large sea-level fluctuations. Phase U5 (12–0 m) appeared under a marine sedimentary environment, with the sea level rising to the modern level.

Author Contributions: Conceptualization, C.H. and L.S.; methodology, C.H., L.D. and Z.L.; software, C.H.; validation, C.H. and D.L.; formal analysis, C.H. and L.S.; investigation, D.Z.; resources, L.S., L.D. and D.Z.; data curation, L.S., L.D., D.L. and D.Z.; writing—original draft preparation, C.H., L.S., Z.L. and D.L.; writing—review and editing, C.H., L.S., Z.L. and D.L.; visualization, J.T. and X.L.; supervision, L.S., D.Z. and D.L.; project administration, L.S., D.Z. and D.L.; funding acquisition, D.L. All authors have read and agreed to the published version of the manuscript.

Funding: This work was supported by the National Natural Science Foundation of China [Grant No. 41776193, 41876215], Science and Technology Planning Project of Fujian Province [Grant No. 2016R1006-2], Ningbo Natural Science Foundation of China [Grant No. 2018A610282] and the K.C. Wong Magna Fund in Ningbo University.

Institutional Review Board Statement: Not applicable.

Informed Consent Statement: Not applicable.

Data Availability Statement: The data presented in this study are available on request from the corresponding author. The data are not publicly available due to privacy.

Acknowledgments: The authors are thankful to Assistant Editor Marijana Ilic for her help. The comments and feedback from two anonymous reviewers significantly improved the manuscript.

Conflicts of Interest: The authors declare no conflict of interest.

Appendix A

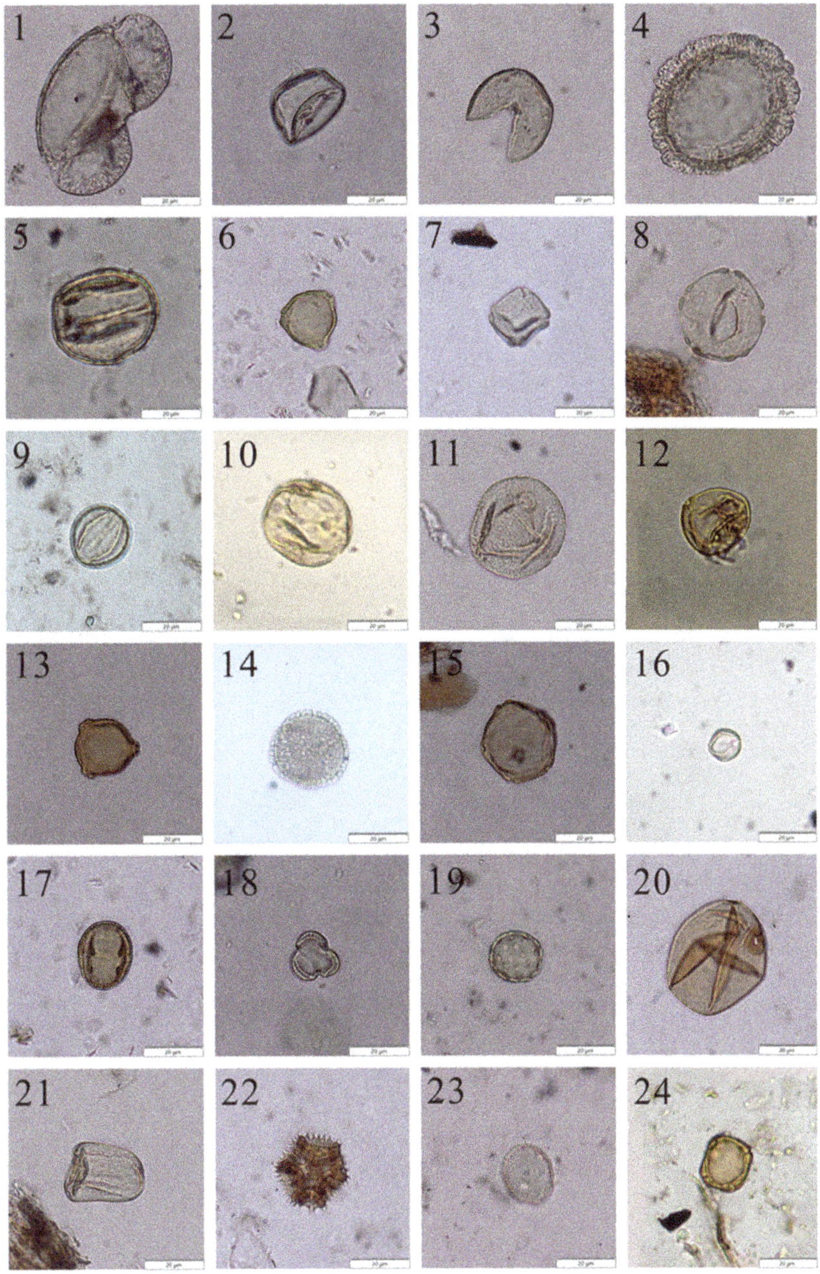

Figure A1. 1: Pinus; 2: Taxodiaceae; 3: Cupressaceae; 4: Tsuga; 5: Deciduous Quercus; 6: Myriccaceae; 7: Aluns; 8: Ulmus; 9: Evergreen Quercus; 10: Altingiaceae; 11: Fagus; 12: Moraceae; 13: Carpinus; 14: Oleaceae; 15: Pterocarya; 16: Eurya; 17: Rutaceae; 18: Artemisia; 19: Chenopodiaceae; 20: Poaceae; 21: Cyperaceae; 22: Compositae; 23: Typha; 24: Myriophyllum spicatum.

References

1. Liao, H.; Yu, H.; Su, C. Morphology and Sedimentation of Sand Bodies in the Tidal Shelf Sea of Eastern Taiwan Strait. *Mar. Geol.* **2008**, *248*, 161–178. [CrossRef]
2. Zhao, D.B.; Sun, R.T.; Jiang, H.Y. Sedimentary Characteristics and Paleoenvironmental Evolution of the Northwestern Taiwan Shoal Since the Late Pleistocene. *Mar. Geol. Quat. Geol.* **2020**, *40*, 1–12.
3. Wu, Q.M. Geochemistry of Rare Earth Elements in Seabed Sediments of Taiwan Shoal in My Country. *Geochemistry* **1983**, *3*, 303–313.
4. Chen, C.H.; Lan, D.Z.; Yu, Y.F.; Zhang, W.L. Late Quaternary Stratigraphy in the Western Taiwan Strait. *Quat. Sci.* **1990**, *10*, 301–307.
5. Lan, D.Z.; Zhang, W.L.; Chen, C.H.; Xie, Z.T. Preliminary Study on Age and Origin of Medium-Coarse Sands in Taiwan Shoal. *Taiwan Strait* **1991**, *10*, 54–59.
6. Lan, D.Z.; Zhang, W.L.; Chen, C.H.; Yu, Y.F. Transgression and Sea Level Change in the Western Taiwan Strait Since the Late Pleistocene. *Acta Oceanol. Sin.* **1993**, *15*, 77–84.
7. Yang, L.J. Distribution Characteristics of Magnetic Susceptibility of Surface Sediments in the Western Taiwan Strait. *Acta Sedimentol. Sin.* **2009**, *27*, 697–703.
8. Dong, D.H.; Deng, G.X.; Liu, M.Z. The Evolution of Paleoenvironment in Southwestern Taiwan Since the Holocene. *J. Subtrop. Resour. Environ.* **2012**, *7*, 1–9.
9. Wang, L.B.; Li, J.; Chen, Z.X.; Zhao, J.T.; Bai, F.L.; Hu, B.Q.; Dou, Y.G.; Zai, B. Late Pleistocene and Holocene Stratigraphy and Paleo-Environmental Evolution in the Western Taiwan Shoal. *Acta Sedimentol. Sin.* **2014**, *32*, 1089–1099.
10. Zheng, Z. Late Quaternary Vegetational and Climatic Changes in the Tropical and Subtropical Areas of China. *Acta Micropalaeontologica Sin.* **2000**, *17*, 125–146.
11. Yue, Y.F.; Zheng, Z.; Huang, K.Y.; Chevalier, M.; Chase, B.M.; Carré, M.; Ledru, M.-P.; Cheddadi, R. A Continuous Record of Vegetation and Climate Change over the Past 50,000 Years in the Fujian Province of Eastern Subtropical China. *Palaeogeogr. Palaeoclimatol. Palaeoecol.* **2012**, *365*, 115–123. [CrossRef]
12. Yang, S.X.; Li, J.; Liu, K.B.; Li, R.H.; Wen, Z.H.; Ye, S.Y.; Yi, S.; Chen, X.H. Pollen-Spore Distribution in the Surface Sediments of the Western Bohai Sea, China. *Quat. Int.* **2016**, *392*, 213–223. [CrossRef]
13. Yang, S.X.; Song, B.; Ye, S.Y.; Laws, E.A.; He, L.; Li, J.; Chen, J.X.; Zhao, G.M.; Zhao, J.T.; Mei, X.; et al. Large-Scale Pollen Distribution in Marine Surface Sediments from the Bohai Sea, China: Insights into Pollen Provenance, Transport, Deposition, and Coastal-Shelf Paleoenvironment. *Prog. Oceanogr.* **2019**, *178*, 102183. [CrossRef]
14. Wang, K.F.; Sun, Y.H.; Zhang, Y.L.; Jiang, H.; Zhang, Y.C. *The Spore-Pollen and Algal Assemblage in the East China Sea Sediments*; China Ocean Press: Beijing, China, 1987; pp. 10–80.
15. Luo, C.X.; Chen, M.H.; Xiang, R.; Liu, J.G.; Zhang, L.L.; Lu, J.; Yang, M.X. Characteristics of Modern Pollen Distribution in Surface Sediment Samples for the Northern South China Sea from Three Transects. *Quat. Int.* **2013**, *286*, 148–158. [CrossRef]
16. Luo, C.X.; Chen, M.H.; Xiang, R.; Liu, J.G.; Zhang, L.L.; Lu, J. Comparison of Modern Pollen Distribution between the Northern and Southern Parts of the South China Sea. *Int. J. Biometeorol.* **2015**, *59*, 397–415. [CrossRef] [PubMed]
17. Luo, C.X.; Lin, G.; Chen, M.H.; Xiang, R.; Zhang, L.L.; Liu, J.G.; Pan, A.D.; Yang, S.X.; Yang, M.X. Characteristics of Pollen in Surface Sediments from the Southern South China Sea and its Paleoclimatic Significance. *Palaeogeogr. Palaeoclimatol. Palaeoecol.* **2016**, *461*, 12–28. [CrossRef]
18. Sun, X.J.; Luo, Y.L.; Huang, F.; Tian, J.; Wang, P.X. Deep-Sea Pollen from the South China Sea: Pleistocene Indicators of East Asian Monsoon. *Mar. Geol.* **2003**, *201*, 97–118. [CrossRef]
19. Sun, X.J.; Li, X. Different Dynamics and Routes of Modern Pollen Transport in the Northern and Southern Parts of the South China Sea. *Sci. China Ser. D* **1998**, *41*, 494–498. [CrossRef]
20. Heusser, L.E. Pollen Distribution in Marine Sediments on the Continental Margin off Northern California. *Mar. Geol.* **1988**, *80*, 131–147. [CrossRef]
21. Sun, X.J.; Li, X. A Pollen Record of the last 37 ka in Deep Sea Core 17940 from the Northern Slope of the South China Sea. *Mar. Geol.* **1999**, *156*, 227–244. [CrossRef]
22. Van der Kaars, S. Pollen Distribution in Marine Sediments from the South-Eastern Indonesian Waters. *Palaeogeogr. Palaeoclimatol. Palaeoecol.* **2001**, *171*, 341–361. [CrossRef]
23. Beaudouin, C.; Suc, J.P.; Escarguel, G.; Arnaud, M.; Charmasson, S. The Significance of Pollen Signal in Present-Day Marine Terrigenous Sediments: The Example of the Gulf of Lions (Western Mediterranean Sea). *Geobios* **2007**, *40*, 159–172. [CrossRef]
24. Jing, X.; Chen, F.; Li, S.; Gui, Z.; Zhong, H. Distribution Pattern of Sporo-Pollen in Surface Sediments of Northwestern Taiwan Strait and Its Geological Significance. *Mar. Geol. Quat. Geol.* **2014**, *34*, 81–88.
25. Liu, Z.X.; Xia, D.X.; Berne, S.; Wang, K.Y.; Marsset, T.; Tang, Y.X.; Bourillet, J.F. Tidal Deposition Systems of China's Continental Shelf, with Special Reference to the Eastern Bohai Sea. *Mar. Geol.* **1998**, *145*, 225–253.
26. Hu, Y.; Chen, J.; Xu, J.; Wang, L.; Li, H.; Liu, H. Sand Wave Deposition in the Taiwan Shoal of China. *Acta Oceanol. Sin.* **2013**, *32*, 26–34. [CrossRef]
27. Shi, Q.; Zhang, J.Y.; Cai, A.Z. Taiwan Shoal-Huge Sand Resource Pool. *J. Nat. Resour.* **2009**, *24*, 507–513.
28. Wang, Y.H.; Jan, S.; Wang, D.P. Transports and Tidal Current Estimates in the Taiwan Strait from Shipboard ADCP Observations (1999–2001). *Estuar. Coast. Shelf Sci.* **2003**, *57*, 193–199. [CrossRef]

29. Takahara, H.Y.; Igarashi, R.; Hayashi, F.; Kumon, P.M.; Liew, M.; Yamamoto, S.; Kawai, T.; Tadamichi, O.; Tomohisa, I. Millennial-Scale Variability in Vegetation Records from the East Asian Islands: Taiwan, Japan and Sakhalin. *Quat. Sci. Rev.* **2010**, *29*, 2900–2917. [CrossRef]
30. Yoneda, M.; Uno, H.; Shibata, Y.; Suzuki, R.; Kumamoto, Y.; Yoshida, K.; Sasaki, T.; Suzuki, A.; Kawahata, H. Radiocarbon Marine Reservoir Ages in the Western Pacific Estimated by Pre-Bomb Molluscan Shells. *Nucl. Instrum. Methods Phys. Res. Sect. B Beam Interact. Mater. At.* **2007**, *259*, 432–437. [CrossRef]
31. Hammer, Ø.; Harper, D.A.T.; Ryan, P.D. PAST: Paleontological statistics software package for education and data analysis. *Palaeontologia electronica.* **2001**, *4*, 9.
32. Qin, J.G.; Wu, G.X.; Zheng, H.B.; Zhou, Q. The Palynology of the First Hard Clay Layer (Late Pleistocene) from the Yangtze Delta, China. *Rev. Palaeobot. Palynol.* **2008**, *149*, 63–72. [CrossRef]
33. Wang, K.F.; Han, X.B. Study on the Cenozoic Fossil Concentricystes of East China. *Acta Palaeontol. Sin.* **1983**, *22*, 468–472.
34. Dai, L.; Weng, C.Y.; Lu, J.; Mao, L.M. Pollen Quantitative Distribution in Marine and Fluvial Surface Sediments from the Northern South China Sea: New Insights into Pollen Transportation and Deposition Mechanisms. *Quat. Int.* **2014**, *325*, 136–149. [CrossRef]
35. Zhao, Y.Y.; Che, C.H.; Yang, H.L.; Jia, F.M. Geochemistry of Fe, Mn, Ti, and P in the Taiwan Bank, China. *Acta Geol. Sin.* **1981**, *18*, 118–126.
36. Bhatt, J.J. Ti/Al Ratio as Chemical Index of Paleoenvironment-A Note. *Chem. Geol.* **1974**, *13*, 75–78. [CrossRef]
37. Schmitz, B. The TiO_2/Al_2O_3 Ratio in the Cenozoic Bengal A byssal Fan Sediments and Its Use as a Paleostream Energy Indicator. *Mar. Geol.* **1987**, *78*, 195–206. [CrossRef]
38. Wei, G.J.; Liu, Y.; Li, X.H.; Shao, L.; Fang, D.Y. Major and Trace Element Variations of the Sediments at ODP Site 1144, South China Sea, during the Last 230 ka and Their Paleoclimate Implications. *Palaegeography Palaeoclimatol. Palaeoecol.* **2004**, *212*, 331–342. [CrossRef]
39. Yang, S.Y.; Lim, D.I.; Jung, H.S.; Oh, B.C. Geochemical Composition and Provenance Discrimination of Coastal Sediments around Cheju Island in the Southeastern Yellow Sea. *Mar. Geol.* **2004**, *26*, 41–53. [CrossRef]
40. Dymond, J.; Suess, E.; Lyle, M. Barium in Deep-Sea Sediment: A Geochemical Proxy for Paleoproductivity. *Paleoceanography* **1992**, *7*, 163–181. [CrossRef]
41. Dean, W.E.; Gardner, J.V.; Piper, D.Z. Inorganic Geochemical Indicators of Glacial-Interglacial Changes in Productivity and Anoxia on the California Continental Margin. *Geochim. Cosmochim. Acta* **1997**, *61*, 4507–4518. [CrossRef]
42. Sun, X.Q.; Liu, S.F.; Li, J.R.; Zhang, H.; Zhu, A.M.; Cao, P.; Chen, M.T.; Zhao, G.T.; Khokiattiwong, S.; Kornkanitnan, N.; et al. Major and Trace Element Compositions of Surface Sediments from the Lower Bengal Fan: Implications for Provenance Discrimination and Sedimentary Environment. *J. Asian Earth Sci.* **2019**, *184*, 104000. [CrossRef]
43. Li, H.J.; Hermann, J. The Effect of Fluorine and Chlorine on Trace Element Partitioning between Apatite and Sediment Melt at Subduction Zone Conditions. *Chem. Geol.* **2017**, *473*, 55–73. [CrossRef]
44. Guo, Z.G.; Yang, Z.S.; Qu, Y.H.; Fan, D.J. Study on Comparison Sedimentary Geochemistry of Mud Area on East China Sea Continental Shelf. *Acta Sedimentol. Sin.* **2000**, *18*, 284–289.
45. Yang, J.W.; Hua, L.; Wu, L.C. Sporopollen, Foram and Diatom Assemblages of Quaternary Strata in the Minjiang River Estuary and Its Paleo-Geographical Significance. *Mar. Geol. Quat. Geol.* **1991**, *11*, 75–83.
46. Wang, K.F.; Lu, J.J.; Zheng, Y.L. Late Quaternary Pollen-Diatom Assemblages along Fujian Seashore and Their Paleoenvironmental Implications. *Acta Micropalaeontologica Sin.* **1995**, *12*, 388–397.
47. Heusser, L.E.; Balsam, W.L. Pollen Sedimentation in the Northwest Atlantic: Effects of the Western Boundary Undercurrent. *Mar. Geol.* **1985**, *69*, 149–153. [CrossRef]
48. Nesbitt, H.W.; Markovics, G.; Price, R.C. Chemical Processes Affecting Alkalis and Alkaline Earths during Continental Weathering. *Geochim. Cosmochim. Acta* **1980**, *44*, 1659–1666. [CrossRef]
49. Nesbitt, H.W.; Young, G.M. Petrogenesis of Sediments in the Absence of Chemical Weathering: Effects of Abrasion and Sorting on Bulk Composition and Mineralogy. *Sedimentology* **1996**, *43*, 341–358. [CrossRef]
50. Tao, J.; Chen, M.T.; Xu, S.Y. A Holocene Environmental Record from the Southern Yangtze River Delta, Eastern China. *Palaeogeogr. Palaeoclimatol. Palaeoecol.* **2006**, *230*, 204–229. [CrossRef]
51. Zou, J.J.; Shi, X.F.; Li, S.L. Distribution of Minor Elements in Near Surface Sediments in north Yellow Sea and the Early Diagenesis. *Mar. Geol. Quat. Geol.* **2007**, *27*, 43–51.
52. Liang, W.J.; Xiao, C.T.; Xiao, K.; Lin, W. The Relationship of Late Jurassic Paleoenvironment and Paleoclimate with Geochemical Elements in Amdo Country of Northern Tibet. *Geol. China* **2015**, *42*, 1079–1091.
53. Lan, D.Z.; Yu, Y.F.; Chen, C.H.; Xie, Z.T. Preliminary Study on the Late Pleistocene Transgression and Holocene Sea-Level Fluctuation in Fuzhou Basin. *Mar. Geol. Quat. Geol.* **1986**, *6*, 103–110.
54. Yang, J.M. The Last Marine Transgression and Sea Level Changes along the Fujian Coast during Late Pleistocene. *Mar. Sci.* **1988**, *5*, 5–9.
55. Zeng, C.S. A Discussion on Problems of Marine Transgression along the Fujian coast during The Late Pleistocene. *Trop. Oceanol.* **1993**, *12*, 39–45.
56. Zheng, Z.; Li, Q. Vegetation, Climate, and Sea Level in the Past 55,000 Years, Hanjiang Delta, Southeastern China. *Quat. Res.* **2000**, *53*, 330–340. [CrossRef]

57. Zheng, Z.; Yang, S.X.; Deng, Y.; Huang, K.Y.; Wei, J.H.; Berne, S.; Suc, J.-P. Pollen Record of the Past 60 ka BP in the Middle Okinawa Trough: Terrestrial Provenance and Reconstruction of the Paleoenvironment. *Palaeogeogr. Palaeoclimatol. Palaeoecol.* **2011**, *307*, 285–300. [CrossRef]
58. Bird, M.I.; Fifield, L.K.; Teh, T.S.; Chang, C.H.; Shirlaw, N.; Lambeck, K. An Inflection in the Rate of Early Mid-Holocene Eustatic Sea-Level Rise: A New Sea-Level Curve from Singapore. *Estuar. Coast. Shelf Sci.* **2007**, *71*, 523–536. [CrossRef]
59. Rolett, B.V.; Zheng, Z.; Yue, Y.F. Holocene Sea-Level Change and the Emergence of Neolithic Seafaring in the Fuzhou Basin (Fujian, China). *Quat. Sci. Rev.* **2011**, *30*, 788–797. [CrossRef]
60. Ge, W.Y.; Li, C.H.; Xing, H.X.; Li, L.; Li, Y.X.; Rittenour, T.; Hu, Z.J. Examining the Chronology of Transgressions Since the Late Pleistocene in the Fujian Coast, Southeastern China. *Quat. Int.* **2019**, *527*, 34–43. [CrossRef]
61. Zong, Y.Q. Mid-Holocene Sea-Level Highstand along the Southeast Coast of China. *Quat. Int.* **2004**, *117*, 55–67. [CrossRef]

Journal of
Marine Science and Engineering

MDPI

Article

Facies Analysis and Sedimentary Architecture of Hybrid Event Beds in Submarine Lobes: Insights from the Crocker Fan, NW Borneo, Malaysia

Muhammad Jamil [1,2,*], Numair Ahmed Siddiqui [1,3], Nisar Ahmed [1], Muhammad Usman [4,5], Muhammad Umar [6], Hamad ur Rahim [7,8] and Qazi Sohail Imran [1]

1 Department of Geosciences, Universiti Teknologi PETRONAS, Seri Iskandar 32610, Malaysia; numair.siddiqui@utp.edu.my (N.A.S.); nisarpu12@gmail.com (N.A.); qazi_17007588@utp.edu.my (Q.S.I.)
2 Department of Earth Sciences, Abbottabad Campus, COMSATS University Islamabad, Abbottabad 22044, Pakistan
3 Shale Gas Research Group, Institute of Hydrocarbon Recovery, Universiti Teknologi PETRONAS, Seri Iskandar 32610, Malaysia
4 Department of Earth and Environmental Sciences, University of Milano-Bicocca, Piazza Della Scienza 4, 20126 Milano, Italy; usman.pu@outlook.com
5 School of Geosciences, China University of Petroleum (East China), Qingdao 266580, China
6 Department of Earth Sciences, The University of Haripur, Haripur 22620, Pakistan; umarkhan09@yahoo.com
7 Earth Sciences Division, Pakistan Museum of Natural History, Islamabad 44000, Pakistan; hamadrahim@gmail.com
8 Department of Earth Sciences, Quaid-i-Azam University, Islamabad 45320, Pakistan
* Correspondence: jamil287@gmail.com

Abstract: Hybrid event beds represent the combined effect of multiple geological processes, which result in complex depositional geometries and distinct facies distribution in marine environments. Previous work on hybrid event beds highlights the classification, origin, and types of hybrid facies. However, in the present study, we discuss the development of hybrid event beds in submarine lobes with an emphasis on the analysis of proximal to distal, frontal to lateral relationships and evolution during lobe progradation. Detailed geological fieldwork was carried out in the classical deep-marine Late Paleogene Crocker Fan to understand the relationship between the character of hybrid bed facies and lobe architecture. The results indicate that hybrid facies of massive or structureless sandstone with mud clasts, clean to muddy sand, and chaotic muddy sand with oversized sand patch alternations (H1–H3) are well developed in proximal to medial lobes, while distal lobes mainly contain parallel to cross-laminated clean to muddy hybrid facies (H3–H5). Furthermore, lateral lobes have less vertical thickness of hybrid beds than frontal lobes. The development of hybrid beds takes place in the lower part of the thickening upward sequence of lobe progradation, while lobe retrogradation contains hybrid facies intervals in the upper part of stratigraphy. Hence, the development of hybrid beds in submarine lobe systems has a significant impact on the characterization of heterogeneities in deep-marine petroleum reservoirs at sub-seismic levels.

Keywords: facies heterogeneity; hybrid event beds; lobe architecture; north-west Borneo; sedimentological characterization; submarine siliciclastic deposition; West Crocker Formation

Citation: Jamil, M.; Siddiqui, N.A.; Ahmed, N.; Usman, M.; Umar, M.; Rahim, H.u.; Imran, Q.S. Facies Analysis and Sedimentary Architecture of Hybrid Event Beds in Submarine Lobes: Insights from the Crocker Fan, NW Borneo, Malaysia. *J. Mar. Sci. Eng.* **2021**, 9, 1133. https://doi.org/10.3390/jmse9101133

Academic Editor: Gemma Aiello

Received: 6 August 2021
Accepted: 17 September 2021
Published: 15 October 2021

Publisher's Note: MDPI stays neutral with regard to jurisdictional claims in published maps and institutional affiliations.

Copyright: © 2021 by the authors. Licensee MDPI, Basel, Switzerland. This article is an open access article distributed under the terms and conditions of the Creative Commons Attribution (CC BY) license (https://creativecommons.org/licenses/by/4.0/).

1. Introduction

Submarine gravity flows are one of the key sedimentary processes responsible for sediment transport, in which a single submarine flow may result in an enormous volume of rock deposition. These sedimentary successions can develop some of the thickest and largest rock accumulations on the planet [1–6]. These density flow deposits have significant petroleum potential and provide valuable information on ancient submarine geological processes in deep-marine environments. The understanding of hybrid bed facies deposited by density flows and their heterogeneities is vital for oil and gas potential, especially in submarine siliciclastic deposition [7–9]. A subaqueous gravity flow carrying sediments may consist of various types of flow mechanisms and the transformation of flow may occur while moving downslope under the action of gravity [10]. The characterization of flow is based on the detailed observations of ancient sediment gravity flow deposits, whereby the basinward variations enhance the probabilities of flow transformation [11–13], resulting in composite beds having abrupt textural breaks [14].

Deep-water depositional systems may contain a combination of sediment gravity flows in a single event termed variably by a number of researchers as bipartite–tripartite deposits [15], linked debrites [16], co-genetic debrite–turbidite (CGDT) beds [17], hybrid event beds HEBs [18], hybrid sediment gravity flow deposits [19], intermediate strength cohesive debris flow deposits [20], transitional flow deposits [21], matrix rich sands [22], sand–mud couplet [23], or hybridite [24]. These hybrid event beds can be recognized on a variable scale (cm to m) in core samples as well as in outcrops [23].

The relative proportion of siliciclastic sediments (grain size and type of sediments) and change in concentration of transporting material in flow determines the evolution of sediment gravity flow processes [25–28]. In addition to this, the fluid turbulence and shear rate significantly influence the flow behavior [12]. A hybrid event bed is considered a type of sediment gravity flow deposit that results from a blend of multiple flow processes including debris flow, turbidity flow, and transitional flows in a single depositional event and generally consisting of basal clean sand followed by muddy sand facies [29–31]. Hybrid event beds are thought to be frequently associated with medial and lateral fringe lobe systems and are further related with prograding lobe settings [16,19,21,32,33] as well as aggradation in basin-plain settings [34]. Hybrid beds are developed when the deceleration of a mud-dominated, highly concentrated sediment flow dispels its energy in medial and distal submarine areas [1,32,35,36].

Recent works on deep-water systems integrate the depositional character of hybrid event beds with the development of lobe systems during the flow transformation in lacustrine and deep-marine basins of Italy [30,34,37,38], the United Kingdom [39], South Africa [40,41], Norway [12], South America [11], Canada [22], Eastern China [36], and Sarawak, Malaysia [42]. In general, these hybrid event beds are present at discrete stratigraphic intervals of lobe systems in a submarine fan environment [38]. Likewise, these alternate clean and muddy sand intervals have considerable impacts on the reservoir characterization and the subsequently, the hydrocarbon production potential of sandy reservoir units [11,30,43–45]. Therefore, it is necessary to understand the effect of flow transformation in stratigraphic records of submarine lobes that will substantially influence the reservoir prediction and limit the uncertainties in the stratigraphic data [23,35,40,43,44]. The present work emphasizes a huge variety of flow transformation where the hybrid event beds are developed in numerous components of lobe system with irregular sedimentary facies.

Hybrid event beds are commonly present in deep-marine environments, especially in submarine lobe systems, and are also termed as linked debrites having an increase in mud content [46]. However, the maximum vertical thickness of individual lobes mainly depends on the origin of sediments, as channelized lobes normally have less vertical thickness than the non-channelized lobes [47]. The base topography of a basin causes the transformation of flow and controls the development of linked debrites in a deep-marine environment [34,46,48]. The development of hybrid event beds in various components of lobes is vital for understanding the paleogeographic reconstruction of submarine fans in deep-water systems and will reduce the ambiguities in the hydrocarbon reservoir potential of deep-marine sandstone intervals [49,50].

The aim of the present study is to associate the various types of hybrid event beds with the characterization of submarine lobes. The key objectives of this work include: (i) the evaluation of hybrid event beds based on submarine lobe processes; (ii) the analysis of spatial distribution of hybrid beds facies with lithological characteristics in several components of lobes; and (iii) to propose a generalized facies model for hybrid event beds for submarine lobes of deep-marine fan systems. Seventeen outcrop sections were selected from the study area that are fresh and recently exposed sections due to infrastructural development of Pan-Borneo Highway (Table 1).

Table 1. Selected locations of onshore Sabah. Hybrid event beds are not commonly present throughout all outcrops. A summary of each section is presented in the table below.

No.	Abbreviation	Location Name	Region	Hybrid Beds Distribution
1	SUL	Jalan Sulaman	NW Sabah	Lateral lobe with complete hybrid facies
2	LPS	Lapasan		Distal lobe with all hybrid facies except H2Frontal lobe with mainly H1 and H3 hybrid facies
3	PQ	Perwira Quarry		No hybrid beds, sandstones, and shale beds
4	UU	University Utama, Telipok		No hybrid facies, mainly massive sandstones
5	SP	Sepangger port bypass		Frontal lobes with H1 and H3 hybrid facies mainly
6	UP	University Prima Condo	W Sabah	Lobe progradation with hybrid beds
7	JK	Jalan UMS behind the KFC		Medial lobes with complete set of hybrid faciesLobe retrogradation with hybrid beds
8	JU	Jalan UMS roadside		Proximal lobe with hybrid beds H1, H2, and H5Lateral lobe with H1, H3, and H5 hybrid facies
9	JBN	Jalan Bantayan		No hybrid beds sandstone and shale intervals
10	KM	Kampung Madpai		No hybrid beds, mainly massive sandstones
11	SUK	Kampung Sukang		No hybrid beds, thick to massive sandstone
12	JLK	Jalan Lok Kawi Pengalat	SW Sabah	No hybrid beds, massive sandstones
13	MK	Kampung Mook		No hybrid beds, massive sandstones, and shales
14	LKW	Lok Kawi wildlife		Distal lobe with hybrid facies H1, H3, and H5
15	KW	Kampung Kawang		No hybrid beds, mainly massive shale intervals
16	PUR	Kampung Purak		No hybrid beds, mainly turbidites only
17	BQ	Benoni Quarry		No hybrid beds, only debrites and turbidites

2. Geological Settings

Borneo exhibits complex geological history, particularly during the Tertiary Period when thousands of meters-thick deep-marine sedimentary successions were deposited in an active tectonic regime [51–56]. Active tectonic subduction in Borneo resulted in the closure of paleo-basins and development of ophiolite rocks [57,58]. During the Late Cretaceous, the island arc and tectonic fragments collided with the continental part of the Sunda Plate, forming the depocenter named Sabah Basin that now represents the northern part of Borneo [52,59,60]. Thus, NW Borneo is located at the complex geological junction of the South China Sea, the Sunda Shelf, the Java Sea, and the Celebes Sea (Figure 1), where convergent tectonic settings resulted in the Sabah orogenic belt exposing the Tertiary deep-water sediments [46,53,61–64]. The NW Sabah Basin mainly consists of the Crocker range or Crocker fold–thrust belt that developed due to the collision of continental plates [65–67].

Tertiary stratigraphy of the Sabah Basin is mainly distributed in two phases, where the first phase of tectonic and sedimentary processes comprised the deposition of the Early Paleogene (Paleocene to Eocene) deep-water sediments of Trusmadi and East Crocker formations. The first phase of deposition was followed by uplifting, erosion, and an unconformable surface termed the Late Eocene Unconformity (LEU). Later, the second Late Paleogene (Late Eocene to Early Miocene) phase was overlain by an unconformity resulting in the deposition of the West Crocker and Temburong formations [62,68]. The upper contact of the West Crocker Formation is marked by the Top Crocker Unconformity (TCU) or the Base Miocene Unconformity (BMU) (Figure 1). The West Crocker Formation is representative of a deep-marine fan system [46,57,62] that mainly comprises thick to massive sandstone, interbedded with siltstone and mudstone, interpreted as various components of submarine lobe systems [46,69].

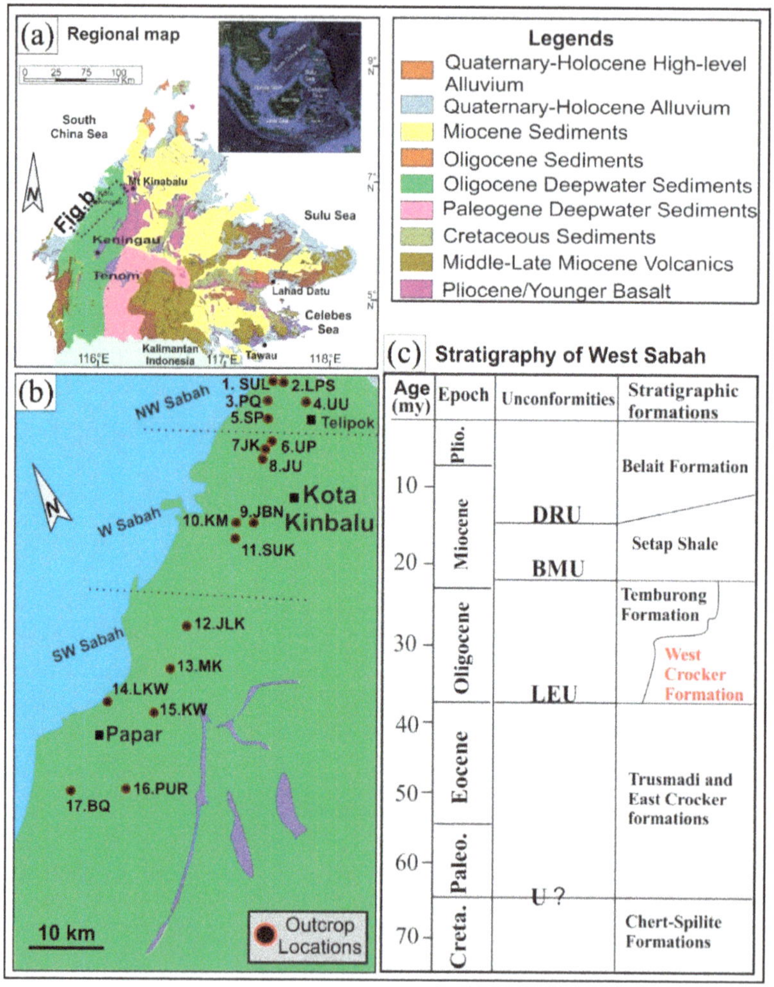

Figure 1. Study area in Sabah, NW Borneo. (**a**) Map of Borneo with surrounding archipelagos with stratigraphic record in the legend outlined with black rectangle showing the study location of Sabah. (**b**) Study area with various regions in Sabah selected for the fieldwork, where locations 1–5 are around Telipok (NW Sabah), locations 6–11 around the Kota Kinabalu (West Sabah), and 12–17 around Papar Town (SW Sabah). (**c**) Stratigraphic position of the sand-rich West Crocker Fan majorly related to the Oligocene age and bound by the Late Eocene Unconformity (LEU) at the base while the upper contact is marked by the Base Miocene Unconformity (BMU).

Previous literature on the Crocker Fan reported that the deep-marine sediments originated from local nearby sources and did not involve long-distance transport. Therefore, these sediments are texturally immature and contain angular to subangular fragments originating from recycled orogen [51,69,70]. The texture of sediments is not diverse; however, they contain a wide variety of deep-marine sedimentary successions including mass transport deposits, debrites, high-density turbidites, and low-density turbidites interpreted to be the parts of outer, middle, and inner fan environments [46,47,66,69–71]. The origin of linked debrites and co-genetic turbidite–debrite intervals was discussed in the West Crocker Formation with respect to their sedimentological and stratigraphic framework and the spatial distribution of these intervals in the Sabah Basin [46,47]. However, this study reveals the development of hybrid event bed facies with various distributive components of submarine lobe system. The outcrops containing these hybrid event beds are part of the Crocker fold–thrust belt that formed due to convergent tectonic regimes. These rocks were initially deposited in a deep-marine basin with a water depth of more than 2000 m and later exposed to the surface by uplifting and erosional processes [72,73].

3. Materials and Methods

The present work includes detailed geological fieldwork in onshore Sabah to observe the characteristics of hybrid event beds in submarine lobe systems. It is pertinent to mention that these hybrid event beds are not present in every outcrop. Therefore, some key sections of outcrops were selected to understand the facies of hybrid event beds in deep-marine sediments.

3.1. Data and Fieldwork

The present work involved detailed geological fieldwork in seventeen locations having more than 1100 m of stratigraphic thickness exposed from NW (Telipok) to SW (Papar) Sabah, NW Borneo (Figure 1), to find excellent examples of outcrops. As most hybrid event beds are vertically extensive but laterally limited in stratigraphic records, they are not expressed in every exposed section (Figure 2) of deep-marine sedimentary successions. Hence, these hybrid event beds are exposed only in a limited number of locations exposed to sedimentary successions. The selected outcrops range in vertical thicknesses from 31 to 192 m and are present over 61.2 km length of transect.

Figure 2. (**a**) Deep-marine sediments contain various types of deposits as we move stratigraphically from base to top of an exposed section in the Benoni Quarry (BQ), SW Sabah. It is possible to find muddy debrite in between two turbidite sandstone units. (**b**) An enlarged image of the contact of turbidites and debrite with lack of transformation of flow and, consequently, no expression of hybrid event intervals in this stratigraphic record.

This extensive field study enabled us to compare various types of hybrid facies in deep-marine sedimentary successions. Sedimentary logs supported the discussion to illustrate the development of hybrid event beds in a lobe system and its stratigraphic framework. Bed-scale heterogeneities in gravity-flow deposits were analyzed, with emphasis on boundary surfaces of individual beds, variation in grain size, and lithological characteristics. Systematic variations in hybrid units bound by stratigraphic intervals were used to determine the flow characterization in submarine lobe deposits.

3.2. Hybrid Bed Facies

Hybrid event beds are classified into five major divisions or facies on the basis of sedimentary features present in each division from base to top and termed as H1 to H5 facies [19] as shown in Figure 3. The base of hybrid event bed succession is massive sandstone occasionally containing dewatering structures at the base and floating mud clasts in the upper part of interval, termed as hybrid event bed facies 1 (H1). This division is overlain by a sandy unit with alternating lighter and darker bands developed due to clean and muddy sand deposition, representing hybrid event bed facies 2 (H2). The third hybrid event bed division is a chaotic muddy sand containing abundant mud clasts while some of the sand patches are outsized particles, collectively denoted H3 facies. A fine sand division comprising parallel or cross laminations is termed hybrid event bed facies 4 (H4). The final depositional unit in the hybrid event bed sequence is a muddy massive unit called hybrid event bed facies 5 or H5 facies.

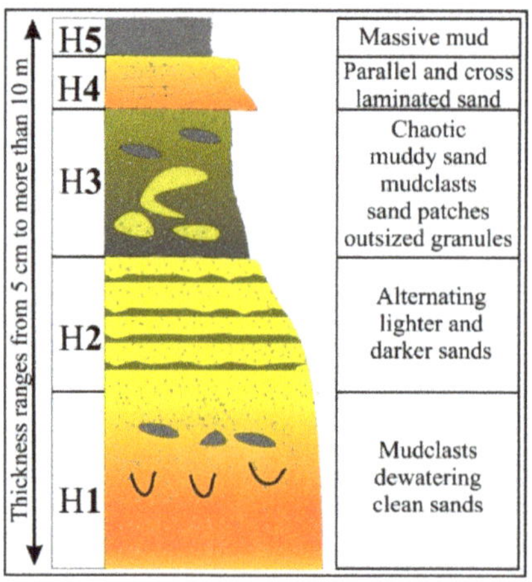

Figure 3. A simple hybrid event bed facies model proposed by Haughton et al. [19], showing five hybrid bed facies or divisions from base H1 to top H5. The basal H1 facies comprises massive sand with dewatering and broken mud clasts, the H2 hybrid event bed facies consists of alternate lighter and darker banded sandstone, H3 has muddy sand with chaotic features both of sand clasts and mud clasts, the H4 facies is formed with fine sand with parallel and cross laminations, and the H5 facies is linked with massive mudstone or shale.

These five hybrid event bed facies vary in scale from a few centimeters to about 10 m in vertical thickness and hence, they can be equally studied in cores as well as in outcrop sections. Moreover, it is quite possible that among these five divisions, one of these divisions only is expressed in less than 10 cm while other hybrid event bed facies may have more than a meter thickness at the same location. However, all these divisions are not necessarily present in every hybrid event bed. In fact, the presence of all five facies in a single hybrid event is relatively uncommon.

3.3. Lobe Architecture

The hierarchy of the lobe system was adopted from Prélat et al. [74] in which the lithological bed or bedset collectively form the lobe element, which is the basic building unit for lobe hierarchy (Figure 4). The beds or bedsets vary in cm to m scale that accurately integrate with the scale of hybrid event bed [33,75]. Therefore, the distribution of hybrid event bed facies at outcrop level represents the lobe element [39,76]. Several lobe elements combine to form an individual lobe at a vertical scale of 5–10 m, which is considered the maximum thickness of a hybrid event bed, and these lobes are grouped together in a single lobe complex, and further lobe complexes combine to form a lobe complex set or lobe complex system [27,77,78].

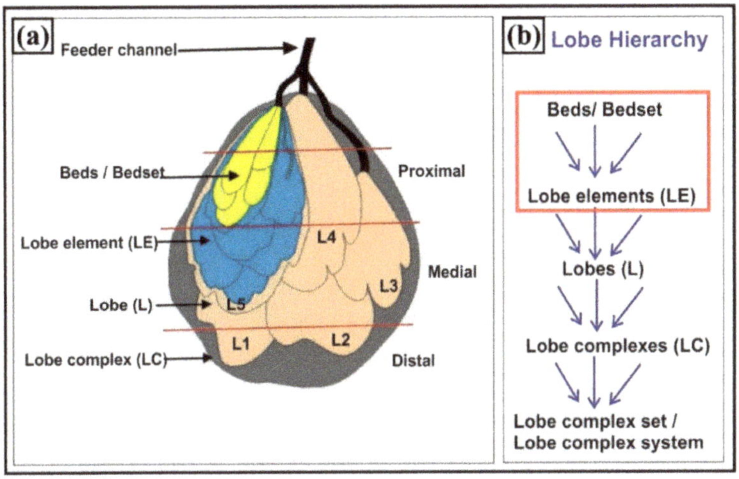

Figure 4. A sketch showing various components of a lobe system from proximal to distal zones. (**a**) Each lobe has the smallest units of bed or bedset which merge into lobe element (LE). Many lobe elements combine to form an individual lobe (L). These lobes are further stacked into a lobe complex [45,79]. (**b**) Lobe hierarchy starting from beds or bedset following to form a lobe. The lobe element is the basic building block of a lobe system while the lobe complex set or lobe complex system is the largest entity in lobe hierarchy.

However, our study has not found the hybrid event bed reaching 10 m on an individual lobe scale. Hence, hybrid event bed divisions are commonly associated with the lobe element and are frequently related with the components of an individual lobe [45]. The occurrence of hybrid beds in a lobe system is common during the transformation of flow from a higher flow regime to a lower flow regime and because of that, these hybrid facies are well-developed in medial lobe settings [18,41]. The dimensions of a lobe in lateral across strike depends on the confined zones in a basin and topographic fluctuations, where confined settings have limited deposition of hybrid beds [12,33,79].

4. Results and Interpretations

4.1. Hybrid Bed Facies in Proximal to Distal Lobes

4.1.1. Proximal Lobes

Proximal lobes having hybrid facies (2–3 m vertical thickness) comprising thick to massive sand vary in grain size from coarse to fine-grained, usually exhibit poor sorting, and may contain water-escape structures (H1 facies). The basal division is clean sand frequently overlain by floating mud clast intervals within massive sand divisions. This massive sand constitutes a considerable part of the base of a hybrid event bed and is often overlain by banded sand and sandy mud deposition (H2 facies) (Figure 5).

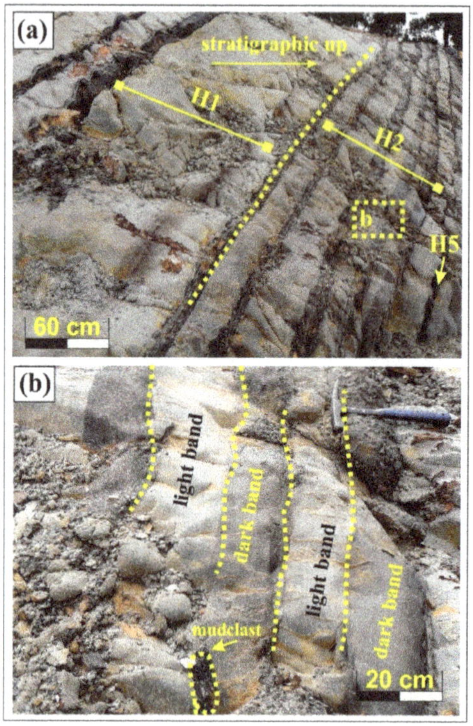

Figure 5. The development of hybrid event bed facies in proximal lobe system. (**a**) An expression of a sandy hybrid event bed (H1 and H2) from the base to the top of a sedimentary unit in the Jalan UMS (JU) section, West Sabah, at multimeter scale where the yellow dashed line indicates the facies boundary between H1 and H2. (**b**) An enlarged image of the H2 hybrid event bed facies containing lighter and darker sand intervals (yellow dashed lines representing the boundary of light and dark bands) with a floating mud clast (labelled in yellow oval-shaped dashed lines) that is relatively uncommon in H2 facies and was not reported previously.

This banded facies develops due to progressive evolution of flow from fluidized to plastic due to reduced turbulency and more cohesion in flow behavior [80]. These hybrid event beds usually occur in axial or proximal lobe settings [38,42] when the flow transformation takes place. Wedge-shaped bed geometries and clastic muddy injections may also occur in chaotic division [32] or in linked debrites, which are mainly related with proximal lobe settings in rock sections of our study area.

4.1.2. Medial Lobes

Medial lobes are ideal avenue for the development of hybrid facies when a sand-rich system is gradually increased in mud content during the phase of deposition. Hence, a complete sequence of hybrid bed facies (H1 to H5) is likely to develop in the medial lobe system (Figure 6). However, the thicknesses of sandy hybrid event beds (H1 and H2) greatly decrease at the expense of increases in the thickness of muddy hybrid event bed intervals (H3 to H5). Hence, the thickness of the basal hybrid event bed facies in medial lobes is less (1–1.5 m) than those in the proximal lobes (2–3 m).

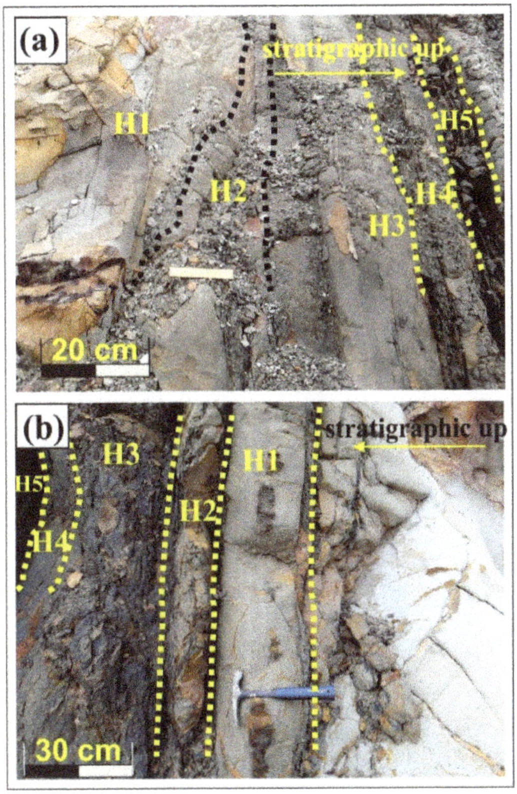

Figure 6. Hybrid bed facies in a medial lobe setting. (**a**) A complete model of hybrid facies is commonly developed in the medial lobe system, where all the facies are distinctively placed in hybrid event beds. One can easily find facies ranging from H1 to H5 from the base to the top, respectively, in the medial lobe exposed in the Sepangger (SP) section, NW Sabah. (**b**) All hybrid event bed facies (H1 to H5) are developed in the medial lobe component exposed in the Jalan UMS behind the KFC (JK) section, West Sabah.

The gradual loss of coarse-grained sediments in the proximal lobes gradually enriches the clay and silt particles, resulting in the transformation of flow character developing the hybrid event beds in medial lobe settings [44]. Moreover, these deep-marine sand–mud couplets are established due to remobilization of sediments during the phase of deposition [14,23,36] which are also termed as the chaotic division of H3 hybrid facies well-developed in the medial lobe settings [14,23,30,36].

4.1.3. Distal Lobes

The vertical thickness of hybrid event beds in distal lobe settings relatively decreases (0.6 to 1.5 m) due to deposited sands in the depositional system. However, they may have formed of thicker residual fine-sand and muddy division of hybrid event beds, while the basal divisions of hybrid event beds (H1 and H2) are less developed (Figure 7). There is no banded hybrid division of H2 in the distal lobe settings.

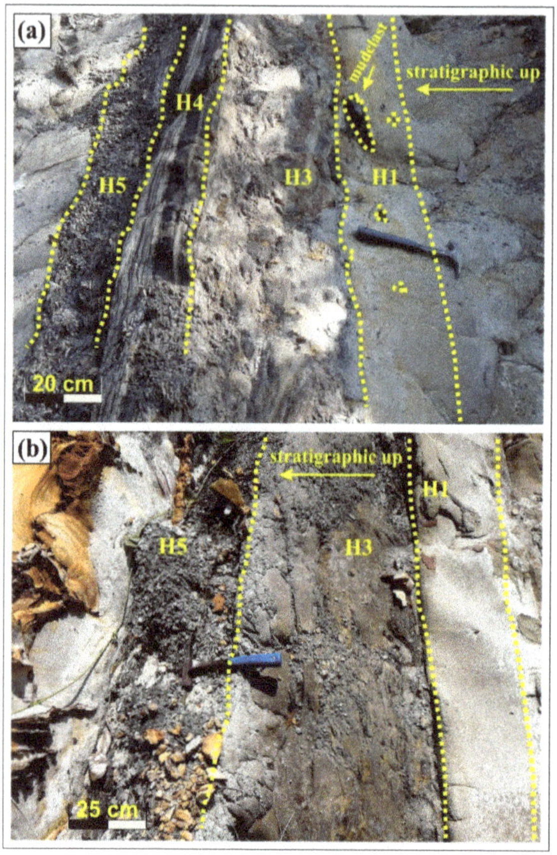

Figure 7. Hybrid event beds interpreted in a distal lobe system. (**a**) Distal lobe contains more vertical thickness of muddy hybrid event bed facies (H3, H4, and H5) than the proximal and medial lobes moving from the base to the top of sequence in the Lapasan (LPS) section, NW Sabah. (**b**) The most distal part of the lobe contains the minimum or least vertical thickness of sandy hybrid facies (H1 or H2) where the facies are mainly composed of chaotic intervals of H3, and mud facies of H5 in the Lok Kawi wildlife (LKW) section, SW Sabah.

Hybrid event beds are frequently reported in the distal lobe system having abrupt pinch out and wedging in beds due to deceleration of flow [2,16,18,21,30,81]. Likewise, linked debrites are usually associated with the proximal and medial settings with higher energy flow, whereas the distal lobes have less vertical thickness of hybrid beds [18,46] as the energy conditions in massive and thick sand packages are already transformed in the proximal and medial lobe settings.

4.2. *Hybrid Bed Facies in Frontal and Lateral Lobes*

4.2.1. Frontal Lobes

Frontal zones of a lobe in the axial component remain with the clay-poor deposition and the lower divisions of hybrid beds are characterized by dewatered sand (H1), overlain by muddy sand intervals mainly in the fringe area (Figure 8). In addition to this, the fringe area of frontal lobes is prone to hybrid events that primarily contain the mud-rich hybrid bed (H3 and H5 facies) because of depositional transition from massive clean sand into muddy sand throughout the deep-marine gravity flow [33].

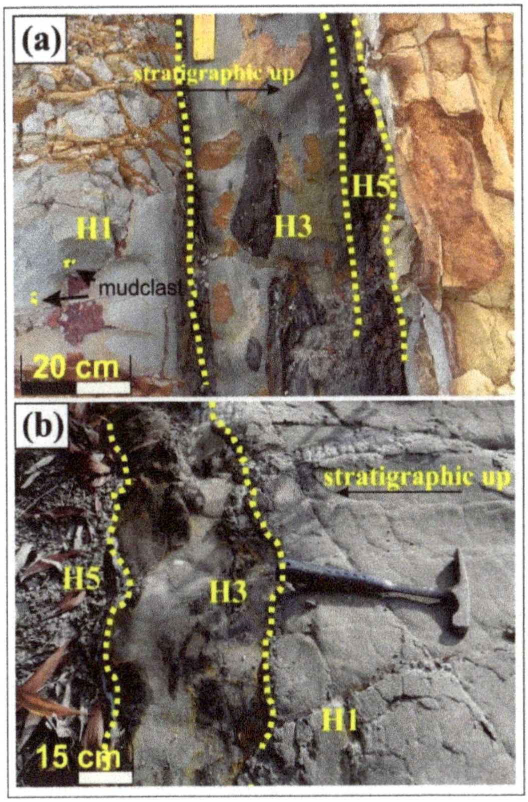

Figure 8. Variable expressions of hybrid event bed facies in frontal lobes. (**a**) Frontal area of the lobe near to the axis has brief hybrid event bed intervals mainly containing H1 and H3 facies with floating mud clasts in the Sepangger (SP) section, NW Sabah. (**b**) The frontal area away from the axial lobe domain has distinct muddy hybrid facies of H3 to H5 and the vertical thickness of the hybrid sequence also increases significantly in the Lapasan (LPS) section, NW Sabah.

4.2.2. Lateral Lobes

Lateral lobes are the frequent sites of flow transformation from the lobe axis sideways. These lateral lobes are abundant in sandy hybrid facies associated with broken fragments or mud clasts (Figure 9) while the upper muddy hybrid facies (H3 to H5) are distinctively developed but have less vertical thickness. Lateral lobes have a slight chaotic effect in the depositional phase, which necessitates the formation of H3 facies in them. The mud clasts of more than 5 cm are often encompassed in the basal hybrid sandy division (H1) but are relatively smaller sized mud clasts in H3 facies.

The distribution of hybrid event bed facies greatly depends on the degree of basin confinement as it determines the lobe stacking and dispersal patterns [33,34,50]. Lateral lobes exhibit abrupt change in thickness and facies, resulting in less development of vertically thick hybrid event bed facies (Figure 9). It is possible to find the hybrid event bed facies in lateral lobes, but the extension of distinct hybrid event bed facies in lateral lobes is less common, especially in a confined sedimentary system where the flow does not have ample space to establish a distinct and well-developed transitional facies architecture [33,49].

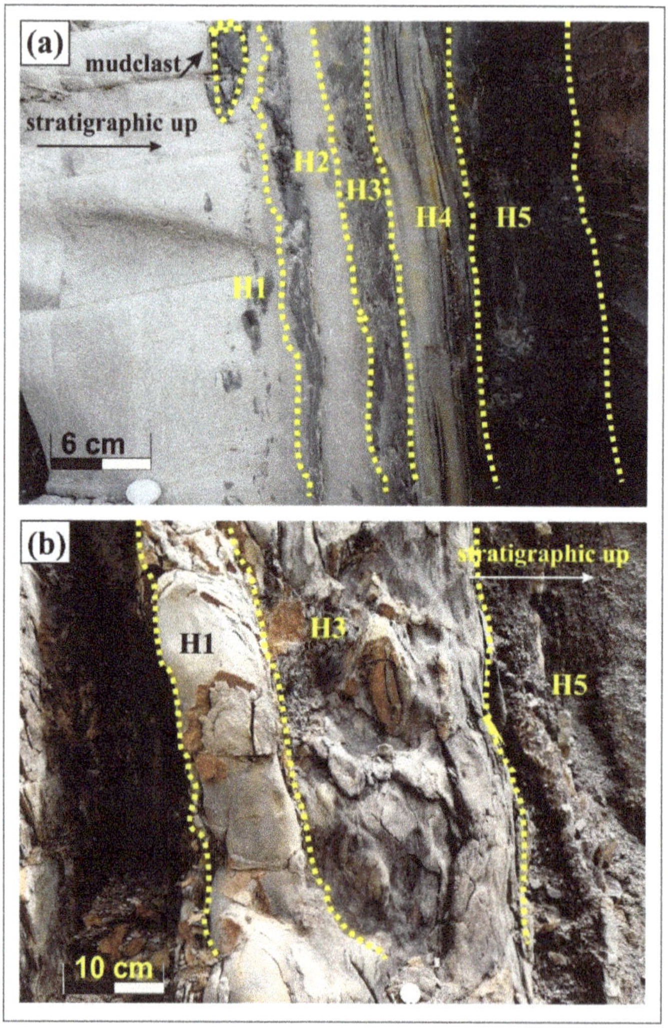

Figure 9. A brief signature of hybrid event beds in lateral lobes. (**a**) Lateral lobes have distinct brief expressions from H1 to H5 facies in the Jalan Sulaman (SUL) section, NW Sabah. (**b**) Lateral lobes predominantly contain H1 sandy hybrid intervals overlying a minor component of H3 and H5 facies, while moving from the base to the top in a stratigraphic succession exposed in the Jalan UMS (JU) section, West Sabah.

4.3. Hybrid Bed Facies in Lobe Progradation

Lobe progradation is a fluctuating geological process that mainly depends on sediment influx, variation in transport distance, and change in relative sea level. Normally, the sand-rich depositional units with a thickening upward cycle are indicative of lobe progradation (Figure 10). These hybrid flows are spatially distributed in the form of complex internal rheology [21,31,40,76,81].

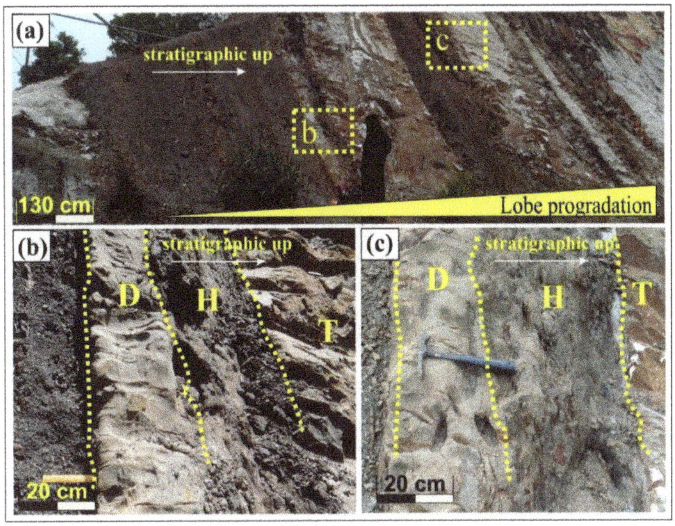

Figure 10. Relative position of hybrid beds in lobe progradation. (**a**) A thickening upward cycle in the Universiti Prima Condo (UP) section, West Sabah, where the vertical thickness of a massive shale is gradually reduced from base to top. Shaded image of a 173 cm tall person is shown as a vertical scale. (**b**) Massive shale overlain by debrite (D), hybrid event bed (H), and then massive sandy turbidite (T) present in the lower part of prograding lobe. (**c**) A second hybrid event bed (H) between debrite (D) and turbidite (T) is present in lobe progradation.

Hybrid event beds with variable facies divisions are more often associated at the base of a prograding lobe pattern, mainly in an unconfined lobe setting [32,41,45]. These hybrid beds are sporadically distributed in a compensational stacking pattern while these hybrid event beds are mainly present at the top of the retrogradational sequence (Figure 11) of a lobe system [41].

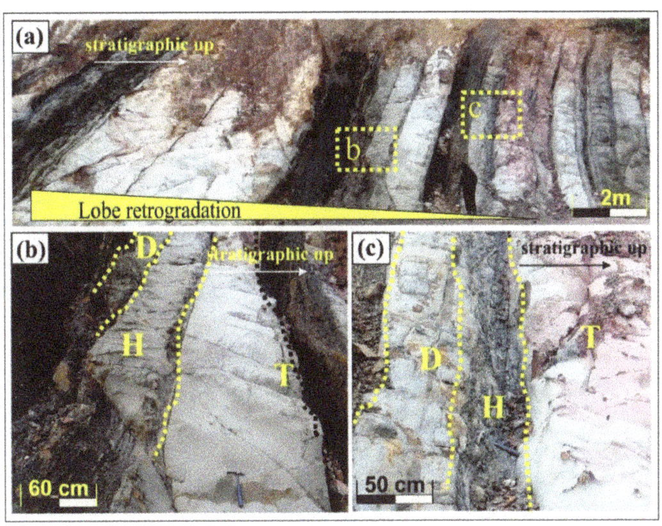

Figure 11. The presence of hybrid event beds during the lobe retrogradation in the Jalan UMS behind

the KFC (JK), West Sabah. (**a**) Massive, amalgamated sandstone (of more than 3 m in thickness) overlain by thick to medium sand units from the base to the top where the hybrid event beds are present in the upper part of sequence. Shaded image of a 169 cm tall person is used as a vertical scale. (**b**) The expression of a hybrid event bed (H) unit bounded by debrite (D) at the base and turbidite (T) at the top of sequence. (**c**) Another chaotic hybrid division (H) present between sandy debrite (D) to turbidite (T) in the upper part of lobe retrogradation.

5. Discussion

5.1. Lithological Heterogenerities in Hybrid Beds and Reservoir Potential

The heterogeneity of hybrid event bed facies adversely hampered the sandy reservoir. The broken clasts revealed in the H1 and H3 divisions would reduce the lateral and vertical connectivity of reservoir [82]. The upper hybrid event bed facies (H3 and H5) are rich in mud which would negatively influence reservoir intervals. The facies of H2 alternate in regular lighter and darker bands due to mud content distribution to less permeable zones in sandy intervals, destroying vertical migrations of fluids in reservoir units.

The scale of heterogeneity should also be considered while addressing the siliciclastic petroleum reservoirs. These hybrid beds are 1–3 m in vertical thickness which is generally not resolved in seismic data. It is quite possible that a massive sandstone interval on a seismic section may contain single or multiple hybrid event beds. According to seismic data, one can easily interpret an interval as a potential reservoir, but actually, the reservoir potential is significantly varied due to the presence of hybrid bed facies. The drilling core data and equivalent outcrop stratigraphy would better give a better insight about the potential of these sandstone reservoirs.

5.2. Distribution of Hybrid Bed Facies

It is vital to understand the distribution of hybrid event bed facies in submarine lobe systems to evaluate the reservoir potential of deep-marine sand intervals and precise input for reservoir modelling. The size of floating mud clasts is reduced during the flow from proximal to distal zones where large mud clasts in H3 are present in the proximal component of a lobe while small mud clasts are present in the distal lobe settings [30,83,84]. Dewatering in the sandy unit may act as lubrication for the mixture of muddy flow, thus forming a sand–mud couplet similar to a turbidite–debrite couplet in a depositional record [23,42].

Generally, the distribution of hybrid event bed facies from the proximal to distal lobes are significantly variable with respect to scale and type of hybrid event bed facies [38]. Proximal lobes have multimeter beds of hybrid events which mainly contain basal facies (H1 and H2) while medial lobes have the maximum distribution of all hybrid event beds (H1 to H5), as shown in Figure 12. However, distal lobes only contain the upper hybrid event bed facies (H3 and H5) with vertical thicknesses of less than one meter. The distal lobes may contain a small proportion of the basal sandy hybrid event beds (H1 and H2) only in the fringe area while distal fringe has the least fraction of basal hybrid facies (H1 and H3).

5.3. Geodynamic Evolution of Hybrid Beds

The development of hybrid bed facies is highly influenced by frontal and lateral lobe settings. The frontal lobes initially deposit coarse sandstones and there is a gradual evolution of flow conditions. The dynamic model of frontal lobes suggests the flow transformation away from the lobe axis in fringe area [39,81,84]. However, the spreading of sediments in lateral lobes starts from an off-axis area that provides a small avenue for the transition of flow and the deposition of hybrid bed facies. It is crucial to understand the role of confinement, especially in lateral lobes where an unconfined basin floor provides ample accommodation space for the well-developed hybrid bed sequence [32]. Hence, hybrid bed facies are frequently associated with boundary conditions of frontal lobes in general and settings of lateral lobes, especially.

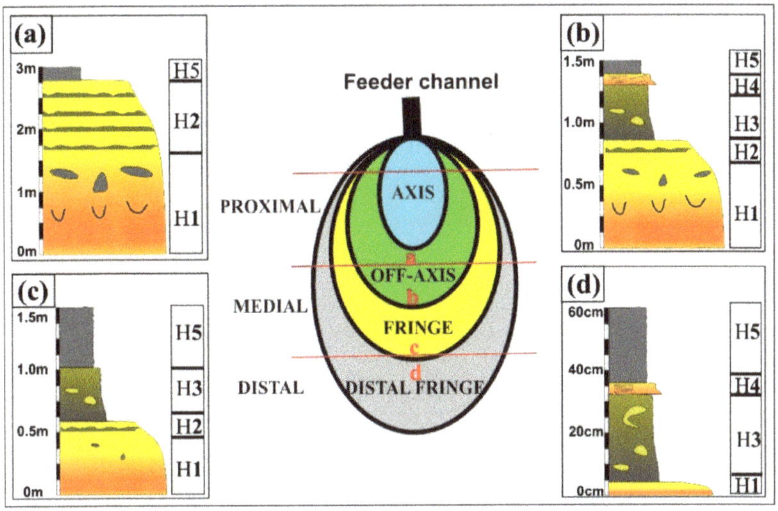

Figure 12. Distribution of hybrid facies from proximal to distal lobes. (**a**) Proximal lobes contain a major chunk of sands contained mainly H1 and H2 facies with thin caps of muds (H5). (**b**) A complete sequence of hybrid beds (H1 to H5) in a medial lobe. (**c**) Distal lobe in the fringe area contains small sandy hybrid event bed facies (H1–H2) but mainly comprises muddy upper facies of hybrid event bed. (**d**) The most distal part of lobe contains muddy hybrid event bed facies with total hybrid sequence less than one meter.

Hybrid event beds are associated with lobe progress as there is an evolution of flow conditions during lobe deposition [38]. The position of a hybrid event bed is the transition zone when a flow changes its velocity either to a higher flow regime or a lower flow regime [39,84]. In the case of lobe progradation, when a sand-rich sedimentary succession gradually gains energy and the shale and thin sandstone units are overlain by a thickening upward sequence, the hybrid beds are generally developed in the lower part of the lobe progradation sequence [40,43,81,85] (Figure 13). However, during lobe retrogradation or cessation, the hybrid beds are frequently present in the upper part of thinning upward stratigraphy due to the loss of energy from massive sandstone deposition to shale and thin sandstone intervals [45].

5.4. Hybrid Event Beds in Sand-Rich Deep-Marine Fan

The Crocker Fan is considered as one of the most classical examples of a sand-rich deep-marine fan environment where the hybrid event beds are less common in the stratigraphic record [47,79]. This sand-dominated deposition is mainly characterized by a debrite–turbidite system while hybrid event beds are only present in fewer parts of the fan lobe architecture [38]. Therefore, this study emphasizes the facies of hybrid event beds in a sand-rich fan system and the distribution of hybrid facies could vary in sand–mud mixed fan and fine-grained shaly fan environments [29]. Based on lobe components, these hybrid beds could possibly be more common in a mixed sand–mud fan deposition where changes in energy conditions and transformation of flow are frequent due to the evolution of sediment supply, resulting in the formation of hybrid beds at discrete stratigraphic intervals.

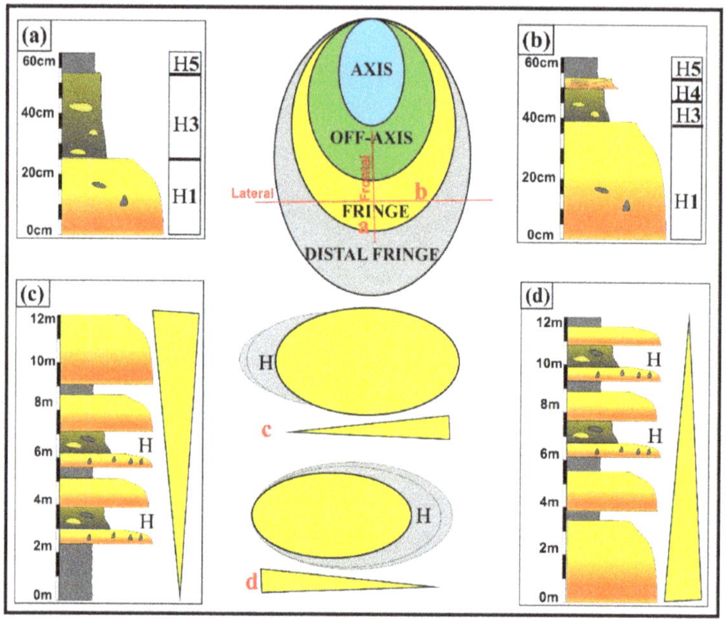

Figure 13. Distribution of hybrid facies. (**a**) Frontal lobes with incomplete hybrid facies. (**b**) clean sand (H1) and muddy sand (H3) facies in lateral lobes. (**c**) Hybrid bed facies in lobe progradation with hybrid beds in the lower part. (**d**) lobe retrogradation with hybrid beds in the upper part of the lobe.

6. Conclusions

The major highlights from this work on facies heterogeneity of hybrid event beds in submarine lobe components distributed along the Crocker Fan System of Sabah are summarized as follows:

1. The sedimentary facies of hybrid events exhibit rapid internal variability in geometry and divisions, characteristic of hybrid event beds in a deep-water system. These variations are typically developed due to numerous forms of basin configuration and transitional flow processes in a depositional environment.
2. It is quite possible that deep-marine sedimentary succession may be devoid of any hybrid event bed facies, especially in the most proximal and the most distal parts of lobes, where the chances of flow transformation are minimum and consequently, no hybrid event beds are formed in these domains.
3. Occasionally, the proximal lobes have multimeter vertical thickness with sandy hybrid event bed facies (H1 and H2), medial lobes have complete sedimentary facies succession of hybrid event bed sequence (H1 to H5), while the distal lobes predominantly contain only muddy hybrid event beds (H3 and H5) ranging in total thickness from 0.6 to 1.5 m of a complete hybrid event bed sequence.
4. Frontal lobes are formed of sediments with variable facies distribution including H1 and H3 in the innermost axial area, while they include a more-developed chaotic division (H3) away from the proximal domain.
5. The development of hybrid event beds is dependent on flow transformation and the zone of transition in a flow event. The gradual transformation of flow favors the deposition of a hybrid sequence while abrupt changes in flow may diminish the chances of hybrid bed deposition.
6. Deep-water sedimentary deposition commonly comprises a fan lobe system and is occasionally associated with hybrid bed facies. Muddy sandstone and clay-rich

hybrid event bed facies adversely affect the reservoir potential of sandy lobe intervals. This will significantly hinder the pore network and connectivity for lateral and vertical migration of fluids from reservoirs.

Author Contributions: Conceptualization, M.J. and N.A.S.; methodology, M.J. and N.A.; software, M.U. (Muhammad Usman); validation, M.J., N.A. and M.U. (Muhammad Umar); formal analysis, M.J. and Q.S.I.; investigation, M.J. and N.A.S.; resources, M.J. and N.A.S.; data curation, H.u.R. and M.U. (Muhammad Usman); writing—original draft preparation, M.J.; writing—review and editing, N.A.S., H.u.R. and M.U. (Muhammad Umar); visualization, M.J., N.A.S. and M.U. (Muhammad Usman); supervision, N.A.S.; project administration, N.A.S. and M.J.; funding acquisition, N.A.S. All authors have read and agreed to the published version of the manuscript.

Funding: The research was funded by the Fundamental Research Grant of the Ministry of Higher Education (MoHE) Malaysia (project ID 16880, reference code FRGS/1/2019/STG09/UTP/03/1), Petroleum Research Fund (PRF) cost number 0153AB-A33 awarded to Eswaran Padmanabhan, YUTP-FRG 1/2021 (RG2021-0927) and YUTP 015LC0-363.

Institutional Review Board Statement: Not applicable.

Informed Consent Statement: Not applicable.

Data Availability Statement: The data presented in this study are available on request from the corresponding author.

Acknowledgments: The authors are thankful to the faculty and staff of Shale Gas Research Group and Department of Geosciences especially Abdul Hadi Abd Rahman, Universiti Teknologi PETRONAS, Malaysia for technical support during geological fieldwork. The comments and feedback from four anonymous reviewers significantly improved the manuscript.

Conflicts of Interest: The authors declare no conflict of interest.

References

1. Li, S.; Zhu, X.; Li, S.; Gong, C.; Pan, R.; Chen, H. Trigger mechanisms of gravity flow deposits in the Lower Cretaceous lacustrine rift basin of Lingshan Island, Eastern China. *Cretac. Res.* **2020**, *107*, 104269. [CrossRef]
2. Talling, P.J.; Masson, D.G.; Sumner, E.J.; Malgesini, G. Subaqueous sediment density flows: Depositional processes and deposit types. *Sedimentology* **2012**, *59*, 1937–2003. [CrossRef]
3. Imran, Q.S.; Siddiqui, N.A.; Latif, A.H.A.; Bashir, Y.; Ali, A.A.A.S.; Jamil, M. Integrated well data and 3D seismic inversion study for reservoir delineation and description. *Bull. Geol. Soc. Malays.* **2020**, *70*, 209–220. [CrossRef]
4. Imran, Q.S.; Siddiqui, N.A.; Latiff, A.H.; Bashir, Y.; Khan, M.; Qureshi, K.; Al-Masgari, A.A.; Ahmed, N.; Jamil, M. Automated Fault Detection and Extraction under Gas Chimneys Using Hybrid Discontinuity Attributes. *Appl. Sci.* **2021**, *11*, 7218. [CrossRef]
5. Usman, M.; Siddiqui, N.A.; Garzanti, E.; Jamil, M.; Imran, Q.S.; Ahmed, L. 3-D seismic interpretation of stratigraphic and structural features in the Upper Jurassic to Lower Cretaceous sequence of the Gullfaks Field, Norwegian North Sea: A case study of reservoir development. *Energy Geosci.* **2021**. [CrossRef]
6. Banerjee, A.; Ahmed Salim, A.M. Stratigraphic evolution of deep-water Dangerous Grounds in the South China Sea, NW Sabah Platform Region, Malaysia. *J. Pet. Sci. Eng.* **2021**, *201*, 108434. [CrossRef]
7. Cullis, S.; Patacci, M.; Colombera, L.; Bührig, L.; McCaffrey, W.D. A database solution for the quantitative characterisation and comparison of deep-marine siliciclastic depositional systems. *Mar. Pet. Geol.* **2019**, *102*, 321–339. [CrossRef]
8. Galy, V.; France-Lanord, C.; Beyssac, O.; Faure, P.; Kudrass, H.; Palhol, F. Efficient organic carbon burial in the Bengal fan sustained by the Himalayan erosional system. *Nature* **2007**, *450*, 407–410. [CrossRef]
9. Hunt, J.E.; Wynn, R.B.; Masson, D.G.; Talling, P.J.; Teagle, D.A.H. Sedimentological and geochemical evidence for multistage failure of volcanic island landslides: A case study from Icod landslide on north Tenerife, Canary Islands. *Geochem. Geophys. Geosyst.* **2011**, *12*. [CrossRef]
10. Zhao, Y.; Wang, H.; Yan, D.; Jiang, P.; Chen, S.; Zhou, J.; Ma, J.; Qin, C.; He, J.; Zhao, Y. Sedimentary characteristics and model of gravity flows in the eocene Liushagang Formation in Weixi'nan depression, South China Sea. *J. Pet. Sci. Eng.* **2020**, *190*, 107082. [CrossRef]
11. Dodd, T.J.H.; McCarthy, D.J.; Richards, P.C. A depositional model for deep-lacustrine, partially confined, turbidite fans: Early Cretaceous, North Falkland Basin. *Sedimentology* **2019**, *66*, 53–80. [CrossRef]
12. Southern, S.J.; Kane, I.A.; Warchoł, M.J.; Porten, K.W.; McCaffrey, W.D. Hybrid event beds dominated by transitional-flow facies: Character, distribution and significance in the Maastrichtian Springar Formation, north-west Vøring Basin, Norwegian Sea. *Sedimentology* **2017**, *64*, 747–776. [CrossRef]

13. Talling, P.J.; Wynn, R.B.; Masson, D.G.; Frenz, M.; Cronin, B.T.; Schiebel, R.; Akhmetzhanov, A.M.; Dallmeier-Tiessen, S.; Benetti, S.; Weaver, P.P.E.; et al. Onset of submarine debris flow deposition far from original giant landslide. *Nature* **2007**, *450*, 541–544. [CrossRef]
14. Stow, D.; Smillie, Z. Distinguishing between Deep-Water Sediment Facies: Turbidites, Contourites and Hemipelagites. *Geosciences* **2020**, *10*, 68. [CrossRef]
15. Sohn, Y.K. Depositional Processes of Submarine Debris Flows in the Miocene Fan Deltas, Pohang Basin, SE Korea with Special Reference to Flow Transformation. *J. Sediment. Res.* **2000**, *70*, 491–503. [CrossRef]
16. Haughton, P.D.W.; Barker, S.P.; McCaffrey, W.D. 'Linked' debrites in sand-rich turbidite systems—Origin and significance. *Sedimentology* **2003**, *50*, 459–482. [CrossRef]
17. Talling, P.J.; Amy, L.A.; Wynn, R.B.; Peakall, J.; Robinson, M. Beds comprising debrite sandwiched within co-genetic turbidite: Origin and widespread occurrence in distal depositional environments. *Sedimentology* **2004**, *51*, 163–194. [CrossRef]
18. Hodgson, D.M. Distribution and origin of hybrid beds in sand-rich submarine fans of the Tanqua depocentre, Karoo Basin, South Africa. *Mar. Pet. Geol.* **2009**, *26*, 1940–1956. [CrossRef]
19. Haughton, P.; Davis, C.; McCaffrey, W.; Barker, S. Hybrid sediment gravity flow deposits—Classification, origin and significance. *Mar. Pet. Geol.* **2009**, *26*, 1900–1918. [CrossRef]
20. Talling, P.J.; Malgesini, G.; Sumner, E.J.; Amy, L.A.; Felletti, F.; Blackbourn, G.; Nutt, C.; Wilcox, C.; Harding, I.C.; Akbari, S. Planform geometry, stacking pattern, and extrabasinal origin of low strength and intermediate strength cohesive debris flow deposits in the Marnoso-arenacea Formation, Italy. *Geosphere* **2012**, *8*, 1207–1230. [CrossRef]
21. Kane, I.A.; Pontén, A.S.M. Submarine transitional flow deposits in the Paleogene Gulf of Mexico. *Geology* **2012**, *40*, 1119–1122. [CrossRef]
22. Terlaky, V.; Arnott, R.W.C. Matrix-rich and associated matrix-poor sandstones: Avulsion splays in slope and basin-floor strata. *Sedimentology* **2014**, *61*, 1175–1197. [CrossRef]
23. Koo, W.M.; Mohrig, D.; Buttles, J.; Sturmer, D.; Pontén, A.; Hess, T. Sand–mud couplets deposited by spontaneous remobilization of subaqueous transitional flows. *Sedimentology* **2020**, *67*, 78–95. [CrossRef]
24. Shanmugam, G. Deep-Water Processes and Deposits. In *Encyclopedia of Geology*; Elsevier: Amsterdam, The Netherlands, 2020.
25. Lowe, D.R.; Guy, M. Slurry-flow deposits in the Britannia Formation (Lower Cretaceous), North Sea: A new perspective on the turbidity current and debris flow problem. *Sedimentology* **2000**, *47*, 31–70. [CrossRef]
26. Baas, J.H.; Best, J.L.; Peakall, J. Depositional processes, bedform development and hybrid bed formation in rapidly decelerated cohesive (mud–sand) sediment flows. *Sedimentology* **2011**, *58*, 1953–1987. [CrossRef]
27. Doughty-Jones, G.; Mayall, M.; Lonergan, L. Stratigraphy, facies, and evolution of deep-water lobe complexes within a salt-controlled intraslope minibasin. *AAPG Bull.* **2017**, *101*, 1879–1904. [CrossRef]
28. Al-Masgari, A.A.; Elsaadany, M.; Siddiqui, N.A.; Latiff, A.H.; Bakar, A.A.; Elkurdy, S.; Hermana, M.; Babikir, I.; Imran, Q.S.; Adeleke, T. Geomorphological Geometries and High-Resolution Seismic Sequence Stratigraphy of Malay Basin's Fluvial Succession. *Appl. Sci.* **2021**, *11*, 5156. [CrossRef]
29. Baker, M.L.; Baas, J.H. Mixed sand–mud bedforms produced by transient turbulent flows in the fringe of submarine fans: Indicators of flow transformation. *Sedimentology* **2020**, *67*, 2645–2671. [CrossRef]
30. Fonnesu, M.; Haughton, P.; Felletti, F.; McCaffrey, W. Short length-scale variability of hybrid event beds and its applied significance. *Mar. Pet. Geol.* **2015**, *67*, 583–603. [CrossRef]
31. Patacci, M.; Marini, M.; Felletti, F.; Di Giulio, A.; Setti, M.; McCaffrey, W. Origin of mud in turbidites and hybrid event beds: Insight from ponded mudstone caps of the Castagnola turbidite system (north-west Italy). *Sedimentology* **2020**, *67*, 2625–2644. [CrossRef]
32. Fonnesu, M.; Felletti, F.; Haughton, P.D.W.; Patacci, M.; McCaffrey, W.D. Hybrid event bed character and distribution linked to turbidite system sub-environments: The North Apennine Gottero Sandstone (north-west Italy). *Sedimentology* **2018**, *65*, 151–190. [CrossRef]
33. Spychala, Y.T.; Hodgson, D.M.; Prélat, A.; Kane, I.A.; Flint, S.S.; Mountney, N.P. Frontal and Lateral Submarine Lobe Fringes: Comparing Sedimentary Facies, Architecture and Flow Processes. *J. Sediment. Res.* **2017**, *87*, 75–96. [CrossRef]
34. Marini, M.; Milli, S.; Ravnås, R.; Moscatelli, M. A comparative study of confined vs. semi-confined turbidite lobes from the Lower Messinian Laga Basin (Central Apennines, Italy): Implications for assessment of reservoir architecture. *Mar. Pet. Geol.* **2015**, *63*, 142–165. [CrossRef]
35. Talling, P.J. Hybrid submarine flows comprising turbidity current and cohesive debris flow: Deposits, theoretical and experimental analyses, and generalized models. *Geosphere* **2013**, *9*, 460–488. [CrossRef]
36. Yang, T.; Cao, Y.; Friis, H.; Liu, K.; Wang, Y. Origin and evolution processes of hybrid event beds in the Lower Cretaceous of the Lingshan Island, Eastern China. *Aust. J. Earth Sci.* **2018**, *65*, 517–534. [CrossRef]
37. Fonnesu, M.; Patacci, M.; Haughton, P.D.W.; Felletti, F.; McCaffrey, W.D. Hybrid Event Beds Generated By Local Substrate Delamination on A Confined-Basin Floor. *J. Sediment. Res.* **2016**, *86*, 929–943. [CrossRef]
38. Mueller, P.; Patacci, M.; Di Giulio, A. Hybrid event beds in the proximal to distal extensive lobe domain of the coarse-grained and sand-rich Bordighera turbidite system (NW Italy). *Mar. Pet. Geol.* **2017**, *86*, 908–931. [CrossRef]

39. Pierce, C.S.; Haughton, P.D.W.; Shannon, P.M.; Pulham, A.J.; Barker, S.P.; Martinsen, O.J. Variable character and diverse origin of hybrid event beds in a sandy submarine fan system, Pennsylvanian Ross Sandstone Formation, western Ireland. *Sedimentology* **2018**, *65*, 952–992. [CrossRef]
40. Kane, I.A.; Pontén, A.S.M.; Vangdal, B.; Eggenhuisen, J.T.; Hodgson, D.M.; Spychala, Y.T. The stratigraphic record and processes of turbidity current transformation across deep-marine lobes. *Sedimentology* **2017**, *64*, 1236–1273. [CrossRef]
41. Spychala, Y.T.; Hodgson, D.M.; Lee, D.R. Autogenic controls on hybrid bed distribution in submarine lobe complexes. *Mar. Pet. Geol.* **2017**, *88*, 1078–1093. [CrossRef]
42. Kuswandaru, G.Y.; Amir Hassan, M.H.; Matenco, L.C.; Taib, N.I.; Mustapha, K.A. Turbidite, debrite, and hybrid event beds in submarine lobe deposits of the Palaeocene to middle Eocene Kapit and Pelagus members, Belaga Formation, Sarawak, Malaysia. *Geol. J.* **2019**, *54*, 3421–3437. [CrossRef]
43. Hofstra, M.; Pontén, A.S.M.; Peakall, J.; Flint, S.S.; Nair, K.N.; Hodgson, D.M. The impact of fine-scale reservoir geometries on streamline flow patterns in submarine lobe deposits using outcrop analogues from the Karoo Basin. *Pet. Geosci.* **2016**, *23*, 159–176. [CrossRef]
44. Porten, K.W.; Kane, I.A.; Warchoł, M.J.; Southern, S.J. A Sedimentological Process-Based Approach to Depositional Reservoir Quality of Deep-Marine Sandstones: An Example From the Springar Formation, Northwestern Vøring Basin, Norwegian Sea. *J. Sediment. Res.* **2016**, *86*, 1269–1286. [CrossRef]
45. Zhang, L.; Pan, M.; Wang, H. Deepwater Turbidite Lobe Deposits: A Review of the Research Frontiers. *Acta Geol. Sin. Engl. Ed.* **2017**, *91*, 283–300. [CrossRef]
46. Jackson, C.A.L.; Zakaria, A.A.; Johnson, H.D.; Tongkul, F.; Crevello, P.D. Sedimentology, stratigraphic occurrence and origin of linked debrites in the West Crocker Formation (Oligo-Miocene), Sabah, NW Borneo. *Mar. Pet. Geol.* **2009**, *26*, 1957–1973. [CrossRef]
47. Zakaria, A.A.; Johnson, H.D.; Jackson, C.A.-L.; Tongkul, F. Sedimentary facies analysis and depositional model of the Palaeogene West Crocker submarine fan system, NW Borneo. *J. Asian Earth Sci.* **2013**, *76*, 283–300. [CrossRef]
48. Bell, D.; Stevenson, C.J.; Kane, I.A.; Hodgson, D.M.; Poyatos-Moré, M. Topographic Controls On the Development of Contemporaneous but Contrasting Basin-Floor Depositional Architectures. *J. Sediment. Res.* **2018**, *88*, 1166–1189. [CrossRef]
49. Brooks, H.L.; Hodgson, D.M.; Brunt, R.L.; Peakall, J.; Hofstra, M.; Flint, S.S. Deep-water channel-lobe transition zone dynamics: Processes and depositional architecture, an example from the Karoo Basin, South Africa. *GSA Bull.* **2018**, *130*, 1723–1746. [CrossRef]
50. Soutter, E.L.; Kane, I.A.; Fuhrmann, A.; Cumberpatch, Z.A.; Huuse, M. The stratigraphic evolution of onlap in siliciclastic deep-water systems: Autogenic modulation of allogenic signals. *J. Sediment. Res.* **2019**, *89*, 890–917. [CrossRef]
51. Hall, R.; van Hattum, M.W.A.; Spakman, W. Impact of India-Asia collision on SE Asia: The record in Borneo. *Tectonophysics* **2008**, *451*, 366–389. [CrossRef]
52. Siddiqui, N.A.; Ramkumar, M.; Rahman, A.H.A.; Mathew, M.J.; Santosh, M.; Sum, C.W.; Menier, D. High resolution facies architecture and digital outcrop modeling of the Sandakan formation sandstone reservoir, Borneo: Implications for reservoir characterization and flow simulation. *Geosci. Front.* **2019**, *10*, 957–971. [CrossRef]
53. Usman, M.; Siddiqui, N.A.; Mathew, M.; Zhang, S.; El-Ghali, M.A.K.; Ramkumar, M.; Jamil, M.; Zhang, Y. Linking the influence of diagenetic properties and clay texture on reservoir quality in sandstones from NW Borneo. *Mar. Pet. Geol.* **2020**, *120*, 104509. [CrossRef]
54. Usman, M.; Siddiqui, N.A.; Zhang, S.-Q.; Mathew, M.J.; Zhang, Y.-X.; Jamil, M.; Liu, X.-L.; Ahmed, N. 3D Geo-Cellular Static Virtual Outcrop Model and its Implications for Reservoir Petro-physical Characteristics and Heterogeneities. *Pet. Sci.* **2021**. [CrossRef]
55. Ahmed, N.; Siddiqui, N.A.; Ramasamy, N.; Ramkumar, M.; Jamil, M.; Usman, M.; Sajid, Z.; Rahman, A.H.B.A. Geochemistry of Eocene Bawang Member turbidites of the Belaga Formation, Borneo: Implications for provenance, palaeoweathering, and tectonic setting. *Geol. J.* **2021**, *56*, 2477–2499. [CrossRef]
56. Sheikh, N.A.; Jamil, M.; Ling Chuan Ching, D.; Khan, I.; Usman, M.; Sooppy Nisar, K. A generalized model for quantitative analysis of sediments loss: A Caputo time fractional model. *J. King Saud Univ. Sci.* **2021**, *33*, 101179. [CrossRef]
57. Hall, R. Contraction and extension in northern Borneo driven by subduction rollback. *J. Asian Earth Sci.* **2013**, *76*, 399–411. [CrossRef]
58. Mathew, M.J.; Menier, D.; Siddiqui, N.; Ramkumar, M.; Santosh, M.; Kumar, S.; Hassaan, M. Drainage basin and topographic analysis of a tropical landscape: Insights into surface and tectonic processes in northern Borneo. *J. Asian Earth Sci.* **2016**, *124*, 14–27. [CrossRef]
59. Siddiqui, N.A.; Mathew, M.J.; Ramkumar, M.; Sautter, B.; Usman, M.; Abdul Rahman, A.H.; El-Ghali, M.A.K.; Menier, D.; Shiqi, Z.; Sum, C.W. Sedimentological characterization, petrophysical properties and reservoir quality assessment of the onshore Sandakan Formation, Borneo. *J. Pet. Sci. Eng.* **2020**, *186*, 106771. [CrossRef]
60. Siddiqui, N.A.; Rahman, A.H.A.; Sum, C.W.; Mathew, M.J.; Hassaan, M.; Menier, D. Generic hierarchy of sandstone facies quality and static connectivity: An example from the Middle-Late Miocene Miri Formation, Sarawak Basin, Borneo. *Arab. J. Geosci.* **2017**, *10*, 237. [CrossRef]

61. Ahmed, N.; Siddiqui, N.A.; Rahman, A.H.B.A.; Jamil, M.; Usman, M.; Sajid, Z.; Kamal Zaidi, F. Evaluation of hydrocarbon source rock potential: Deep marine shales of Belaga Formation of Late Cretaceous-Late Eocene, Sarawak, Malaysia. *J. King Saud Univ. Sci.* **2021**, *33*, 101268. [CrossRef]
62. Hall, R. Cenozoic geological and plate tectonic evolution of SE Asia and the SW Pacific: Computer-based reconstructions, model and animations. *J. Asian Earth Sci.* **2002**, *20*, 353–431. [CrossRef]
63. Jamil, M.; Rahman, A.H.A.; Siddiqui, N.A.; Ahmed, N. Deep marine Paleogene sedimentary sequence of West Sabah: Contemporary opinions and ambiguities. *War. Geol.* **2019**, *45*, 198–200.
64. Jamil, M.; Rahman, A.H.A.; Siddiqui, N.A.; Ibrahim, N.A.; Ahmed, N. A contemporary review of sedimentological and stratigraphic framework of the Late Paleogene deep marine sedimentary successions of West Sabah, North-West Borneo. *Bull. Geol. Soc. Malays.* **2020**, *69*, 53–65. [CrossRef]
65. Clennell, B. Far-field and gravity tectonics in Miocene basins of Sabah, Malaysia. *Geol. Soc. Lond. Spec. Publ.* **1996**, *106*, 307–320. [CrossRef]
66. Cullen, A.B.; Zechmeister, M.S.; Elmore, R.D.; Pannalal, S.J. Paleomagnetism of the Crocker Formation, northwest Borneo: Implications for late Cenozoic tectonics. *Geosphere* **2012**, *8*, 1146–1169. [CrossRef]
67. Wang, P.; Li, S.; Guo, L.; Jiang, S.; Somerville, I.; Zhao, S.; Zhu, B.; Chen, J.; Dai, L.; Suo, Y. Mesozoic and Cenozoic accretionary orogenic processes in Borneo and their mechanisms. *Geol. J.* **2016**, *51*, 464–489. [CrossRef]
68. Jamil, M.; Siddiqui, N.A.; Umar, M.; Usman, M.; Ahmed, N.; Rahman, A.H.A.; Zaidi, F.K. Aseismic and seismic impact on development of soft-sediment deformation structures in deep-marine sand-shaly Crocker fan in Sabah, NW Borneo. *J. King Saud Univ. Sci.* **2021**, *33*, 101522. [CrossRef]
69. Lambiase, J.J.; Tzong, T.Y.; William, A.G.; Bidgood, M.D.; Brenac, P.; Cullen, A.B. The West Crocker formation of northwest Borneo: A Paleogene accretionary prism. *Spec. Pap. Geol. Soc. Am.* **2008**, *436*, 171–184.
70. van Hattum, M.W.A.; Hall, R.; Pickard, A.L.; Nichols, G.J. Provenance and geochronology of Cenozoic sandstones of northern Borneo. *J. Asian Earth Sci.* **2013**, *76*, 266–282. [CrossRef]
71. Abdullah, W.H.; Togunwa, O.S.; Makeen, Y.M.; Hakimi, M.H.; Mustapha, K.A.; Baharuddin, M.H.; Sia, S.-G.; Tongkul, F. Hydrocarbon source potential of Eocene-Miocene sequence of Western Sabah, Malaysia. *Mar. Pet. Geol.* **2017**, *83*, 345–361. [CrossRef]
72. Madon, M. Deep-sea trace fossils in the West Crocker Formation, Sabah (Malaysia), and their palaeoenvironmental significance. *Bull. Geol. Soc. Malays.* **2021**, *71*, 23–46. [CrossRef]
73. Usman, M.; Siddiqui, N.A.; Zhang, S.; Ramkumar, M.; Mathew, M.; Sautter, B.; Beg, M.A. Ichnofacies and sedimentary structures: A passive relationship with permeability of a sandstone reservoir from NW Borneo. *J. Asian Earth Sci.* **2020**, *192*, 103992. [CrossRef]
74. Prélat, A.; Hodgson, D.M.; Flint, S.S. Evolution, architecture and hierarchy of distributary deep-water deposits: A high-resolution outcrop investigation from the Permian Karoo Basin, South Africa. *Sedimentology* **2009**, *56*, 2132–2154. [CrossRef]
75. Shanmugam, G. The hyperpycnite problem. *J. Palaeogeogr.* **2018**, *7*, 6. [CrossRef]
76. Brunt, R.L.; Di Celma, C.N.; Hodgson, D.M.; Flint, S.S.; Kavanagh, J.P.; van der Merwe, W.C. Driving a channel through a levee when the levee is high: An outcrop example of submarine down-dip entrenchment. *Mar. Pet. Geol.* **2013**, *41*, 134–145. [CrossRef]
77. Groenenberg, R.M.; Hodgson, D.M.; Prelat, A.; Luthi, S.M.; Flint, S.S. Flow–deposit interaction in submarine lobes: Insights from outcrop observations and realizations of a process-based numerical model. *J. Sediment. Res.* **2010**, *80*, 252–267. [CrossRef]
78. Zhang, L.-F.; Pan, M.; Li, Z.-L. 3D modeling of deepwater turbidite lobes: A review of the research status and progress. *Pet. Sci.* **2020**, *17*, 317–333. [CrossRef]
79. Jamil, M.; Siddiqui, N.A.; Rahman, A.H.B.A.; Ibrahim, N.A.; Ismail, M.S.B.; Ahmed, N.; Usman, M.; Gul, Z.; Imran, Q.S. Facies Heterogeneity and Lobe Facies Multiscale Analysis of Deep-Marine Sand-Shale Complexity in the West Crocker Formation of Sabah Basin, NW Borneo. *Appl. Sci.* **2021**, *11*, 5513. [CrossRef]
80. Davis, C.; Haughton, P.; McCaffrey, W.; Scott, E.; Hogg, N.; Kitching, D. Character and distribution of hybrid sediment gravity flow deposits from the outer Forties Fan, Palaeocene Central North Sea, UKCS. *Mar. Pet. Geol.* **2009**, *26*, 1919–1939. [CrossRef]
81. Zhang, L.-F.; Dong, D.-Z. Thickening-upward cycles in deep-marine and deep-lacustrine turbidite lobes: Examples from the Clare Basin and the Ordos Basin. *J. Palaeogeogr.* **2020**, *9*, 1–16. [CrossRef]
82. Southern, S.J.; Patacci, M.; Felletti, F.; McCaffrey, W.D. Influence of flow containment and substrate entrainment upon sandy hybrid event beds containing a co-genetic mud-clast-rich division. *Sediment. Geol.* **2015**, *321*, 105–122. [CrossRef]
83. Lee, S.H.; Jung, W.-Y.; Bahk, J.J.; Gardner, J.M.; Kim, J.K.; Lee, S.H. Depositional features of co-genetic turbidite–debrite beds and possible mechanisms for their formation in distal lobated bodies beyond the base-of-slope, Ulleung Basin, East Sea (Japan Sea). *Mar. Geol.* **2013**, *346*, 124–140. [CrossRef]
84. Giacomone, G.; Olariu, C.; Steel, R.; Shin, M. A coarse-grained basin floor turbidite system—the Jurassic Los Molles Formation, Neuquen Basin, Argentina. *Sedimentology* **2020**, *67*, 3809–3843. [CrossRef]
85. Hussain, A.; Haughton, P.D.W.; Shannon, P.M.; Turner, J.N.; Pierce, C.S.; Obradors-Latre, A.; Barker, S.P.; Martinsen, O.J. High-resolution X-ray fluorescence profiling of hybrid event beds: Implications for sediment gravity flow behaviour and deposit structure. *Sedimentology* **2020**, *67*, 2850–2882. [CrossRef]

Journal of
Marine Science
and Engineering

MDPI

Article

The Depositional Environments in the Cilento Offshore (Southern Tyrrhenian Sea, Italy) Based on Marine Geological Data

Gemma Aiello [1,*] and Mauro Caccavale [1,2]

[1] Instituto di Scienze Marine (ISMAR), Consiglio Nazionale delle Ricerche (CNR), 80133 Napoli, Italy; mauro.caccavale@cnr.it
[2] Instituto Nazionale di Geofisica e Vulcanologia, Osservatorio Vesuviano, 80125 Napoli, Italy
* Correspondence: gemma.aiello@cnr.it

Abstract: The depositional environments offshore of the Cilento Promontory have been reconstructed based on the geological studies performed in the frame of the marine geological mapping of the geological sheet n. 502 "Agropoli". The littoral environment (toe-of-coastal cliff deposits and submerged beach deposits), the inner continental shelf environment (inner shelf deposits and bioclastic deposits), the outer continental shelf environment (outer shelf deposits and bioclastic deposits), the lowstand system tract and the Pleistocene relict marine units have been singled out. The littoral, inner shelf and outer shelf environments have been interpreted as the highstand system tract of the Late Quaternary depositional sequence. This sequence overlies the Cenozoic substratum (ssi unit), composed of Cenozoic siliciclastic rocks, genetically related with the Cilento Flysch. On the inner shelf four main seismo-stratigraphic units, overlying the undifferentiated acoustic basement have been recognized based on the geological interpretation of seismic profiles. On the outer shelf, palimpsest deposits of emerged to submerged beach and forming elongated dunes have been recognized on sub-bottom profiles and calibrated with gravity core data collected in previous papers. The sedimentological analysis of sea bottom samples has shown the occurrence of several grain sizes occurring in this portion of the Cilento offshore.

Keywords: littoral deposits; continental shelf deposits; palimpsest deposits; marine geological maps; Cilento Promontory; Southern Italy

Citation: Aiello, G.; Caccavale, M. The Depositional Environments in the Cilento Offshore (Southern Tyrrhenian Sea, Italy) Based on Marine Geological Data. *J. Mar. Sci. Eng.* **2021**, *9*, 1083. https://doi.org/10.3390/jmse9101083

Academic Editor: Antoni Calafat

Received: 5 September 2021
Accepted: 29 September 2021
Published: 4 October 2021

Publisher's Note: MDPI stays neutral with regard to jurisdictional claims in published maps and institutional affiliations.

Copyright: © 2021 by the authors. Licensee MDPI, Basel, Switzerland. This article is an open access article distributed under the terms and conditions of the Creative Commons Attribution (CC BY) license (https://creativecommons.org/licenses/by/4.0/).

1. Introduction

The aim of this paper is to present some new geological, seismo-stratigraphic and sedimentological data on the depositional environments recognized in the coastal area surrounding the Cilento Promontory, located in the Southern Tyrrhenian Sea, as derived by the marine geological mapping of the geological sheet n. 502 "Agropoli" [1–3]. The geological sheet n. 502 "Agropoli" (1:50,000 scale) has shown the distribution of several lithostratigraphic units, cropping out at the sea bottom and of the main morphological lineaments, accordingly to the CARG (CARta Geologica) set of rules [4,5]. The main stratigraphic units individuated through the analysis of the sediments cropping out at the sea bottom belong to the Late Quaternary depositional sequence (Figure 1). The spatial and temporal evolution and the lateral and vertical migration of the depositional environments belonging to the Late Quaternary depositional sequence, i.e., the coastal setting, the continental shelf setting and the slope setting, have been previously discussed [6–10]. The variations of the accommodation space of the Late Quaternary deposits during the last fourth-order glacio-eustatic cycle, ranging in age between 128 ky B.P. (Tyrrhenian stage) and the isotopic stage 5 have been recorded by the stratigraphic succession investigated through the marine geological survey. Catuneanu et al. [9] studied the sequence stratigraphic concepts in detail, particularly referring to the depositional sequences, bounded by subaerial unconformities and their marine correlative conformities. These unconformities have been used as sequence boundaries, marking hiatuses in the stratigraphic record. All the genetic models of the depositional sequences are based on the distinction of forced regressive deposits, normal regressive (lowstand and highstand) deposits and transgressive deposits as distinct genetic units [9]. Forced regressive deposits, normal regressive (lowstand and highstand) deposits and transgressive deposits can be identified when they can be genetically related to the changes in shoreline trajectory [8]. Catuneanu [10] further specified the definition of sequence as a rock-stratigraphic unit bounded by interregional unconformities in the 1940s with a resolution of 10^2–10^3 m [11–16], as a relatively conformable succession of genetically related strata bounded by unconformities or their correlative conformities in the 1970s with a resolution of 10^1–10^2 m and, finally, as a stratigraphic cycle defined by the recurrence of the same type of sequence stratigraphic surface in the rock record in the 2010 s with a resolution of 1–10 m.

The facies analysis criteria and the schematic representation of the depositional environments are herein briefly resumed. The system tracts of the Late Quaternary depositional sequence consist of deposits typified by facies genetically related to continental, coastal, shelf and deep-sea depositional environments. Here, we will concentrate on the coastal and continental shelf depositional environments, since the other kinds of deposits have not been recognized in the study area. The coastal depositional systems are characterized by a great variability, both in the morphology and in the depositional style. This variability has suggested different budgets between the available sediments, coupled with the genetic control of the oceanographic regimes ("wave-dominated", "tide-dominated" or "mixed") [17–22]. The Mediterranean Sea is typified by a micro-tidal regime. In particular, on the Italian continental margins, the coastal systems are wave-dominated. The coastal deposits grow during each phase of a relative sea level cycle, but with different facies. Regressive systems arise during the sea-level falls ("forced regressions") [23,24] or when the siliciclastic supply counterbalances the relative sea-level rate. If we consider the present day continental shelf deposits, three main types have been summarized, including the relict sediments, deposited during a seawards advancement of the shoreline and then drowned, the palimpsest sediments, which are relict sediments reworked by currents, storm waves and tides and the Late Quaternary highstand deposits, in equilibrium with the present day depositional processes [25–27]. The highstand deposits are younger than the maximum marine flooding occurred at the end of the last sea-level rise (about 4 to 5 ky B.P.). Offshore Italy, they exhibit their maximum thickness on the inner shelf in correspondence to the main deltas (Po, Tiber and Arno), reduced to a few meters on the outer shelf (Figure 2). In Naples Bay, the highstand deposits have been mapped offshore Campania [28–31].

Maërl layers (Maërl is a hard seaweed with a purple-pink color that forms reef-like carpets, known as maërl beds, in the dim light conditions of the shallow seas along the European coasts) have been recognized on the continental shelf offshore Cilento through the geological interpretation of sub-bottom profiles [32]. The maërl facies has been recognized based on the component analysis of 32 grab samples and is preferentially concentrated on the submerged depositional terraces, located at water depths ranging between 42 m and 52 m. This preferential distribution has been probably controlled by strong bottom currents occurred in this area due to the local oceanographic circulation, preventing for the deposition of siliciclastic deposits. The coralline algae have controlled the carbonate deposition offshore the Cilento Promontory at water depths ranging between 40 m and 60 m [32]. Another significant contribution to the depositional environments deals with the searching of relict sandy deposits on the continental shelf offshore the Cilento Promontory aimed at beach nourishment [33]. The performed analyses have allowed to individuate the submarine relict sands suitable for beach nourishment. Moreover, the terraced landforms occurring both onshore and offshore the Cilento Promontory have been recently analyzed and aimed at reconstructing the meaning of the terraced surfaces as Quaternary records of the sea level changes [34]. Two main types of terraced surfaces have been recognized, including the erosional terraces ("wave-cut platforms" or "abrasion platforms") [35–37] and the depositional terraces (STDs) [38–40]. Two types of landforms have been controlled by the interaction of several geological processes, including the type of bedrock, the geodynamic setting, the sedimentary input and the relative sea level changes [34]. In the Cilento offshore, the submarine marine terraces were formed when the rocky outcrops were exposed on the continental shelf and were predominantly generated during the interglacial periods [34].

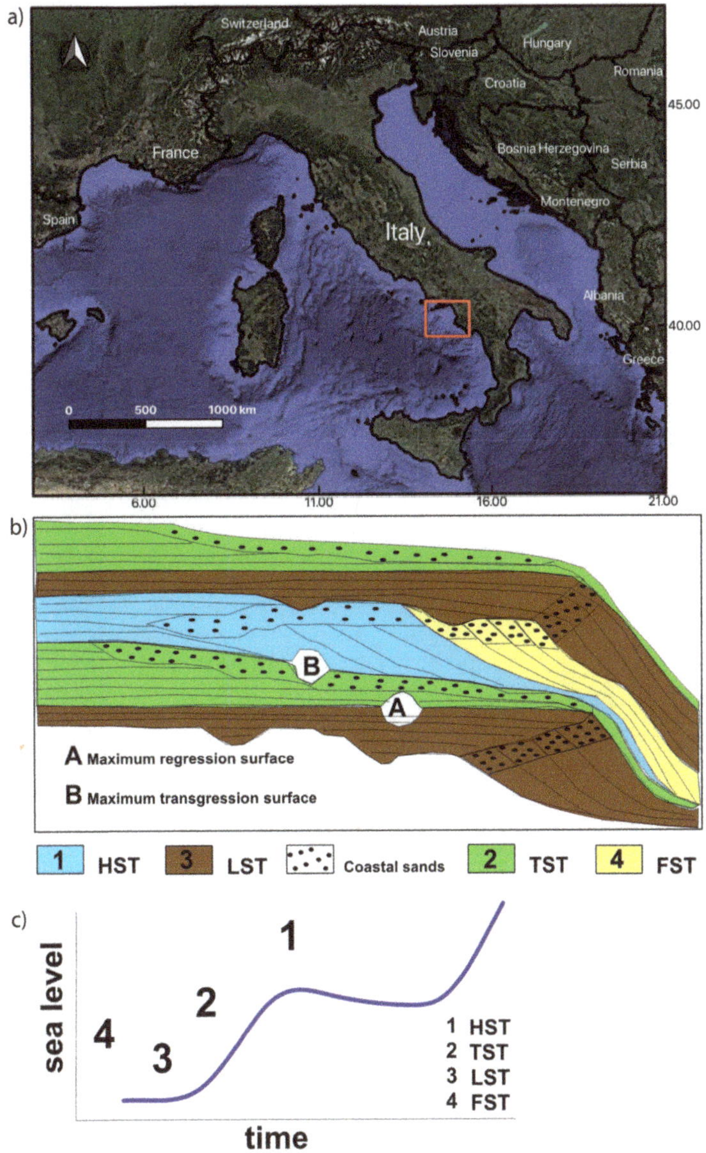

Figure 1. (**a**) Location map of the study area (red inset). (**b**) Sketch stratigraphic diagram as a function of depth (**c**) and time, showing the geometric relationships between the system tracts and the distribution of siliciclastic facies in unconformity-bounded depositional sequences.HST: Highstand system tract. TST: Transgressive system tract. LST: Lowstand system tract. FST: Forced regression system tract.

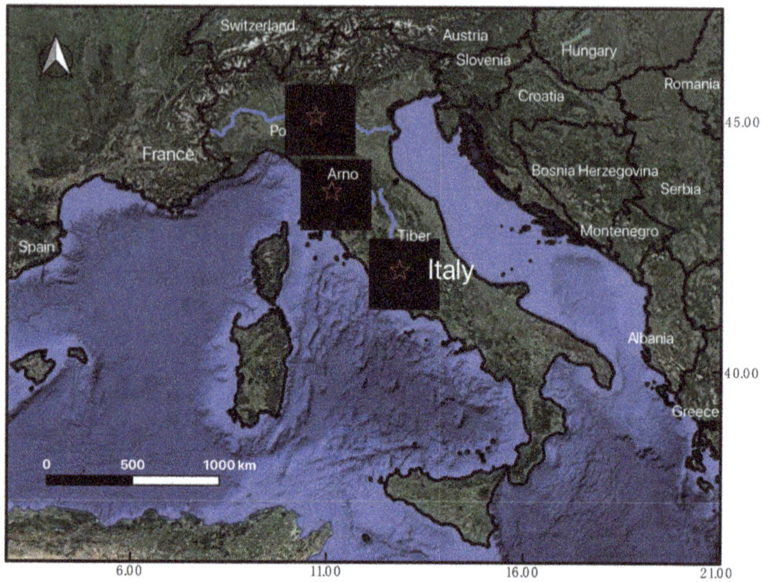

Figure 2. Satellite map of Italy, showing the main river systems (Po, Tiber and Arno) whose deltas have fed the highstand deposits of the Italian continental shelf.

In this paper, the marine geological maps have allowed to show the depositional environments occurring offshore of the Cilento Promontory and to interpret these environments in terms of system tracts of the Late Quaternary depositional sequence [41–43]. Moreover, the sedimentological data of sea bottom samples have been analyzed in order to show the main grain sizes occurring at the sea bottom in this portion of the Cilento offshore. Ternary plots have been constructed in order to evaluate the different grain sizes, considering as variables shale, sand and silt and gravel and sand and silt, respectively. These plots allowed to analyze the grain sizes occurring at the sea bottom in this portion of the Cilento offshore. Sub-bottom profiles have been interpreted based on the criteria of seismic stratigraphy in order to reconstruct the stratigraphic setting of the area and to complement the cartographic representation.

2. Geologic Setting

The Southeastern Tyrrhenian margin is a passive-type continental margin involved by listric faults, with blocks dipping both seawards and landwards. Along the Tyrrhenian margin, this tectonic style has controlled the formation of half-graben basins on the continental shelf and slope, alternating with structural highs [44,45].

The marine area surrounding the Cilento Promontory represents a structural high resulting from the seawards prolongation of the Licosa Cape structural high, bounded northwards and southwards by two half-graben basins: the Salerno Valley and the Policastro Gulf. The Salerno Valley is a half-graben basin whose individuation has been controlled during the Early Pleistocene by the master fault Capri-Sorrento Peninsula, with average throws in the order of 1500 meters [46,47]. Previous seismo-stratigraphic data have shown a main regional unconformity, located at depths ranging between 2000 and 2500 meters, correlated with the top of the Meso-Cenozoic carbonates and marking the base of the Plio-Pleistocene filling of the Salerno Valley [46,47].

The Cilento structural high has been deeply investigated based on its geologic and seismo-stratigraphic characteristics. The interpretation of multichannel profiles has shown the occurrence of wide structural highs, characterized by an acoustically transparent seismic facies, corresponding to the acoustic basement, alternating with parallel seismic reflectors corresponding to the Quaternary marine filling of sedimentary basins [48]. More recently, the geological interpretation of newly acquired deep multichannel seismic lines along the Tyrrhenian margin has confirmed this structural framework, showing that the geological structure of the Cilento high is locally complicated by folding, reverse faults and basin inversions [49–51].

In the Cilento Promontory, the siliciclastic successions of the Cilento Flysch crop out, which have involved the deformation of the Apenninic chain during the Cenozoic and have been then deformed by the Plio-Quaternary extensional tectonic phases [52–54]. These units have been deeply revised from a stratigraphic and structural point of view [53]. The revision was focused both on the Cilento Group, composed of the "Pollica" Sandstones and of the "S. Mauro" Formation and on the Northern Calabria Unit, constituted by the "Crete Nere" Formation, by the "Saraceno" Formation, by the "Cannicchio" Sandstones and by the "Sicilide" units [53]. These stratigraphic-structural units represent the rocky acoustic basement of the Plio-Pleistocene and Holocene marine deposits of the continental shelf between the Licosa Cape and the Palinuro Cape—namely, the ssi unit described in this paper (see the section on the results).

3. Materials and Methods

The research has been developed through the acquisition, the processing and the geologic interpretation of a densely spaced grid of high-resolution seismic profiles (sub-bottom Chirp) collected by the CNR-ISMAR onboard of the R/V Urania (National Research Council of Italy). The grid of sub-bottom Chirp profiles superimposed to the onshore–offshore DEM of the Cilento Promontory is reported in Figure 3. The acoustic profiling system CAP-6600 Chirp II has been used. Its linear frequency modulated signals of 2–7 kHz or 8–23 kHz and 4-kW provide high-resolution sounding reaching a 10–30-cm resolution within uppermost marine sediments. The obtained data, recorded onboard with a SEG-Y format, was processed by using the software Seisprho [55], allowing to plot the sub-bottom profiles as bitmap images. The geological interpretation of sub-bottom profiles was carried out based on the criteria of seismic stratigraphy in order to identify the main seismic sequences and the related unconformities [56].

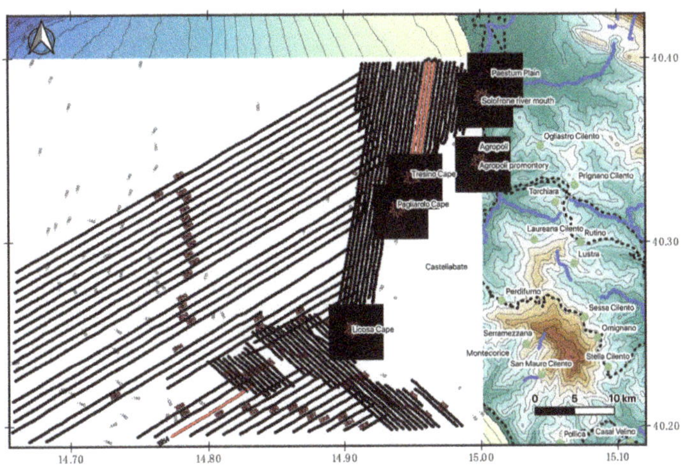

Figure 3. Location map of sub-bottom Chirp profiles superimposed to onshore–offshore DEM of the Cilento Promontory. Red lines indicate the seismic profiles shown in the text.

The sea bottom samples were collected during the GMS 03_01 cruise (R/V Urania, CNR) in November 2003 by a Van Veen grab and immediately described in terms of visible features and grain size. The location of samples superimposed on the onshore–offshore Digital Elevation Model (DEM) of the Cilento offshore is shown in Figure 4. The grain size analyses were performed at the CNR-ISMAR (Naples, Italy) sedimentological laboratory using a grain size laser analyzer (SYMPATEC Laser Particle Size Analyzer).

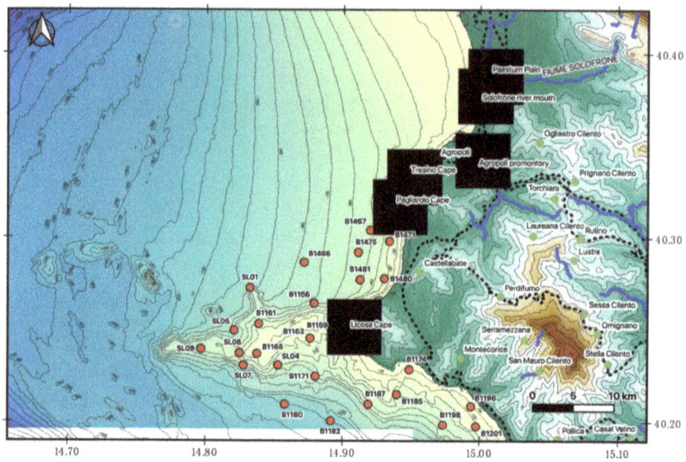

Figure 4. Location of sea bottom samples superimposed on onshore–offshore DEM of the Cilento offshore.

The particle size analysis involved a phase of preparation and pre-treatment of the sample. The collected samples were dried in thermostatic ovens at 105 °C for 24 h, until the weight stabilization was obtained. Then, the pretreatment was carried out with a solution of hydrogen peroxide and distilled water for 24/48 h. Subsequently, the sample was stirred with a mechanical stirrer at 600 rpm for about 2 h and then wet-separated with a 63-μm sieve into a coarse component (>63 μm) and a fine component (<63 μm). The coarse fraction was dry screened with a stack of ASTM sieves with mesh sizes ranging from 4000 μm to 63 μm, with intermediate sieves and a subsequent determination of the weights of the obtained fractions. The fine fraction was first analyzed dry, until a sub-sample was obtained, dispersed in an aqueous solution and subsequently analyzed with the laser granulometer. The processed sedimentological data were used for the construction of cumulative and frequency curves and were classified according to the classification of Shepard (1954), classical or modified.

4. Results

Marine geological mapping (Figure 5) coupled with sedimentological and seismo-stratigraphic data has allowed to reconstruct the depositional environments offshore of Northern Cilento. Four geological maps show the distribution of the Late Quaternary deposits at the seafloor in the Cilento offshore and have been superimposed to the onshore–offshore DEM of the Cilento Promontory (Figure 5).

The map n. I NW (scale 1: 25,000) covers a continental shelf area, dipping with low gradients up to water depths of 95 m (Figure 5). The northeastern extremity of this map, including the farthest southern sector of the Paestum Plain, includes little ramps up to water depths of 35 m. The rocky shoreline and the outcrops of acoustic basement alongside the Agropoli promontory control the convex bathymetric trend up to 25 m. Similarly, the rocky promontory between the Tresino Cape and the Pagliarolo Cape conditioned the physiography of the submerged area, with a convex bathymetric trend surrounding high coastlines. The physiographic unit of the Licosa Cape high, represented by an E-W trending ridge, is the most representative morpho-structural lineament of both maps n. II SW and n. III SE (scale 1: 25,000; Figure 5). An articulated E–W bathymetric trend occurs, with a wide area showing remnants of terraced surfaces located at different water depths [3,34]. The concave bathymetric trend highlights the occurrence of slide scars, incised by drainage axes, in the southern sector of the structural high. Three main morpho-structural highs of the acoustic basement occur, respectively, N–S, E–W and NNW–SSE trending (Figure 5). A continental shelf, ranging at water depths between 105 and 185 m, is covered by the map n. IV NE (scale 1: 25,000; Figure 5).

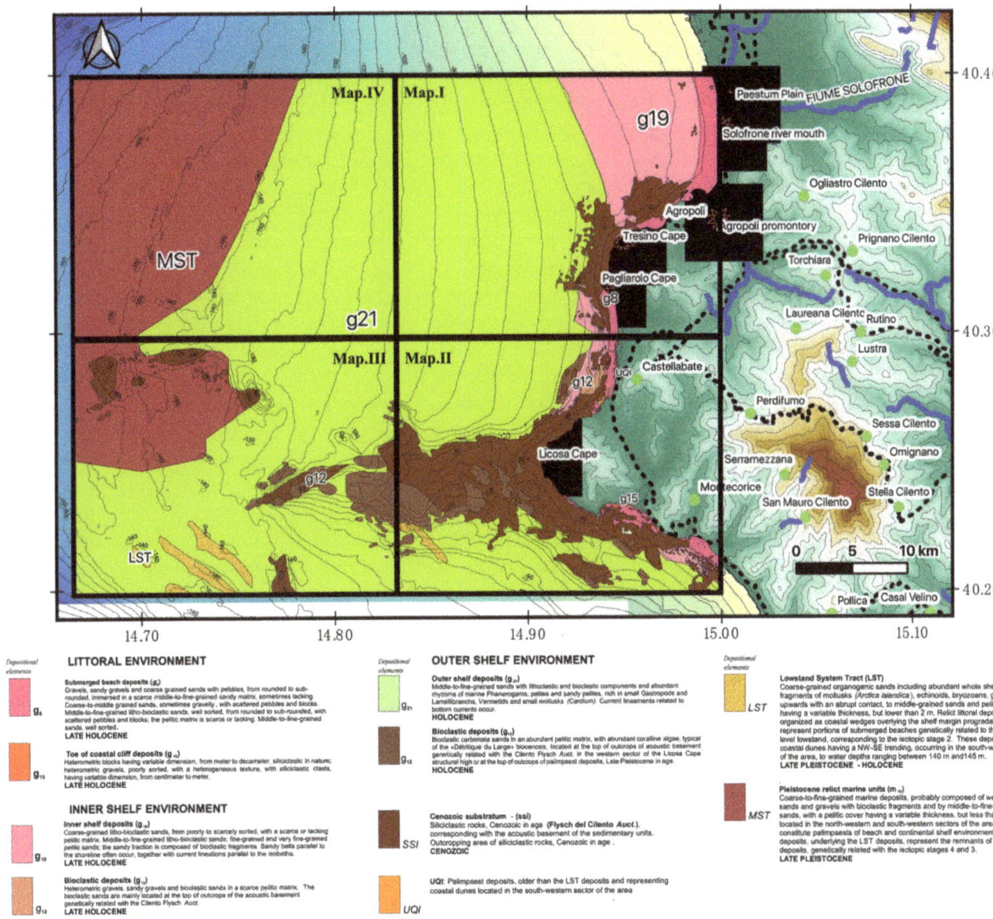

Figure 5. Onshore–offshore DEM of the Cilento Promontory with superimposed the results of the marine geological survey at the 1:25,000 scale (map n. I NW, map n. II SW, map n. III SE and map n. IV NE).

The submerged beach deposits and the toe-of-coastal cliff deposits, particularly abundant at the foot of the present day coastal cliffs, incised in Pollica formation, characterizing the littoral environment. Next to Licosa Cape, the toe-of-coastal cliff deposits surrounded a wide terrace of marine abrasion, located at water depths ranging between 4 m and 10 m, extending from the S. Marco Plain to Ogliastro Marina Bay. Poorly sorted blocks and gravels constitute the toe-of-coastal cliff deposits. The submerged beach deposits are composed of gravels, sandy gravels and coarse-grained sands with rounded to subrounded pebbles, immersed in a scarce middle-to-fine-grained sandy matrix. The inner shelf deposits and the bioclastic deposits characterize the inner shelf environment. The inner shelf deposits are composed of poorly sorted coarse-grained litho-bioclastic sands, middle-to-fine-grained litho-bioclastic sands and fine-grained pelitic sands. The bioclastic deposits are composed of bioclastic gravels, gravelly sands and bioclastic sands immersed in a scarce pelitic matrix. The bioclastic sands often represent the base of meadows of marine Phanerogams and are located at the top of wide outcrops of the Cenozoic substratum (ssi unit), genetically related with the Cilento Flysch, based on criteria of stratigraphic correlation of the geological units cropping out in the adjacent emerged coastal belt with the corresponding geological units recognized offshore. The outer shelf deposits and the bioclastic deposits characterize the outer shelf environment. The outer shelf deposits are composed of middle-to-fine-grained sands with lithoclastic and bioclastic fragments, including abundant rhizomes of marine Phanerogams, and of pelites and sandy pelites. Bioclastic sands in a pelitic matrix, with abundant calcareous algae, constitute the bioclastic deposits.

The sedimentological analyses disclose the grain size distribution of the sediments at the seafloor offshore the Cilento Promontory. The recognized grain sizes include sandy gravels, gravelly sands, sands, silty sands, muddy sands, sandy silts, silts and muds. The results of the sedimentological investigation showed that the analyzed samples mainly consist of fine-grained lithologies. Fine-grained sands are widespread along the coast in the northern sector of the study area. Offshore of Licosa Cape, both fine-grained and coarse-grained sands were recognized. Ternary plots of sea bottom samples (Figure 6) were constructed (shales–sands–silts) and (gravels–sands–silts) in order to improve the processing of the sedimentological data and to elaborate the sedimentological results.

The seismic analysis of the sub-bottom Chirp profiles has allowed us to study the seismo-stratigraphic characteristics of the Cilento offshore between the Solofrone river mouth and the Licosa Cape. The main outcrops at the sea bottom of the rocky acoustic basement were bounded, relatively to the sedimentary covers, composed of coarse-grained sands grading towards middle-fine-grained sands and fine-grained sands. The sandy facies are prevalent in the sector of continental shelf located between the Solofrone river mouth and Agropoli Town, where they form N–S sandy belts parallel to the isobaths and located at water depths ranging between 10 and 17 m. In the same area, the rocky outcrops have a limited extension and occur at water depths ranging between 15 m and 20 m.

The rocky acoustic basement (ssi) widely crops out next to the shoreline from Agropoli Town to Tresino Cape and from Tresino Cape to Pagliarolo Cape, where it represents the seawards prolongation of the coastal cliffs incised in the deposits of the Cilento Group [52–54]. From Agropoli to Tresino Cape, the ssi unit is represented by a terraced surface having a low gradient, dipping seawards and outcropping at the sea bottom at water depths ranging between 5 m and 25 m. In this area, the basement is colonized by Posidonia meadows. From Tresino Cape and Pagliarolo Cape, the rocky substratum forms a terraced surface having a low gradient between the emerged sea cliff and the isobath of 20 m, physically continuous, lacking the adjoining sandy facies of a submerged beach. This outcrop strongly reshapes the trending of the two rocky promontories onshore, showing the strong control played by the land geology on the marine geology. In the whole sector, the acoustic basement is downthrown below the recent sedimentary cover through normal faults, whose upper part is sometimes evident on seismic profiles.

The seismo-stratigraphic analysis evidenced that the recent sedimentary cover, ranging in age between the Late Pleistocene and the Holocene, is organized into four main seismo-stratigraphic units (Figure 7), overlying the undifferentiated acoustic basement (ssi). The first unit (seismo-stratigraphic unit 1) is characterized by an acoustically transparent seismic facies and ranges in thickness between 7 m and 10 m. Unit 1, probably composed of sands, unconformably overlies the undifferentiated acoustic basement (ssi). Its top is strongly eroded, forming palaeo-channels, in which the seismo-stratigraphic unit 2 was deposited.

The second unit (seismo-stratigraphic unit 2; Figure 7) is distinguished from alternating acoustically transparent intervals and continuous intervals, probably corresponding with alternating sands and shales for a whole thickness of about 10 meters. The unit forms the filling of depressions or intra-platform basins individuated at the top of unit 1 and of erosional depressions located at the top of the unit ssi.

The third unit (seismo-stratigraphic unit 3; Figure 7) is characterized by acoustic facies with parallel and discontinuous reflectors of high amplitude and represents a first phase of the recent filling, Holocene in age.

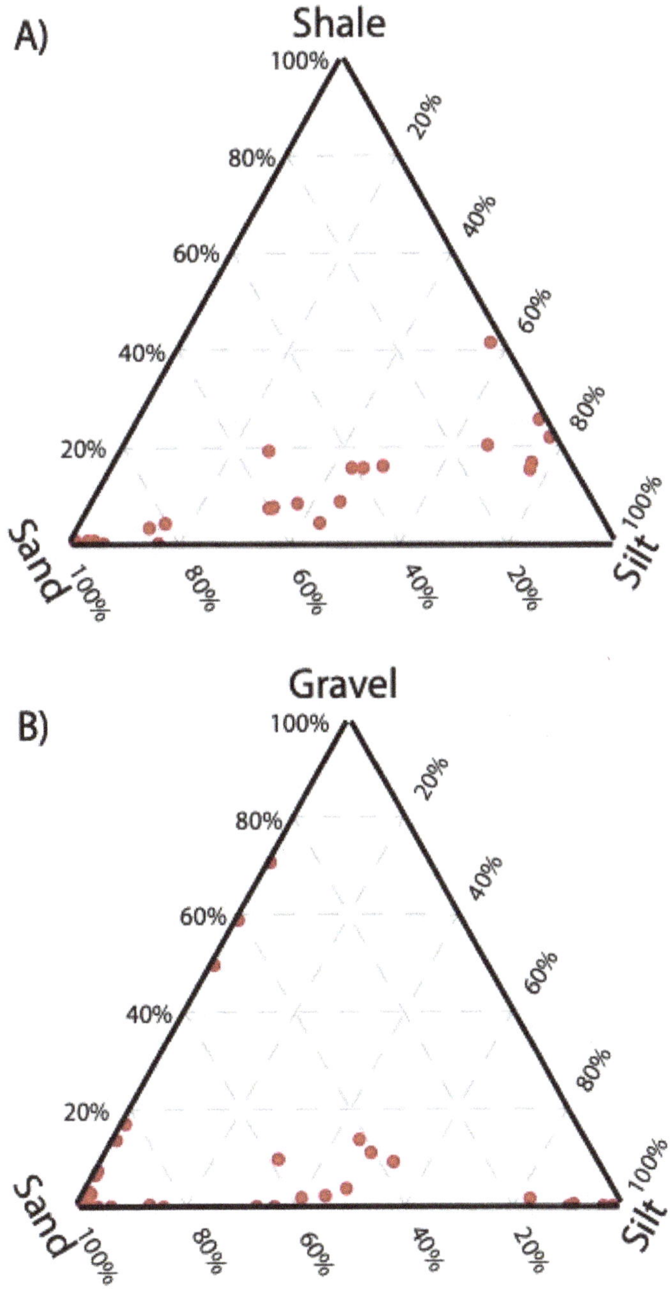

Figure 6. Ternary plots of the sea bottom samples. Inset (**A**) ternary plot shale–sand–silt. Inset (B) ternary plot gravel–sand–silt.

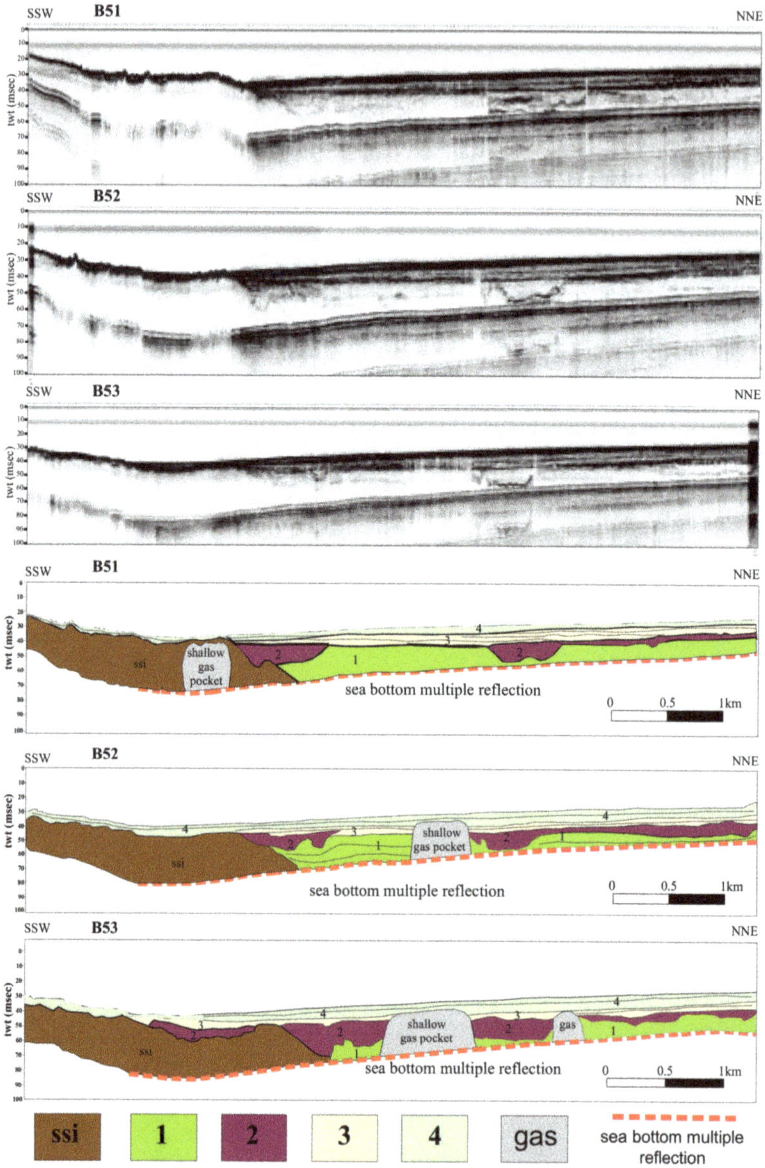

Figure 7. Sub-bottom profiles B51, B52 and B53 and the corresponding geological interpretations (see Figure 3 for location).

The fourth unit (seismo-stratigraphic unit 4; Figure 7) is distinguished from seismic facies with parallel continuous-to-discontinuous reflectors of high amplitude and represents a second phase of the recent filling, Holocene in age. The geological interpretation of sub-bottom profiles has shown the occurrence of shallow gas pockets (Figure 7).

The seismo-stratigraphic interpretation of the sub-bottom profile Bl14 was calibrated with the lithostratigraphic data of a gravity core, herein named the Licosa gravity core, previously published (Figure 8) [57]. On the left in the seismic profile, the rocky acoustic basement genetically related with the Cilento Flysch (ssi unit) was recognized. This seismic unit is overlain by a thick progradational unit, interpreted as the beach deposits of the isotopic stages 4 and 5 [58,59], Late Pleistocene in age. Based on the calibration with the Licosa gravity core, the uppermost part of this unit has been interpreted as coarse-grained sands, very rich in Mollusk shells, including Arctica islandica, a cold host of the Pleistocene (Sg unit; Figure 8). This represents a key seismic unit in the seismo-stratigraphic setting of this area and is composed of coarse-grained organogenic sands. Being located at higher water depths with respect to the sedimentary distribution model, which locates this kind of organogenic sand in the coastal belt [60,61], this unit can be interpreted as relict sands [62], which can be compared with the organogenic sands described by Péres and Picard (1964) [63], including cold hosts attributed to the end of the Würm.

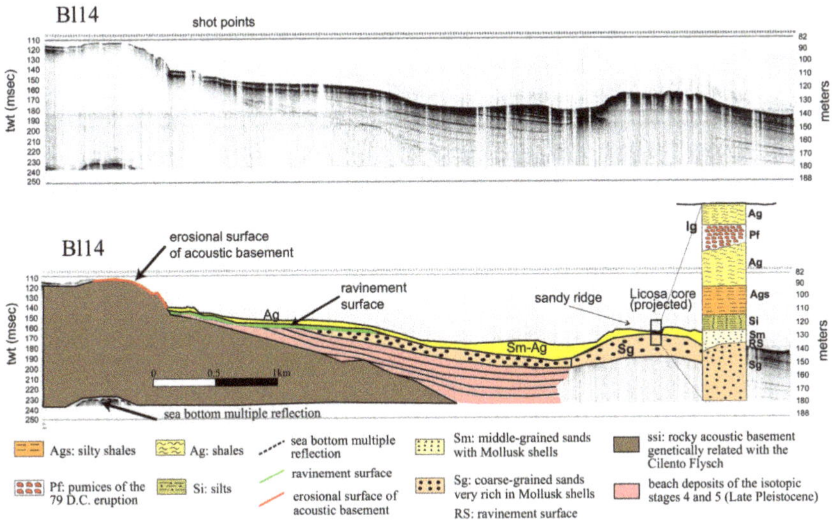

Figure 8. Sub-bottom profile Bl14 and corresponding geologic interpretation, calibrated with the lithostratigraphic data of the Licosa gravity core (see Figure 3 for location). The subfigure on the right side of the seismic profile represents the stratigraphy of the Licosa core (see the key to the seismic profile for the explanation).

The Sg unit forms a sandy ridge, which is located at water depths ranging between 130 m and 140 m of the water depth (Figure 8). This sandy ridge has been interpreted as a part of the submerged beach (shoreface) and could be related to the last lowstand phase, corresponding to the isotopic stage 2 [58,59]. The underlying seismic unit, characterized by pro-grading clinoforms, could represent the remnants of older beach systems, genetically related with the isotopic stages 4 and 3 (Late Pleistocene; Figure 8).

The abrupt contact between the coarse-grained sands with *Arctica islandica* and the overlying finer-grained deposits (Sm to Ag deposits in the Licosa core; Figure 8) is represented by an erosional surface ("ravinement surface"; Figure 8) [64]. The ravinement surface is a time-transgressive or diachronous subaqueous erosional surface resulting from nearshore marine and shoreline erosion associated with a sea level rise [64]. Proceeding landwards, this erosional surface laterally grades into the erosional surface located at the top of the rocky acoustic basement (Figure 8).

5. Discussion

The seismic and sequence stratigraphy are techniques of the analysis of seismic profiles that have undergone a rapid evolution, starting from the basic concepts of Vail et al. 1977 [6]. In this paper, a definition of the depositional sequence was given, coupled with the criteria for its determination, based on the lateral terminations of the inner strata of sequences, both upwards (erosional truncation) and downwards (onlap and downlap). These criteria were and still remain basic for the identification of regional unconformities bounding the depositional sequences in their lower part and in their upper part. Other basic concepts of this technique have been developed by Christie-Blick 1991 [7], who defined, for the first time, the "unconformity-bounded" depositional sequences based on their stratal patterns (onlap and offlap). In particular, the "unconformity-bounded" depositional sequences show onlapping strata at their base and offlapping strata at their top [7]. Despite how many sequence stratigraphic papers have been further developed, reaching the modern applications and concepts of sequence stratigraphy of Catuneanu [9,10], we believe that some basic concepts of the first sequence stratigraphic papers are still valid [6–8], as the identification of regional unconformities for the identification of the depositional sequences.

Based on marine geological mapping, coupled with seismo-stratigraphic interpretation and confirmed by the sedimentological data, all the described deposits (littoral, inner shelf and outer shelf environments) pertain to the highstand system tract of the Late Quaternary depositional sequence (Figure 5). The lowstand system tract is composed of organogenic coarse-grained sands, including an abundant bioclastic component (Mollusks, Echinoids, Bryozoans), passing upwards through an abrupt contact to middle-grained sands and thin pelitic drapes, whose thickness does not exceed 2 m (Figure 5). They are relict littoral deposits, organized as coastal wedges overlying the shelf margin prograding clinoforms, which represent portions of submerged beaches genetically related with the last sea-level lowstand, corresponding with the isotopic stage 2 [58,59]. The lowstand deposits form dune strips with NW-SE elongation, which occur in the southwestern sector of the area, at water depths ranging between 140 m and 145 m (Figure 5). The Pleistocene relict marine units are composed of coarse-to-fine-grained marine deposits and are probably constituted by well-sorted sands and gravels with bioclastic fragments and by middle-to-fine-grained sands overlain by thin pelitic drapes not exceeding a thickness of 2 m. These deposits, which constitute palimpsests of beach and continental shelf environments, are mainly located in the northwestern and southwestern quadrants of the marine area covered by the geological sheet n. 502 "Agropoli". These deposits, underlying the lowstand system tract represent the remaining parts of older beach systems correlated to isotopic stages 4 and 5 [58,59] (Figure 8).

The bioclastic deposits are typical of the biocenosis of "Détritique Du Large" [63], developed on hard sea bottoms and are located both at the top of outcrops of Cenozoic acoustic basement, genetically related with the Cilento Flysch (ssi unit) in the western sector of the morpho-structural high of the Licosa Cape and at the top of palimpsest deposits, Late Pleistocene in age (Figure 5). In the Mediterranean Sea, the zonation of benthic assemblages carried out by Péres and Picard [63] represents a basic tool in order to know the lithology and to interpret the facies of the bioclastic deposits (Figure 9). In the Mediterranean Sea, the bioclastic deposits occur at water depths ranging between 40 m and 100 m ("Détritique Cotier" of Péres and Picard) [63] (Figure 9). The main components of this biocoenosis are produced after the reworking and the deposition of the benthic communities on both mobile sea bottoms (biocoenosis of the "Détritique Cotier") and on hard sea bottoms (biocoenosis of the "Détritique Du Large"). After the Holocene, sea level rise relict and drowned sediments, characterized by low rates of sedimentation and by the occurrence of glauconite, were deposited on the seafloor ("Détritique Du Large"; Figure 9) [64,65].

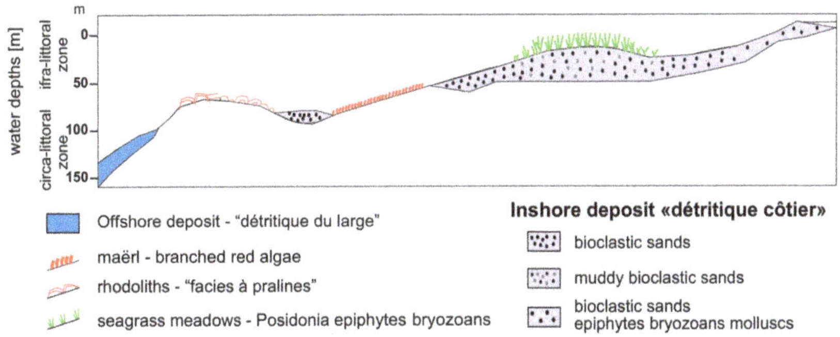

Figure 9. Sketch diagram showing the main biocenosis of the mobile sea bottoms.

6. Conclusions

The geological evolution of the Cilento continental shelf during the Late Pleistocene-Holocene has been reconstructed based on the marine geological data (Figure 5), coupled with the sedimentological data (Figure 6) and with the seismo-stratigraphic data (Figures 7 and 8).

Following a general climate warming [66–69], during the Upper Pleistocene Holocene, there was a rapid rise of sea level on a global scale [70–74]. Transgressive and highstand deposits have been individuated on the continental shelves all around the world [26,75–80]. During the transgression, the high rate of the sea level rise and the low gradient of the Cilento continental shelf led to the almost simultaneous submersion of large areas and to the drastic landwards shift of the coastal facies. As a consequence, the geological interpretation of the Chirp sub-bottom profiles did not allow the identification of retrogradational seismo-stratigraphic units, which can be interpreted as beach systems deposited during the transgression. On the contrary, highstand and lowstand deposits have been well-documented based on seismo-stratigraphic data (Figures 7 and 8).

The ravinement surface, representing an erosional truncation involving the upper part of the offlap succession, is characterized by an irregular topography (Figure 8). The Licosa core, calibrating the sub-bottom profile Bl14 (Figure 8), has shown that the sediments overlying the ravinement surface are characterized by middle-grained sands grading upwards into silty shales (Sm, Si, Ags; Figure 8), with an overall decrease of the grain size upwards. This decrease in grain size indicates a deepening of the continental platform. These sediments have been interpreted as deposited during the transgressive phase started at the end of the glacial stage of the isotopic stage 2 [57]. These deposits are overlain, in turn, by shales (Ag), including a pumiceous ash layer of Vesuvius, and are located on the ravinement surface (Figure 8). In the sub-bottom profile Bl14, this succession (transgressive and highstand deposits) constitutes a thin drape overlying the ravinement surface (Figure 8). Unfortunately, tephrostratigraphic data on this layer are not still available, allowing it to function as a good chronostratigraphic layer in this sector of the Tyrrhenian margin.

From the end of the isotopic stage 5a, the sea, despite the cyclical oscillations, is constantly lowering up to the isotopic stage 2, when it was located at a depth of about 120 m in the Mediterranean Sea [16,81–90]. During this phase of "forced regression" [24,91–94], the progradational wedges of the Cilento offshore were deposited, allowing for a platform widening in the order of several kilometers. The forced regression is controlled by the seaward migration of the coastline as a reaction to the relative sea-level fall. This kind of regression occurs during periods of falling of the sea level, since the coastline is forced to regress due to the falling of the base level, without taking into account the sediment supply [24]. During the forced regression, the fluvial incision is accompanied by the deposition of pro-gradational deposits, in a shoreface setting.

Author Contributions: Writing, original draft preparation; data interpretation and writing, review and editing, G.A.; Methodology, software and data curation, M.C. All authors have read and agreed to the published version of the manuscript.

Funding: This research received no external funding, but the analyzed data were acquired by the CNR ISMAR of Naples (Italy) in the frame of the realization of the geological sheet n. 502 "Agropoli" (marine sector). The CARG Project was previously funded by the National Research Council of Italy (CNR-IAMC, now CNR-ISMAR) and by the Campania Region-Sector of Defence Soil, Geothermics and Geotechnics, Naples Italy.

Institutional Review Board Statement: Not applicable.

Informed Consent Statement: Not applicable.

Acknowledgments: We thank the Academic Editor Dott. Antoni Calafat and two anonymous reviewers, whose comments greatly improved this manuscript. We thank Dott.Lucia Monti for making available the data shown in this paper.

Conflicts of Interest: The authors declare no conflict of interest.

References

1. Aiello, G.; Marsella, E. The contribution of marine geology to the knowledge of marine coastal environment off the Campania region (southern Italy): The geological map n. 502 "Agropoli" (southern Campania). *Mar. Geophys. Res.* **2013**, *34*, 89–113. [CrossRef]
2. Martelli, L.; Nardi, G.; Cammarosano, A.; Cavuoto, G.; Aiello, G.; D'Argenio, B.; Marsella, E.; Ferraro, L. *Note Illustrative Della Carta Geologica d'Italia Alla Scala 1:50,000*; Regione Campania, ISPRA (Servizio Geologico d'Italia): Rome, Italy, 2016; pp. 1–110. Available online: http://www.isprambiente.gov.it/Media/carg/noteillustrative/502_Agropoli.pdf (accessed on 28 September 2021).
3. Aiello, G. Elaborazione ed interpretazione geologica di sismica di altissima risoluzione nell'offshore del promontorio del Cilento (Tirreno meridionale, Italia). *Quad. Geofis.* **2019**, *155*, 1–24. [CrossRef]
4. Catalano, R.; Fabbri, A.; Argnani, A.; Bortoluzzi, G.; Correggiari, A.; Gamberi, F.; Ligi, M.; Marani, M.; Penitenti, D.; Roveri, M.; et al. *Linee Guida al Rilevamento Delle Aree Marine del Servizio Geologico Nazionale*; Bozza n. 1; APAT, Servizio Geologico d'Italia: Rome, Italy, 1996.
5. Fabbri, A.; Argnani, A.; Bortoluzzi, G.; Correggiari, A.; Gamberi, F.; Ligi, M.; Marani, M.; Penitenti, D.; Roveri, M.; Trincardi, F. *Carta Geologica Dei Mari Italiani Alla Scala 1:250,000. Guida Al Rilevamento; Quaderni Serie III*; Presidenza del Consiglio dei Ministri, Dipartimento per i Servizi Tecnici Nazionali, Servizio Geologico d'Italia: Rome, Italy, 2002; Volume 8, pp. 1–93.
6. Vail, P.R.; Mitchum, R.M.; Todd, R.G.; Widmier, J.M.; Thomson, S.; Sangree, J.B.; Bubb, J.N.; Hatlelid, W.J. Seismic stratigraphy and global changes of sea level. In *Seismic Stratigraphy—Applications to Hydrocarbon Exploration*; Payton, C.E., Ed.; Mem. 26; American Association of Petroleum Geologists: Tulsa, OK, USA, 1977; pp. 49–212.
7. Christie-Blick, N. Onlap, offlap, and the origin of unconformity-bounded depositional sequences. *Mar. Geol.* **1991**, *97*, 35–56. [CrossRef]
8. Helland-Hansen, W.; Gjelberg, G.J. Conceptual basis and variability in sequence stratigraphy: A different perspective. *Sediment. Geol.* **1994**, *92*, 31–52. [CrossRef]
9. Catuneanu, O.; Abreu, V.; Bhattacharya, J.P.; Blum, M.D.; Dalrymple, R.W.; Eriksson, P.G.; Fielding, C.R.; Fisher, W.L.; Galloway, W.E.; Gibling, M.R.; et al. Towards the standardization of sequence stratigraphy. *Earth Sci. Rev.* **2009**, *92*, 1–33. [CrossRef]
10. Catuneanu, O. Model-independent sequence stratigraphy. *Earth Sci. Rev.* **2018**, *188*, 312–388. [CrossRef]
11. Di Capua, A.; Scasso, R.A. Sedimentological and petrographical evolution of a fluvio-lacustrine environment during the onset of volcanism:Volcanically-induced forcing of sedimentation and environmental responses. *Sedimentology* **2020**, *67*, 1879–1913. [CrossRef]
12. Giannetti, B.; De Casa, G. Stratigraphy, chronology and sedimentology of ignimbrites from the white trachytic tuff, Roccamonfina Volcano, Italy. *J. Volcanol. Geotherm. Res.* **2000**, *96*, 243–295. [CrossRef]
13. Lucchi, F. On the use of unconformities in volcanic stratigraphy and mapping: Insights from the Aeolian Islands (Southern Italy). *J. Volcanol. Geotherm. Res.* **2019**, *385*, 3–26. [CrossRef]
14. International Commission on Stratigraphy. Stratigraphic Guide. Available online: https://stratigraphy.org/guide/uncon (accessed on 28 September 2021).
15. Martinson, D.G.; Pisias, N.J.; Hays, J.D.; Imbrie, J.; Moore, T.C.; Shackleton, N.J. Age dating and the orbital theory of the ice ages: Development of a high-resolution 0 to 300,000-year chronostratigraphy. *Quat. Res.* **1987**, *27*, 1–29. [CrossRef]
16. Lisiecki, L.E.; Raymo, M.E. A Pliocene-Pleistocene stack of 57 globally distributed benthic $\delta^{18}O$ records. *Paleoceanography* **2005**, *20*, PA1003. [CrossRef]
17. Reading, H. *Sedimentary Environments and Facies*, 1st ed.; Elsevier: New York, NY, USA, 1978; pp. 1–557. [CrossRef]

18. Nummedal, D.; Swift, D.J.P. Transgressive stratigraphy at sequence-bounding unconformities: Some principles derived from Holocene and Cretaceous examples. In *Sea Level Fluctuation and Coastal Evolution*, 1st ed.; Nummedal, D., Pilkey, O.H., Howard, J.D., Eds.; SEPM: Tulsa, OK, USA, 1987; Volume 41, pp. 241–260.
19. Trincardi, F.; Correggiari, A.; Roveri, M. Late Quaternary transgressive erosion and deposition in a modern epicontinental shelf: The Adriatic Semienclosed Basin. *GeoMar. Lett.* **1994**, *14*, 41–51. [CrossRef]
20. Chiocci, F.L.; Ercilla, G.; Torre, J. Stratal architecture of Western Mediterranean Margins as the result of the stacking of Quaternary lowstand deposits below glacio-eustatic fluctuation base-level. *Sediment. Geol.* **1997**, *112*, 195–217. [CrossRef]
21. Cattaneo, A.; Steel, R. Transgressive deposits: A review of their variability. *Earth Sci. Rev.* **2003**, *62*, 187–228. [CrossRef]
22. Lobo, F.J.; Ridente, D. Stratigraphic architecture and spatio-temporal variability of high frequency (Milankovitch) depositional cycles on modern continental margins: An overview. *Mar. Geol.* **2014**, *352*, 215–247. [CrossRef]
23. Posamentier, H.W.; Allen, G.P.; James, D.P.; Tesson, M. Forced regressions in a sequence stratigraphic framework: Concepts, examples, and exploration significance. *AAPG Bull.* **1992**, *76*, 1687–1709.
24. Catuneanu, O. Sequence stratigraphy of clastic systems: Concepts, merits, and pitfalls. *J. Afr. Earth Sci.* **2002**, *35*, 1–43. [CrossRef]
25. Bozzano, A.; Corradi, N.; Fanucci, F.; Ivaldi, R. Late Quaternary deposits from the Ligurian continental shelf (NW Mediterranean): A response to problems of coastal erosion. *Chem. Ecol.* **2006**, *22* (Suppl. S1), S349–S359. [CrossRef]
26. Sulli, A.; Agate, M.; Mancuso, M.; Pepe, F.; Pennino, V.; Polizzi, S.; Lo Presti, V.; Gargano, F.; Interbartolo, F. Variability of depositional setting along the north-western Sicily continental shelf (Italy) during Late Quaternary: Effects of sea level changes and tectonic evolution. *Alp. Mediterr. Quat* **2012**, *25*, 141–156.
27. Martorelli, E.; Falese, F.; Chiocci, F.L. Chapter 12 Overview of the variability of Late Quaternary continental shelf deposits of the Italian peninsula. In *Continental Shelves of the World: Their Evolution during the Last Glacio-Eustatic Cycle*, 1st ed.; Memoirs; Chiocci, F.L., Chivas, A.R., Eds.; Geological Society of London: London, UK, 2014; Volume 41, pp. 171–186. [CrossRef]
28. Aiello, G.; Budillon, F.; Cristofalo, G.; D'Argenio, B.; de Alteriis, G.; De Lauro, M.; Ferraro, L.; Marsella, E.; Pelosi, N.; Sacchi, M.; et al. Marine Geology and Morphobathymetry in the Bay of Naples (South-Eastern Tyrrhenian Sea, Italy). In *Mediterranean Ecosystems*, 1st ed.; Faranda, F.M., Guglielmo, L., Spezie, G., Eds.; Springer: Milano, Italy, 2001; pp. 1–8.
29. D'Argenio, B.; Aiello, G.; de Alteriis, G.; Milia, A.; Sacchi, M.; Tonielli, R.; Budillon, F.; Chiocci, F.L.; Conforti, A.; De Lauro, M.; et al. *Digital Elevation Model of the Naples Bay and Adjacent Areas, Eastern Tyrrhenian Sea*, 1st ed.; Editore De Agostini: Rome, Italy, 2004; pp. 1–10.
30. Sacchi, M.; Insinga, D.D.; Milia, A.; Molisso, F.; Torrente, M.M.; Conforti, A. Stratigraphic signature of the Vesuvius 79 A.D. event off the Sarno prodelta system, Naples Bay. *Mar. Geol.* **2005**, *222*, 443–469. [CrossRef]
31. Insinga, D.D.; Molisso, F.; Lubritto, C.; Sacchi, M.; Passariello, I.; Morra, V. The proximal marine record of Somma–Vesuvius volcanic activity in the Naples and Salerno bays, Eastern Tyrrhenian Sea, during the last 3 kyrs. *J. Volcanol. Geotherm. Res.* **2008**, *177*, 170–186. [CrossRef]
32. Savini, A.; Basso, D.; Bracchi, V.A.; Corselli, C.; Pennetta, M. Maërl-bed mapping and carbonate quantification on submerged terraces offshore the Cilento peninsula (Tyrrhenian Sea, Italy). *Geodiversitas* **2012**, *34*, 77–98. [CrossRef]
33. Pennetta, M.; Bifulco, A.; Savini, A. Ricerca di depositi di sabbia sottomarina relitta sulla piattaforma continentale del Cilento (SA) utilizzabile per interventi di ripascimento artificiale dei litorali. *Geol. Dell'ambiente* **2013**, *1*, 1–22.
34. Savini, A.; Bracchi, V.A.; Cammarosano, A.; Pennetta, M.; Russo, F. Terraced landforms onshore and offshore the Cilento Promontory (south-eastern Tyrrhenian margin) and their significance as Quaternary records of sea level changes. *Water* **2021**, *13*, 566. [CrossRef]
35. Trenhaile, A.S. *The Geomorphology of Rocky Coasts*, 1st ed.; Oxford University Press: Oxford, UK, 1987; pp. 1–384. [CrossRef]
36. Retallack, G.J.; Roering, J.J. Wave-cut or water-table platforms of rocky coasts and rivers? *GSA Today* **2012**, *22*, 4–10. [CrossRef]
37. Guida, D.; Valente, A. Terrestrial and marine landforms along the Cilento coastland (Southern Italy): A framework for landslide hazard assessment and environmental conservation. *Water* **2019**, *11*, 2618. [CrossRef]
38. De Pippo, T.; Pennetta, M. Terrazzi deposizionali sommersi nel Golfo di Policastro (Campania). *Mem. Descr. Carta Geol.* **2004**, *LVIII*, 57–62.
39. Casalbore, D.; Romagnoli, C.; Adami, C.; Bosman, A.; Falese, F.; Ricchi, A.; Chiocci, F.L. Submarine depositional terraces at Salina island (Southern Tyrrhenian sea) and implications on the Late-Quaternary evolution of the insular shelf. *Geosciences* **2018**, *8*, 20. [CrossRef]
40. Budillon, F.; Amodio, S.; Contestabile, P.; Alberico, I.; Innangi, S.; Molisso, F. *The Present-Day Nearshore Submarine Depositional Terraces off the Campania Coast (South-Eastern Tyrrhenian Sea): An Analysis of Their Morpho-Bathymetric Variability*; Extended Abstract; IMEKO TC-19 International Workshop on Metrology for the Sea: Naples, Italy, 2020; pp. 132–138.
41. Milia, A.; Torrente, M.M. Late-Quaternary volcanism and transtensional tectonics in the Bay of Naples, Campanian continental margin, Italy. *Mineral. Petrol.* **2003**, *79*, 49–65. [CrossRef]
42. Chiocci, F.L.; Chivas, A.L. Chapter 1 An overview of the continental shelves of the world. In *Continental Shelves of the World: Their Evolution during the Last Glacio-Eustatic Cycle*, 1st ed.; Chiocci, F.L., Chivas, A.R., Eds.; Memoirs; Geological Society of London: London, UK, 2014; Volume 41, pp. 1–5. [CrossRef]
43. Aiello, G.; Marsella, E. Interactions between Late Quaternary volcanic and sedimentary processes in the Naples Bay, Southern Tyrrhenian sea. *Ital. J. Geosci.* **2015**, *134*, 367–382. [CrossRef]
44. Bartole, R. Tectonic structure of the Latian Campanian shelf (Tyrrhenian Sea). *Boll. Oceanol. Teor. Appl.* **1984**, *2*, 197–230.

45. Bartole, R.; Savelli, C.; Tramontana, M.; Wezel, F.C. Structural and sedimentary features in the Tyrrhenian margin off Campania, southern Italy. *Mar. Geol.* **1984**, *55*, 163–180. [CrossRef]
46. Aiello, G.; Di Fiore, V.; Marsella, E.; D'Isanto, C. Stratigrafia sismica e morfobatimetria della Valle di Salerno. In Proceedings of the 26th National Congress GNGTS (Gruppo Nazionale di Geofisica della Terra Solida), Rome, Italy, 24–27 November 2007; Extended Abstract; pp. 495–498.
47. Aiello, G.; Marsella, E.; Di Fiore, V.; D'Isanto, C. Stratigraphic and structural styles of half graben offshore basins in Southern Italy: Multichannel seismic and Multibeam morpho bathymetric evidences on the Salerno Valley (Southern Campania continental margin, Italy). *Quad. Geofis.* **2009**, *77*, 1–33.
48. Sacchi, M.; Infuso, S.; Marsella, E. Late Pliocene early Pleistocene compressional tectonics in offshore Campania (eastern Tyrrhenian sea). *Boll. Geof. Teor. Appl.* **1994**, *36*, 469–482.
49. Aiello, G.; Marsella, E.; Cicchella, A.G.; Di Fiore, V. New insights on morpho structures and seismic stratigraphy along the Campania continental margin (Southern Italy) based on deep multichannel seismic profiles. *Rend. Lincei* **2011**, *22*, 349–373. [CrossRef]
50. Aiello, G.; Cicchella, A.G. Dati sismostratigrafici sul margine continentale della Campania tra Ischia, Capri ed il Bacino del Volturno (Tirreno meridionale, Italia) in base al processing sismico ed all'interpretazione geologica di profili sismici a riflessione multicanale. *Quad. Geof.* **2019**, *149*, 1–52.
51. Zitellini, N.; Ranero, C.; Loreto, M.F.; Ligi, M.; Pastore, M.; D'Oriano, F.; Sallares, V.; Grevemeyer, I.; Moeller, S.; Prada, M. Recent inversion of the Tyrrhenian Basin. *Geology* **2020**, *48*, 123–127. [CrossRef]
52. Bonardi, G.; Amore, F.O.; Ciampo, G.; De Capoa, P.; Miconnet, P.; Perrone, V. Il Complesso Liguride Auct: Stato delle conoscenze e problemi aperti sull'evoluzione pre appenninica ed i suoi rapporti con l'Arco Calabro. *Mem. Soc. Geol. Ital.* **1988**, *41*, 17–35.
53. Cammarosano, A.; Cavuoto, G.; Martelli, L.; Nardi, G.; Toccaceli, R.M.; Valente, A. Il Progetto CARG nell'area silentina (area interna Appennino meridionale): Il nuovo assetto stratigrafico strutturale derivante dal rilevamento dei fogli 503, 502, e519 (Vallo della Lucania, Agropoli e Capo Palinuro). *Rend. Online Soc. Geol. Ital.* **2011**, *12*, 19–21.
54. Vitale, S.; Ciarcia, S. Tectono stratigraphic setting of the Campania region (southern Italy). *J. Maps* **2018**, *14*, 9–21. [CrossRef]
55. Gasperini, L.; Stanghellini, G. SEISPRHO: An interactive computer program for processing and interpretation of high-resolution seismic reflection profiles. *Comp. Geosci.* **2009**, *35*, 1497–1507. [CrossRef]
56. Vail, P.R.; Hardenbol, J.; Todd, R.G. Jurassic unconformities, chronostratigraphy, and sea-level changes from seismic stratigraphy and biostratigraphy. In *Interregional Unconformities and Hydrocarbon Accumulation*, 1st ed.; Memoirs; Schlee, J.S., Ed.; American Association of Petroleum Geologists: Tulsa, OK, USA, 1984; Volume 36, pp. 129–144.
57. Ferraro, L.; Pescatore, T.; Russo, B.; Senatore, M.R.; Vecchione, C.; Coppa, M.G.; Di Tuoro, A. Studi di geologia marina del margine tirrenico: La piattaforma continentale tra Punta Licosa e Capo Palinuro (Tirreno meridionale). *Boll. Soc. Geol. Ital.* **1997**, *116*, 473–485.
58. Shackleton, N.J.; Opdyke, N.D. Oxygen isotope and paleomagnetic stratigraphy of equatorial Pacific Core V28-238: Oxygen isotope temperatures and ice volume on a 105 year and 106 year scale. *Quat. Res.* **1973**, *3*, 39–55. [CrossRef]
59. Zazo, C. Interglacial sea levels. *Quat. Intern.* **1999**, *55*, 101–113. [CrossRef]
60. Emery, K.O. Relict Sediments on Continental Shelves of World. *AAPG Bull.* **1968**, *52*, 445–464.
61. Swift, D.J.P. Continental shelf sedimentation. In *Marine Sediment Transport and Environmental Management*, 1st ed.; Stanley, D.J., Swift, D.J.P., Eds.; John Wiley: New York, NY, USA, 1976; pp. 311–350.
62. Orme, G.R. Relict sediment. In *Beaches and Coastal Geology. Encyclopedia of Earth Sciences Series*; Springer: New York, NY, USA, 1982. [CrossRef]
63. Péres, J.M.; Picard, J. Noveau manuel de bionomie marine benthique de la Mer Méditerranee. *Recl. Trav. Stn. Mar. D'endoume* **1964**, *31*, 5–137.
64. Thorne, J.A.; Swift, D.J.P. Sedimentation on continental margins: VI. A regime model for depositional sequences, their component systems tracts, and bounding surfaces. In *Shelf Sand and Sandstone Bodies—Geometry, Facies and Sequence Stratigraphy*, 1st ed.; Special Publication; Swift, D.J.P., Oertel, G.F., Tillman, R.W., Thorne, J.A., Eds.; International Association of Sedimentologists, Wiley: New York, NY, USA, 1991; Volume 14, pp. 189–255. [CrossRef]
65. Carannante, G.; Esteban, M.; Milliman, J.D.; Simone, L. Carbonate lithofacies as a paleolatitude indicators: Problems and limitations. *Sedim. Geol.* **1988**, *60*, 333–346. [CrossRef]
66. Ruddiman, W.F.; McIntyre, A. The mode and mechanism of the last deglaciation: Oceanic evidence. *Quat. Res.* **1981**, *16*, 125–134. [CrossRef]
67. Jansen, E.; Overpeck, J.; Briffa, K.R.; Duplessy, J.C.; Joos, F.; Masson-Delmotte, V.; Olago, D.; Otto-Bliesner, B.; Peltier, W.R.; Rahmstorf, S.; et al. Palaeoclimate. In *Climate Change: The Physical Science Basis. Contribution of Working Group I to the Fourth Assessment Report of the Intergovernmental Panel on Climate Change*; Solomon, S., Qin, D., Manning, M., Chen, Z., Marquis, M., Averyt, K.B., Tignor, M., Miller, H.L., Eds.; Cambridge University Press: Cambridge, UK; New York, NY, USA, 2007.
68. Ruddiman, W.F. The early anthropogenic hypothesis: Challenges and responses. *Rev. Geophys.* **2007**, *45*, RG4001. [CrossRef]
69. Murray-Wallace, C.; Woodroffe, C.D. Sea-level changes since the Last Glacial Maximum. In *Quaternary Sea-Level Changes: A Global Perspective*, 1st ed.; Murray-Wallace, C., Woodroffe, C.D., Eds.; Cambridge University Press: Cambridge, UK, 2014; pp. 320–368.
70. Fairbridge, R.W. Eustatic changes in sea level. *Phys. Chem. Earth* **1961**, *4*, 99–185. [CrossRef]

71. Fairbanks, R.G. A 17,000-year glacio-eustatic sea level record: Influence of glacial melting rates on the Younger Dryas event and deep-ocean circulation. *Nature* **1989**, *342*, 637–642. [CrossRef]
72. Varekamp, J.C.; Thomas, E.; Van de Plassche, O. Relative sea-level rise and climate change over the last 1500 years. *Terra Nova* **1992**, *4*, 293–304. [CrossRef]
73. Benjamin, J.; Rovere, A.; Fontana, A.; Furlani, S.; Vacchi, M.; Inglis, R.H.; Galili, E.; Antonioli, F.; Sivan, D.; Miko, S.; et al. Late Quaternary sea-level changes and early human societies in the central and eastern Mediterranean Basin: An interdisciplinary review. *Quat. Int.* **2017**, *449*, 29–57. [CrossRef]
74. Antonioli, F.; De Falco, G.; Lo Presti, V.; Moretti, L.; Scardino, G.; Anzidei, M.; Bonaldo, D.; Carniel, S.; Leoni, G.; Furlani, S.; et al. Relative Sea-Level Rise and Potential Submersion Risk for 2100 on 16 Coastal Plains of the Mediterranean Sea. *Water* **2020**, *12*, 2173. [CrossRef]
75. Chiocci, F.L.; D'Angelo, S.; Orlando, L.; Pantaleone, E.A. Evolution of the Holocene shelf sedimentation defined by high-resolution seismic stratigraphy and sequence analysis (Calabro-tyrrhenian continental shelf). *Mem. Soc. Geol. Ital.* **1989**, *48*, 359–380.
76. Díaz, J.I.; Maldonado, A. Transgressive sand bodies on the Maresme continental-shelf, western Mediterranean-sea. *Mar. Geol.* **1990**, *91*, 53–72. [CrossRef]
77. Trincardi, F.; Campiani, E.; Correggiari, A.; Foglini, F.; Maselli, V.; Remia, A. Bathymetry of the Adriatic Sea: The legacy of the last eustatic cycle and the impact of modern sediment dispersal. *J. Maps* **2014**, *10*, 151–158. [CrossRef]
78. Ercilla, G.; Estrada, F.; Casas, D.; Duran, R.; Nuez, M.; Alonso, B.; Farran, M.L. The El Masnou infralittoral sedimentary environment (Barcelona province, NW Mediterranean Sea): Morphology and Holocene seismic stratigraphy. *Sci. Mar.* **2010**, *74*, 179–196. [CrossRef]
79. Ridente, D. Late Pleistocene Post-Glacial Sea Level Rise and Differential Preservation of Transgressive "Sand Ridge" Deposits in the Adriatic Sea. *Geosciences* **2018**, *8*, 61. [CrossRef]
80. Distefano, S.; Gamberi, F.; Baldassini, N.; Di Stefano, A. Quaternary Evolution of Coastal Plain in Response to Sea-Level Changes: Example from South-East Sicily (Southern Italy). *Water* **2021**, *13*, 1524. [CrossRef]
81. Bonifay, E. L'Ere Quaternaire: Definition, limites and subdivision sur la base de la chronologie mediterraneénne. *Bull. Soc. Geol. Fr.* **1975**, *17*, 380–393. [CrossRef]
82. Yokoyama, Y.; Purcell, A. On the geophysical processes impacting palaeo-sea-level observations. *Geosci. Lett.* **2021**, *8*, 13. [CrossRef]
83. Lambeck, K.; Yokoyama, Y.; Purcell, T. Into and out of the Last Glacial Maximum: Sea-level change during Oxygen Isotope Stages 3 and 2. *Quat. Sci. Rev.* **2002**, *21*, 343–360. [CrossRef]
84. Lambeck, K.; Antonioli, F.; Purcell, A.; Silenzi, S. Sea-level change along the Italian coast for the past 10,000 yr. *Quat. Sci. Rev.* **2004**, *23*, 1567–1598. [CrossRef]
85. Ferranti, L.; Antonioli, F.; Mauz, B.; Amorosi, A.; Dai Pra, G.; Mastronuzzi, G.; Monaco, C.; Orrù, P.; Pappalardo, M.; Radtke, U.; et al. Markers of the last interglacial sea-level high stand along the coast of Italy: Tectonic implications. *Quat. Int.* **2006**, *145–146*, 30–54. [CrossRef]
86. Peltier, W.R.; Fairbanks, R.G. Global glacial ice volume and Last Glacial Maximum duration from an extended Barbados sea level record. *Quat. Sci. Rev.* **2006**, *25*, 3322–3337. [CrossRef]
87. Lambeck, K.; Antonioli, F.; Anzidei, M.; Ferranti, L.; Leoni, G.; Scicchitano, G.; Silenzi, S. Sea level change along the Italian coast during the Holocene and projections for the future. *Quat. Int.* **2011**, *232*, 250–257. [CrossRef]
88. Ishiwa, T.; Yokoyama, Y.; Okuno, J.; Obrochta, S.; Uehara, K.; Ikehara, M.; Miyairi, Y. A sea-level plateau preceding the Marine Isotope Stage 2 minima revealed by Australian sediments. *Sci. Rep.* **2019**, *9*, 6449. [CrossRef]
89. Cerrone, C.; Vacchi, M.; Fontana, A.; Rovere, A. Last Interglacial sea-level proxies in the Western Mediterranean. *Earth Syst. Sci. Data Discuss.* **2021**, *13*, 4485–4527. [CrossRef]
90. Thompson, S.B.; Creveling, J.R. A Global Database of Marine Isotope Stage 5a and 5c Marine Terraces and Paleoshoreline Indicators. *Earth Syst. Sci. Data Discuss.* **2021**, *13*, 3467–3490. [CrossRef]
91. Mitchum, R.M. Seismic stratigraphy. In *Seismic Stratigraphy—Applications to Hydrocarbon Exploration*, 1st ed.; Memoirs; Payton, C.E., Ed.; American Association Petroleum Geologists: Tulsa, OK, USA, 1977; Volume 26, pp. 205–212.
92. Hunt, D.; Tucker, M.E. Stranded parasequences and the forced regressive wedge systems Tract: Deposition during base-level fall. *Sediment. Geol.* **1992**, *81*, 1–9. [CrossRef]
93. Posamentier, H.W.; Allen, G.P. Siliciclastic Sequence Stratigraphy: Concepts and applications. *SEPM Concepts Sedimentol. Paleontol.* **1999**, *7*, 1–210. [CrossRef]
94. Plint, A.G.; Nummedal, D. The Falling Stage Systems Tract: Recognition and importance in sequence stratigraphic analysis. In *Sedimentary Response to Forced Regression*, 1st ed.; Special Publication; Hunt, D., Gawthorpe, R.L., Eds.; Geological Society of London: London, UK, 2000; Volume 172, pp. 1–17. [CrossRef]

Journal of *Marine Science and Engineering*

Article

Depositional Setting and Cementation Pattern of Al-Mejarma Beachrocks, Saudi Arabia: A Proxy for the Late Quaternary Red Sea Coastal Evolution

Ibrahim M. Ghandour [1,2,*], Hamad A. Al-Washmi [1], Athar A. Khan [1], Ammar A. Mannaa [1], Mohammed H. Aljahdali [1] and Brian G. Jones [3]

1 Marine Geology Department, Faculty of Marine Science, King Abdulaziz University, P.O. Box 80200, Jeddah 21589, Saudi Arabia; halwashmi@kau.edu.sa (H.A.A.-W.); geoathar@gmail.com (A.A.K.); amannaa@kau.edu.sa (A.A.M.); maljahdli@kau.edu.sa (M.H.A.)
2 Geology Department, Faculty of Science, Tanta University, Tanta 31527, Egypt
3 School of Earth, Atmospheric and Life Sciences, University of Wollongong, Wollongong, NSW 2522, Australia; briangj@uow.edu.au
* Correspondence: ighandour@kau.edu.sa

Abstract: This study utilizes lithofacies characteristics, petrographic, XRD, and stable isotope data of Al-Mejarma beachrocks, Red Sea, Saudi Arabia, to interpret its depositional setting, origin of cement, and coastal evolution. The beachrock is 1.15 m thick, medium to very coarse-grained sandstone with scattered granules. It shows massive to graded bedding, horizontal, ripple, and shore parallel to slightly oblique planar cross-laminations, with a remarkable absence of bioturbation. It was deposited by shore-parallel longshore currents in a relatively high-energy beach environment. The framework comprises quartz, feldspars, and lithic fragments admixed with biogenic remains of algae, mollusca, foraminifera, corals, and echinoids. They are cemented by high magnesium calcite in the form of isopachous rims and pore-filling blades, and rarely, as a meniscus bridge. The mean values of $\delta^{18}O_{VPDB}$ and $\delta^{13}C_{VPDB}$ are 0.44‰ and 3.65‰, respectively, suggesting a seawater origin for the cement. The framework composition, facies geometry, and association with back-barrier lagoon impose a deposition as a shoreface-beach barrier through two stages corresponding to the middle and late Holocene. The first stage attests landward migrating sediment accumulation and rapid marine cementation. The sediments stored offshore during the early and middle Holocene humid periods migrated landward from offshore and alongshore by onshore waves and longshore drift during the middle and late Holocene sea-level highstand. They were cemented to form beachrock and subsequently emerged as the late Holocene sea-level fell.

Keywords: stable isotopes; coastal sand barrier; Red Sea coastal evolution; late Holocene climate and sea level; beachrock petrography

Citation: Ghandour, I.M.; Al-Washmi, H.A.; Khan, A.A.; Mannaa, A.A.; Aljahdali, M.H.; Jones, B.G. Depositional Setting and Cementation Pattern of Al-Mejarma Beachrocks, Saudi Arabia: A Proxy for the Late Quaternary Red Sea Coastal Evolution. *J. Mar. Sci. Eng.* **2021**, *9*, 1012. https://doi.org/10.3390/jmse9091012

Academic Editor: Gemma Aiello

Received: 16 August 2021
Accepted: 13 September 2021
Published: 15 September 2021

Publisher's Note: MDPI stays neutral with regard to jurisdictional claims in published maps and institutional affiliations.

Copyright: © 2021 by the authors. Licensee MDPI, Basel, Switzerland. This article is an open access article distributed under the terms and conditions of the Creative Commons Attribution (CC BY) license (https://creativecommons.org/licenses/by/4.0/).

1. Introduction

Marine processes play an important role in shaping coastal landforms through the dynamic interplay between geomorphic setting, climate, hydrodynamics, sediment transport, and biogeochemistry [1–5]. The Holocene climate and sea-level changes are the primary drivers of the world's coastal evolution [6]. These changes determine the trends of emergence and submergence, and the formation of several coastal geomorphological features [7–10]. Understanding the past changes in sea level and climate is crucial for effective future long-term and safe coastal development planning [11]. Several proxies including vertical variations in sedimentary facies, upcore fossil and isotopic changes, marine notches, coral reef terraces, and geoarchaeological records have been applied to infer coastal evolution [8,12–14] and sea-level changes [7,10,13,15–17].

Beachrocks are carbonate-cemented sediments of variable composition and grain sizes typically found within the supratidal, intertidal, and upper subtidal zones mostly along tropical and subtropical coastlines [18–24]. They form during sea-level stillstands or minor regression [7,12]. Various biotic and abiotic processes have been put forward to account for beachrock cementation [20,25,26]. These processes include the mixing of marine and meteoric waters [26], direct precipitation of calcium carbonate in pore spaces as a result of evaporation [27], CO_2 degassing of intertidal groundwater during tidal pumping and spraying [28], and direct or indirect activity of organisms [29,30]. The cement style of beachrocks is affected by coastal zone morphodynamics [31]. Meniscus high Mg-calcite cements grow during the beach phase accretion, when the locus of beachrock formation occurs in the undersaturated sediment in the upper intertidal to supratidal zone, whereas acicular cements form during the landward shift of the beachface, as the beachrocks are exposed to saturated conditions [31]. The texture of the beachrock cement can then reveal short-term cycles of beach progradation and retrogradation [31]. However, such changes may not be apparent where beachrock forms under very arid conditions, such as along the Saudi Red Sea coast.

Beachrocks are widely distributed along the Saudi Red Sea and Gulf of Aqaba coasts [32–36]. Despite their considerable extent, the research on beachrock along the Saudi Red Sea coast is still in earlier stages and is underrepresented. A single published study focused on the cementation patterns, cement composition, and microfabrics of Al-Shuaiba beachrocks [33]. This study showed that the cement consists mainly of aragonite and high Mg-calcite in the form of micritic coatings, isopachous to asymmetric aragonite rim, and cryptocrystalline partial pore-fillings. The cement of the beachrocks of Al-Shuaiba grows selectively around carbonate grains and does not grow around siliciclastic grains. Cement nucleation on carbonate grains was easier as they provide good "seed crystals" for carbonate cement growth [33].

Another interesting beachrock occurrence crops out along the eastern shoreline of the back-barrier Al-Mejarma Lagoon on the Saudi Red Sea coast (Figure 1). It displays sedimentary structures and textural attributes that indicate a high-energy environment and are therefore inconsistent with the low-energy lagoonal system. The present study aims to interpret the depositional setting and the origin of cement in beachrock that is exposed along the eastern coast of Al-Mejarma Lagoon, Central Red Sea, Saudi Arabia, based on field observations, petrography, and carbon and oxygen isotopic data.

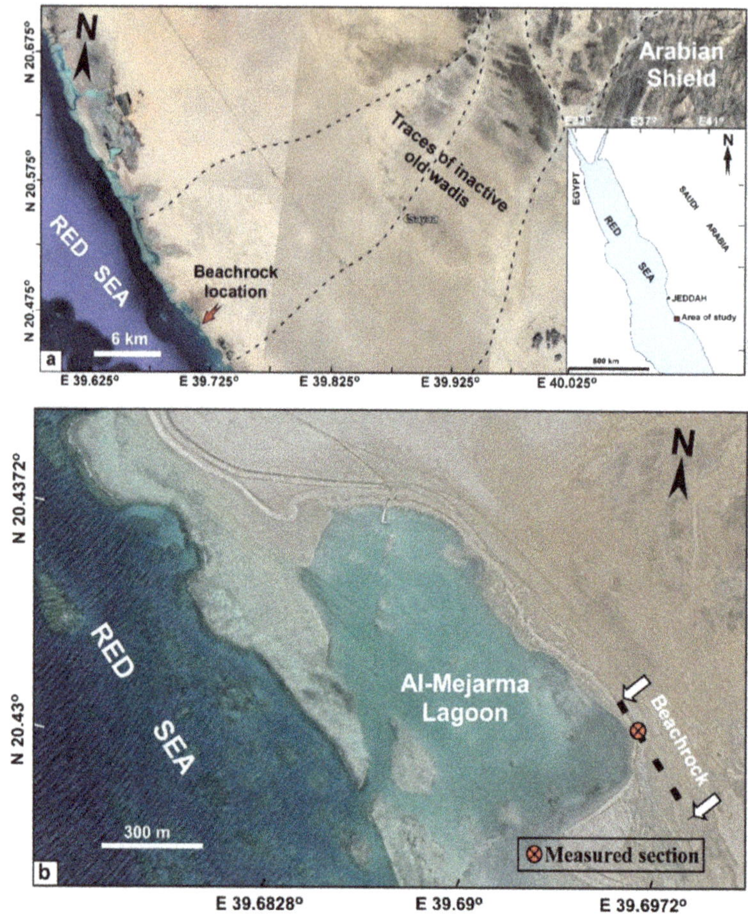

Figure 1. Landsat image showing the location of the area of study: (**a**) Location map of the study area; (**b**) Close up of Al-Mejarma Lagoon showing the extension of the beachrock and the location of the measured section.

2. Holocene Climate and Sea Level in the Red Sea Region

The equatorial regions have experienced a shift in climate and sea level during the Holocene [37,38]. Several unequivocal geological and archaeological archives indicate increased precipitation rates during the early-middle Holocene in the eastern Sahara and Saudi Arabian regions [39–42]. The humid climate was associated with the northward and eastward expansion of the East African Summer Monsoon [42,43]. The humid interval was followed by a progressive increase in aridity during the late Holocene associated with the southward shift of the Intertropical Convergence Zone (ITCZ) [44,45].

The Red Sea region has a generally hot and arid climate throughout the year with low precipitation (<63 mm/y) and high evaporation (2.06 ± 0.22 m/y) rates [46,47]. The Red Sea surface water temperature ranges between 22 and 32 °C [48] and increases from north to the south showing wide variability. In winter, it is 22 °C at the entrance of the Gulf of Suez, about 27 °C at the southern end of the Red Sea and the highest temperature is 28.2 °C recorded at a latitude of N 20°. In contrast, the Red Sea surface water temperature during summer varies from 28 °C at the entrance of the Gulf of Suez to 31 °C at the southern end of the sea, whereas the highest temperature of 32.5 °C is shifted a little to the south at N 16° [49].

The wave characteristics in the Red Sea area are strongly controlled by the persistent winds that increase the longshore transport [50]. The wind systems of the Red Sea area are dominated by the north-westerly (NNW) wind that blows year-round extending to the southern end of the Red Sea [51]. This wind system generates the most energetic waves in the Red Sea basin. These waves propagate to the southern end of the Red Sea, in the form of swell and frequently as wind sea [52]. In winter (October to April), the NNW winds are restricted to the northern region and the most regular Indian monsoonal winds blow much stronger from SSE to the NNW over the southern and central parts of the Red Sea resulting in the development of a convergence zone at N 20°–25° [46,51,53]. This wind system generates waves that propagate northward and attenuate when they approach the Red Sea's northern end [51]. The Red Sea coast is oriented NNW-SSE, roughly parallel to the prevailing winds, a situation likely to favor longshore currents and littoral drift.

Relative sea-level highstands with spatial, temporal, and amplitude variabilities are recorded during the mid-Holocene (between 7.0 and 4.5 ka), followed by sea-level fall to the present-day level [13,54–59]. Following the last post-glacial sea-level rise, sea-level fluctuations during the Holocene have had a profound influence on the Red Sea's coastal evolution. Evidence for the highstand along the Red Sea and the Gulf of Aqaba coasts is well documented by coral terraces, palaeoshoreline notches, and erosional benches [60–62], shallow subsurface coastal sedimentary facies [63–66] and upcore change in benthic foraminiferal distribution [67,68]. Ghandour and Haredy [64] recognized transgressive-regressive successions from the Red Sea coastal plain of Al-Kharrar Lagoon, Saudi Arabia. The lower transgressive part consisting of gravel-rich fluvial sediments that show an upward increase in marine influence and the overlying lagoonal sediments was attributed to the mid-Holocene relative sea-level rise/highstand. These deposits correspond to the mid-Holocene humid interval [68]. The upper regressive part consists of shallow lagoonal and tidal flat deposits showing features of pedogenesis formed during a phase of sea-level fall under arid climate.

3. Area of Study

Al-Mejarma Lagoon (Figure 1b) is a shallow back-barrier coastal lagoon covering an area of 0.88 km^2 with a maximum water depth of 6 m. It is located between latitudes N 20°25′29″ and N 20°26′4.02″, and longitudes E 39°41′23″ and E 39°42′4″. It is connected to the Red Sea through a shallow (7 m deep) and narrow (70 m width) inlet channel [69,70]. The water dynamic is variable, energetic along the channel, and very calm on both the northern and southern parts of the lagoon. The lagoon has a semidiurnal low tidal range of about 0.3 m that is similar to that of the central Red Sea [71]. The area has no source of freshwater except rare intermittent supply during catastrophic flash floods through occasionally active wadis to the north and south of the area of study.

A low-lying intertidal flat borders the lagoon to the south, to the north and north-east, it is surrounded by coastal dunes, and to the east by beachrocks (Figure 1b). The recent bottom sediments of Al-Mejarma Lagoon are dominated by argillaceous sand and sandy mud; however, coarse bioclastic sediments are abundant along the inlet channel [69,70]. In an ongoing research, a short (2.5 m long) core collected from the intertidal flat shows stacking of two facies; lagoonal gray mud containing skeletal remains of bivalves, gastropods and corals, sharply overlain by intertidal flat yellowish-brown medium-grained sand. This facies organization is consistent with the shallow subsurface coastal facies organization at Al-Shuaiba [67], Al-Kharrar [63,64,68], north Al-Wajh [65], and Al Lith [66].

The terrigenous influx into the lagoon was derived from the volcanic and sedimentary rocks of the Al Lith belt [72]. The belt generally consists of basalt, andesite, and dacitic tuff interbedded with and overlain by volcaniclastic and epiclastic rocks, and thin marble and chert beds [73]. Since extreme aridity developed in the late Holocene, most wadis eventually became inactive and the sediment supply to the coast ceased (Figure 1a).

4. Materials and Methods

Beachrocks are best exposed on the eastern coast of the Al-Mejarma Lagoon, Saudi Arabia (Figure 1). Laterally, they are poorly exposed, with only the upper surface exposed while the rest is hidden within intertidal flat deposits. A sedimentary log (Figure 2) was measured, described, and samples were collected from the exposure on the lagoonal coast. The elevation above the mean sea level (MSL) was determined using a Leica Viva GNSS GS15 receiver. Different lithofacies were identified based on texture and sedimentary structures. Seven thin sections were prepared in the Laboratory of Petrology, Geology Department, Cairo University, Egypt. The original samples were cut into 3 × 2 cm chips perpendicular to the bedding plane. The chips were then mounted on a glass slide using Canada balsam, and after drying it, is polished to a thickness of 0.03 mm. The thin sections were investigated by a polarizing microscope to determine framework composition, texture, and cement morphology. The bulk mineralogical composition was determined using X-ray diffraction (XRD) analysis. A gram of each dry sample was powdered using an agate mortar and pestle, and the powdered sediment was packed into a cavity-bearing slide, which was scanned from 2–40° 2θ at speed of 1°/min. To convert X-ray peak angles to percentage of magnesium (in mol% of $MgCO_3$) in the calcite lattice, the chart suggested by Goldsmith et al. [74] was used. Stable isotopic values of $\delta^{18}O$ and $\delta^{13}C$ in the carbonate cement were measured using an automated carbonate preparation device (KIEL-III) coupled to a gas-ratio mass spectrometer (Finnigan MAT 252) at the Environmental Isotope Laboratory, Geosciences Department, University of Arizona. The cement was separated using a micro-drill to avoid contamination. Powdered samples were reacted with dehydrated phosphoric acid under vacuum at 70 °C. The isotope ratio measurement is calibrated based on repeated measurements of international reference materials (NBS-19 and NBS-18) and precision is ±0.10‰ for $\delta^{18}O$ and ±0.08‰ for $\delta^{13}C$ (1 sigma). The isotopic values are reported relative to the Vienna Pee Dee Belemnite (VPDB) international standard [75]. To distinguish the type of porewater from which the carbonate cement is precipitated, the Z value [76] is calculated as follows:

$$Z = a\,(\delta^{13}C + 50) + b\,(\delta^{18}O + 50)$$

where a and b are constant values equal to 2.048 and 0.498, respectively. Z values > 120 indicate a cement of marine water origin, whereas values <120 suggest a cement of freshwater origin. The morphology and microstructures of the cement were examined using scanning electron microscopy (5 kV accelerating voltage, SEM-EDS, JSM-6360 LA, JEOL) at Kafrelsheikh University, Egypt. The Mg mass % of the cement was determined for two samples (HMD3 and HMD7) using EDX.

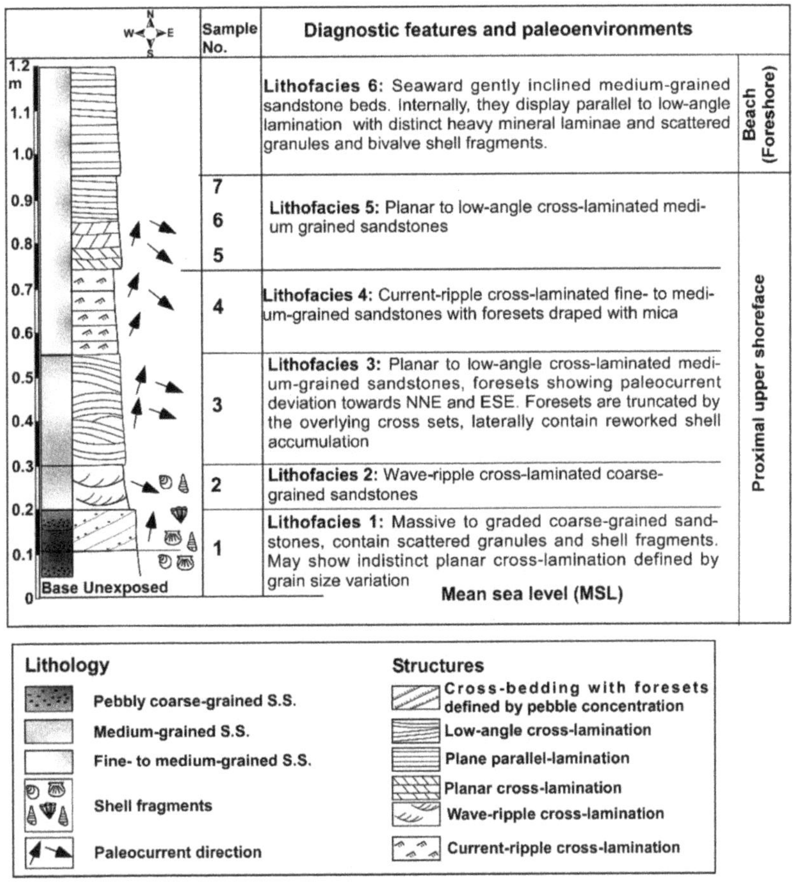

Figure 2. The sedimentary log, location of analyzed samples, and the lithofacies characteristics that were recognized from the Al-Mejarma beachrock (see Figure 1 for location).

5. Results

An interesting beachrock crops out along the eastern shoreline of Al-Mejarma Lagoon displaying relatively high-energy sedimentary structural and textural attributes. It extends for 400 m in a NNW-SSE direction parallel to the Red Sea shoreline and about 7 m inland with a thickness of 1.15 m (Figures 2 and 3a). The composition and texture of the beachrock are texturally and compositionally different from the surrounding lagoonal and tidal flat unconsolidated sediments.

Figure 3. Field photographs showing the characteristic features of the beachrock of the Al-Mejarma: (**a**) An overview of strike and inland extension of Al-Mejarma beachrock. This part shows the seaward inclined foreshore sandstones (lithofacies 6); (**b**) General overview showing breakdown and sliding of beachrocks due to modern wave erosion. The person is 1.7 m tall; (**c**) Coarse-grained pebbly sandstones of lithofacies 1 overlain by wave ripple cross-laminated sandstones of lithofacies 2; (**d**) Low-angle cross-lamination of lithofacies 3; (**e**) Ripple cross-laminated (lithofacies 4) showing foresets draped by mica and foresets dipping toward N and NNE, and low-angle planar cross-laminated medium-grained sandstones of lithofacies 5 with foresets trending SSE, and (**f**) Low-angle tabular cross-laminated medium-grained sandstones of lithofacies 5. Note the current directions towards SSE and NNE. The scale is hammer = 0.33 m, pen in (**d**–**f**) = 0.14.

5.1. Morphology and Lithofacies

The beachrock occurs from 1.2 m above the mean sea level (MSL) in the form of tabular horizontal to slightly inclined cemented horizons. The cemented beachrock horizons consist of massive, planar, and ripple cross-laminated sand with foresets trending parallel to slightly oblique to the Red Sea coastline (Figure 3a,b). Textural characteristics and sedimentary structures were used to subdivide the studied beachrocks into six vertically stacked lithofacies (Figure 2). Lithofacies 1 is defined from the base of the beachrock successions. It is about 0.15 m thick and consists of sharp-based massive to normal graded, poorly sorted, coarse- to very coarse-grained sandstone containing dispersed granules of lithic fragments and skeletal remains (Figure 3c). The basal part is substantially coarser. Laterally, it shows indistinct planar cross-laminations defined by variations in grain size. Lithofacies 2 consists of a 0.1 m thick sharp-based, wave ripple cross-laminated, moderately sorted, coarse-grained sandstone (Figure 3c). Lithofacies 3 is about 0.25 m thick, planar-laminated to low angle cross-laminated, moderately sorted, medium- to coarse-grained sandstone (Figure 3d).

Lithofacies 4 consists of a 0.2 m thick, moderately well-sorted, medium-grained sandstone, displaying current ripple cross-laminations with foresets dipping to N and NNE defined by mica concentrations (Figure 3e). Lithofacies 5 consists of a 0.2 m thick, moderately well-sorted, medium-grained sandstone displaying small scale planar tabular to low-angle cross-stratification with foresets showing paleo flows towards NNE and SSE (Figure 3e,f). Lithofacies 6 occupies the upper part of the beachrock exposure showing gently inclined geometry dipping about 9° towards the west (seaward). It consists of a 0.25 m thick, plane to low-angle parallel-laminated, moderately well-sorted, medium- to coarse-grained, occasionally pebbly sandstone with foresets defined by heavy mineral laminae (Figure 3a).

5.2. Framework Composition and Cement

The framework composition includes mixed siliciclastic and biogenic calcareous grains of variable sizes ranging from silt to granule size (Figure 4). Siliciclastic grains are dominated by quartz: mono- and polycrystalline, silt to gravel size, angular to well-rounded grains. Feldspars are the second most abundant siliciclastic grains including both K-feldspars and plagioclase (Figure 4a,b). In addition, relatively rare biotite, amphiboles, heavy minerals, and basaltic and chert lithic fragments are recorded. The skeletal remains are dominated by coralline algae and benthic foraminifera (Figure 4a,b). Remains of bivalves, gastropods, echinoids, corals, and ostracods occur with relatively low abundance. The $CaCO_3$ content of the bulk samples (Table 1) varies between 31 and 68% (mean 48%). The XRD analysis shows that the bulk mineral composition is dominated by quartz, plagioclase, K-feldspars and high Mg-calcite (mol% of $MgCO_3$ varies between 13 and 16%), and traces of amphiboles. EDX analysis of the cement shows that the Mg mass % in samples HMD3 and HMD7 are 3.55 and 0.72%, respectively. High Mg-calcite (HMC) cement occurs dominantly as pore-filling and isopachous rims around grains (Figure 4c,d), and rarely as microcrystalline meniscus bridges linking grains (Figure 4b). It occurs as subhedral to euhedral crystals, which may have a blade-shape and scalenohedral forms (Figure 5a–d). The scalenohedral crystals in the upper part have a morphology that exhibits dissolution features such as corroded or cracked surfaces (Figure 5e,f).

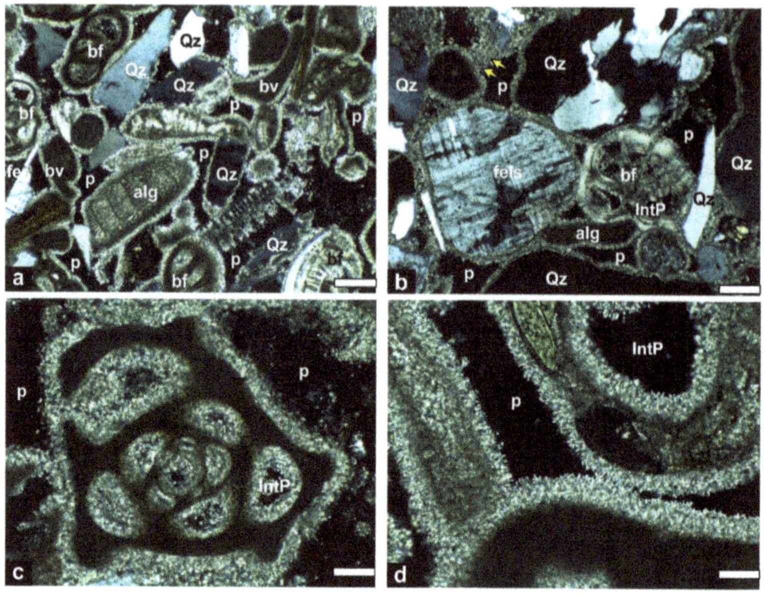

Figure 4. Photomicrographs showing the framework composition and the cement in the beachrock of Al-Mejarma: (**a**,**b**) Varieties of siliciclastic grains dominated by mono- and polycrystalline quartz (Qz) and feldspar (fels) mixed with skeletal remains dominated by benthic foraminifera (bf), coralline red algae (alg), bivalves (bv), and intergranular pores (p); the grains are cemented by HMC mainly in the form of isopachous rims and rarely as microcrystalline meniscus bridge between grains (yellow arrow); (**c**,**d**) HMC cement grows as isopachous fringes with elongated bladed crystals regularly surrounding skeletal remains and filling the intragranular pore (IntP) in the foraminifera test. (The scale bar in (**a**,**b**) = 200 μm and in (**c**,**d**) = 40 μm).

Table 1. The $CaCO_3$ content of the bulk beachrock samples and stable isotope, and Z values of carbonate cement in the Al-Mejarma beachrock. The isotopic values are normalized to Vienna Pee Dee Belemnite (VPDB).

Sample	$CaCO_3$ %	$\delta^{13}C_{VPDB}$	$\delta^{13}O_{VPDB}$	Z
7	68	3.64	0.54	135.0
6	31	3.75	0.31	135.1
5	54	3.78	0.10	135.1
4	52	3.88	0.68	135.6
3	51	3.48	0.37	134.6
2	48	3.32	0.33	134.3
1	36	3.65	0.73	135.1
Min	31	3.32	0.10	134.26
Max	68	3.88	0.73	135.58
Mean	48	3.65	0.44	134.98

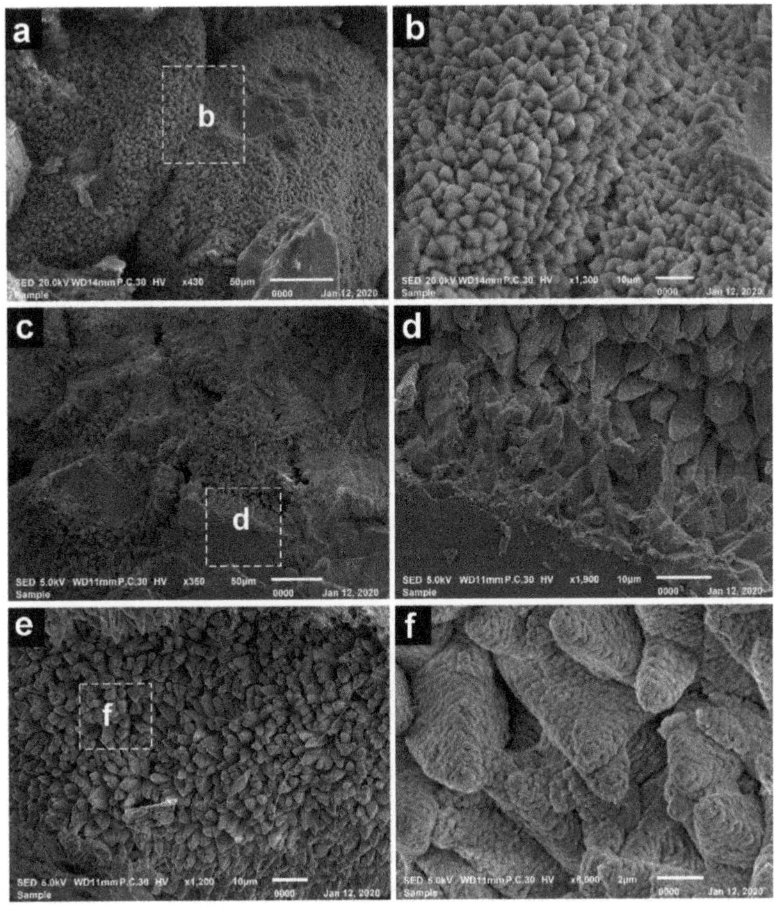

Figure 5. SEM photomicrographs showing the characteristic features of the HMC cement in the beachrock of Al-Mejarma: (**a**,**b**) Isopachous rim with scalenohedral HMC crystals; (**c**–**f**) HMC cement with bladed morphology, with crystal showing dissolution features of rounded corroded crystal outlines in HMD7 (**e**,**f**).

5.3. Oxygen and Carbon Isotopic Data

Stable isotope ($\delta^{13}C$ and $\delta^{18}O$) data for the seven studied beachrock samples are shown in Table 1. Generally, the $\delta^{13}C_{VPDB}$ values of calcite cement vary between 3.32‰ and 3.88‰ (mean 3.65‰), and the $\delta^{18}O_{VPDB}$ values range from 0.10‰ to 0.73‰ (mean 0.44‰). The calculated Z values for the Al-Mejarma beachrocks range from 134.3 to 135.6 (Table 1) indicating a marine origin for the cement.

6. Discussion

The calcareous sandbody bordering the eastern shoreline of Al-Mejarma Lagoon displays cross-lamination, a medium to coarse grain size, with granules and shell fragments indicating a relatively high-energy marginal marine depositional setting [77], which contrast with the adjacent low-energy lagoonal environment. It suggests that the geometry of the paleo-coastline at the time of the sandbody formation was different compared to the current situation. As discussed in the following sections, the elongated, coast-parallel geometry, texture, framework composition, and the bladed and cryptocrystalline high Mg-calcite cements of marine origin, all point to a beachrock origin [19,20].

6.1. Depositional Setting

The shoreline aligned sandbody, the lithofacies characteristics, the absence of fine-grained sediments, and the presence of reworked shell fragments suggest deposition either as prograding upper shoreface-beach deposits of a strandplain or a coastal barrier spit/island [2,4–6,78–80]. Coastal barriers and strandplains are prominent depositional features supplied and molded almost entirely by marine processes. They consist of sandbodies of similar origin and facies characteristics. They are highly dynamic and show significant facies variability even within minimal thicknesses [1,80]. Contrary to strandplains, coastal barriers are elongate sandy islands, of different width, parallel to the shore and separated from the mainland by lagoons [81]. The association of the present beachrock with a back-barrier lagoon (Al-Zubeiri, pers. comm. 2021) excludes a strandplain origin [82]. Beachrocks on barrier island/spits are not recorded from the Red Sea, whereas beachrocks associated with a strandplain are recognized from Al-Shuaiba [33].

The formation and evolution of coastal barriers/spits are largely controlled by availability of mobile sediments, hydraulic (wave and tide) energy, sea-level (SL) changes and a low-gradient profile to provide space for a back-barrier lagoon to form [4,5,78,80,83–85].

6.2. Source and Availability of Sediments

The composition and the availability of sediments for building coastal barriers depend largely on the regional geology and climate. Modern-day siliciclastic barriers are recognized from humid regions covering about 12 to 13% of the world's coasts [6,78,80], whereas carbonate barriers are developed under arid to semi-arid climate comprising only 1% [4]. Barriers develop when the supply of bulk sediments from alongshore or offshore sources are plentiful. The source of sediments needed for building barrier island/spits vary widely. The sediments of some barriers, such as Provincelands Spit in northern Cape Cod, Massachusetts, and Lawrencetown barrier along the northeast shore of Nova Scotia, may be derived from eroding headlands [86,87]. The late-Holocene carbonate spits at Al Dakhirah, Qatar, Arabian Gulf, were derived from the erosion of shore-parallel reefal materials seaward and from the downdrift 'headland' [4]. The sediments for barrier systems may also be derived from the inner continental shelf or derived directly from rivers. For example, Sloss et al. [88] suggested that the sediments which built the barrier estuary system in the Lake Illawarra, New South Wales, Australia, were derived from sediments stored on the continental shelf during the last glacial sea-level fall. In addition, an inner shelf source of reworked cool-water biogenic sediments was recognized for barrier construction in South Australia [89].

The framework composition of the Al-Mejarma beachrock consists of a mixture of calcareous skeletal remains, representing communities living in the inner shelf, and land-derived siliciclastic terrigenous sediments that were discharged offshore from the surrounding catchment. In the area of study, there is no preserved headland on the downdrift side and siliciclastic detritus was derived from the Al-Lith volcanic belt in the hinterland to the east. Thus, the beachrock composition reflects a period of sediment availability, coastal zone morphology, lagoonal extension, and hydrodynamic forcings different from the current situation.

The modern nearshore area receives very little or negligible terrigenous sediments because of the extremely arid climate and the very low precipitation rates and runoff, which are unable to activate wadis [90]. Therefore, the coarse siliciclastic sediments in the beachrocks must have been discharged into the coastal area during a period of relatively higher precipitation rates and greater runoff compared to the present-day climatic conditions. This humid interval may be correlated with the early-middle Holocene period [39,90]. The mid-Holocene humid climate with an annual monsoonal precipitation rate of 400 mm/y was recorded by geological and geoarchaeological features in Arabia and North Africa [40,61]. During this humid interval, floods may have been more frequent than they are today, and were perhaps capable of carrying a huge amount of siliciclastic sediment to be discharged into the offshore. The sediments that were delivered are medium- to very coarse-grained, occasionally pebbly, and mineralogically and texturally submature. Therefore, the heterogeneous composition of the beachrock suggests that the terrigenous siliciclastic sediments spent significant time on the inner shelf and mixed with autochthonous calcareous biogenic sediments. These sediments were later transported shoreward by energetic flows from offshore and by onshore and longshore drifts, and they migrated landward forming a shoreface-beach barrier bordering the Al-Mejarma Paleolagoon during the middle and late Holocene sea-level highstand and fall.

6.3. Sediment Transportation and Deposition

Onshore and alongshore sediment transport that is necessary for the formation of barrier systems is significantly controlled by hydrodynamic wave energy and wave-induced currents [91]. Although rare, barrier systems are recognized from hypertidal coasts [91], whereas they typically characterize microtidal coasts with wave heights less than 2 m [92]. The Red Sea coast is suitable for the formation of coastal barriers because of its limited tidal range and average wave height (<2 m). The sediments of the Al-Mejarma beachrock lack sedimentological size alternation reflecting stability in sediment supply and almost continuous sediment transport. In such an arid region, the study area lacks perennial streams and, therefore, contemporaneous supply by longshore drift from deltas at the mouths of wadis is also excluded. Instead, the stored offshore sediments were moved onshore from the inner shelf during long-term accretionary wave conditions. In contrast, the formation of coastal barriers in humid regions suggest a continuous longshore supply of river mouth coarse siliciclastic sediments [78,80]. In this setting, the onshore movement of sediments that come from the inner shelf, is much less efficient than alongshore movement because normal fair-weather wave action is thought to preferentially move sediment onshore from relatively shallow depths (i.e., the upper shoreface [89,93,94]).

Sedimentary structures and bedding style in the Al-Mejarma beachrock indicate deposition in an upper shoreface to beach environment with the role of wind-generated waves and currents as driving forces for sediment transport and deposition. Graded beds with sharp erosive base together with wave rippled, horizontal to slightly inclined planar laminated coarse-grained sandstones (lithofacies 1–3) strongly indicate high-energy storm events. Cross-lamination with foresets trending parallel to or slightly oblique to the Red Sea shoreline (lithofacies 4 and 5) were formed by the migration of straight-crested megaripples in the surf (upper shoreface) zone by longshore flow [95]. The gently seaward inclined strata, planar, and low-angle stratifications with foresets defined by heavy minerals suggest deposition of lithofacies 6 by wave swash and backwash in the foreshore [77,96].

The major depositional processes operating on barrier systems are wind-driven waves [97]. When waves approach shorelines obliquely, they generate longshore currents, which transport sediments parallel to the shoreline, and form spits and barrier islands [78,97]. As the wind systems change seasonally, the direction of wave approach and resulting longshore current systems also change seasonally [97,98]. Measured paleo flow trends in the Al-Mejarma beachrock are largely towards NNE and SSE parallel to slightly oblique the Red Sea shoreline. This trend reflects the direction of the longshore flow following the direction of wind stresses. The prevailing winds from the NNW in summer would have generated appreciable and significant SSE longshore drift that was capable of transporting sediments towards SSE. The wind direction is reversed in winter and the southern and central Red Sea regions up to latitudes of N 20–25° are strongly influenced by the strong Indian monsoon-related SSE winds [53]. As a result of the southerly monsoon-related winds, there would have also been a strong net sediment transport northward producing NNE-trending cross-beds.

6.4. Sea Level and Climate Controls

The Holocene sea level and climate changes played a significant role in the formation of the barrier system and the coastal evolution at Al-Mejarma. Sea-level changes regulated sediment dispersion, transportation, and shoreline trajectory, whereas climate controlled sedimentation rate. The Al-Mejarma area was subaerially exposed during the Last Glacial Maximum (LGM), and was subsequently inundated by the post glacial rapid sea-level rise as the sea level rose to its present level or slightly higher during middle Holocene [56,61,99]. The humid mid-Holocene period coincided with a relative sea-level highstand as reported in many localities in the Red Sea area [61,64,90]. The sea level was 0.5–1.0 m higher than its present level [61,99], which accounts for the raised elevation of the Al-Mejarma beachrock. Hein et al. [61] reported marine notches and terraces as well as coral reef terraces of the mid-Holocene age along the Egyptian Red Sea coast suggesting a highstand of 1.1 to 1.8 m above mean sea level. Huge amounts of siliciclastic sediments of different grain sizes would have been transported to the inner shelf through wadis during this early to middle Holocene humid period.

Durgaprasada Rao and Behairy [90] attributed the abrupt cessation of coarse siliciclastic input into the Red Sea coast and subaerial exposure of reef flats to the change of climate from humid to arid and a drop of sea level by about one meter. Prograding barrier systems form in a regime of abundant sand supply during a period of sea-level highstand and fall [79,88,100,101]. The late Holocene climate became extremely arid and the wadis turned inactive. The sediments that were initially dispersed in the proximal continental shelf were transferred toward the shoreline by wave action during the transgressive phase, in what was termed a cannibalistic process [85,102].

The framework composition, lithofacies characteristics, and cementation suggest that the formation of Al-Mejarma beachrock has been accomplished through two evolutionary stages (Figure 6a–c) in response to the interplay of the Holocene climate and sea-level changes. During the middle Holocene sea-level highstand and late Holocene sea-level fall, continued onshore migration of stored inner shelf sediments forced coastal progradation, provided a nucleation site for the paleo-barrier system, and caused seaward shoreline movement (Figure 6a,b). Al-Mejarma Paleolagoon formed and developed simultaneously with or after barrier formation. Continuous coastal progradation and sea-level fall during the late Holocene led to a major rearrangement of the depositional system, a gradual reduction and abandonment of the back-barrier lagoon leaving a lowland area that is filled with the present-day tidal flat deposits, while a new barrier and lagoon developed seaward of the old barrier coast and beachrock (Figure 6c).

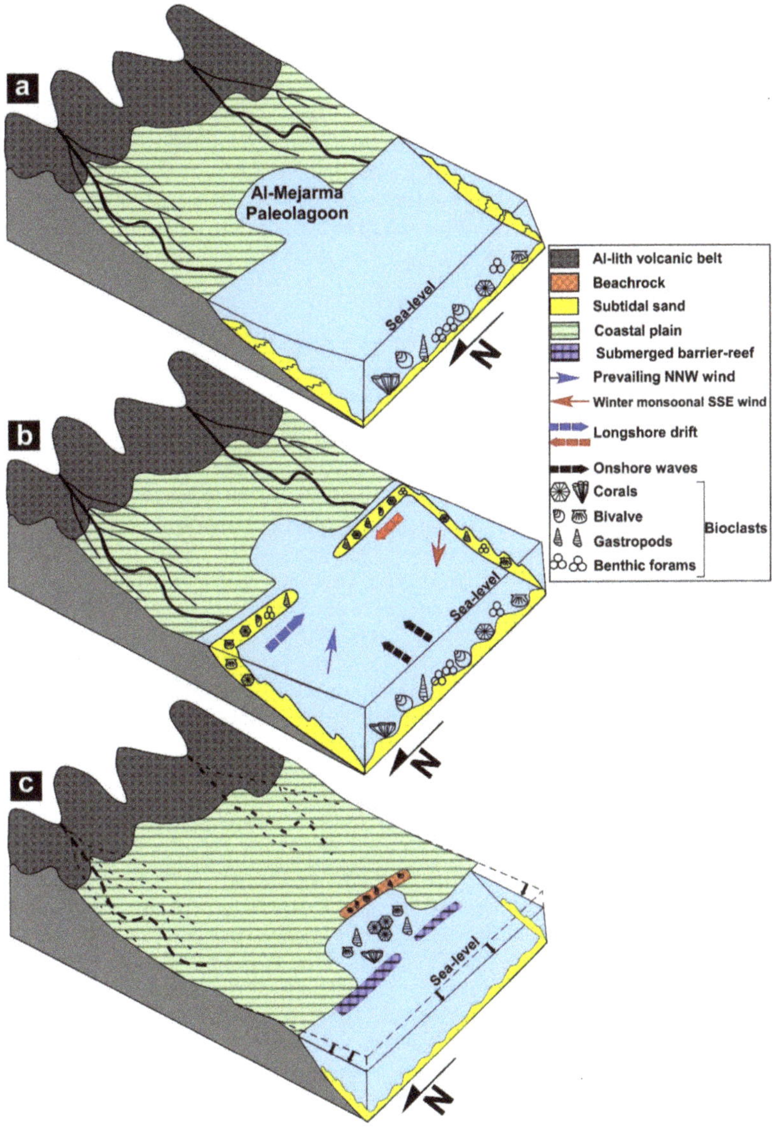

Figure 6. Al-Mejarma coastal evolution and depositional model of beachrock formation: (**a**) A landward extension of Al-Mejarma Lagoon following the post-glacial sea-level rise, where lowland coastal areas were flooded; (**b**) A huge amount of siliciclastic terrigenous sediments were injected through wadis to the north and south into the Red Sea coastal area and offshore. The sediments were prograded landward by the action of onshore waves and longshore drift building a shoreface-beach barrier during mid-Holocene sea-level highstand; (**c**) Finally, the shoreface-barrier sediments were emerged, cemented, and the lagoon shrunk to it its current position following the late Holocene relative sea-level fall and intensive arid climate.

6.5. Origin of Cement

Rapid cementation processes enhance the preservation of beachrock [103]. High-energy wave and current conditions and the relatively coarse, porous, and permeable sediments facilitate the free circulation of seawater allowing supersaturated marine water to move through the pore system, and helped to precipitate the cement [104].

Petrographic and SEM examinations, XRD data, and the isotopic values of the cement in the Al-Mejarma beachrock suggest that the regular distribution of isopachous HMC cement showing prismatic fringes and small equant crystals suggest precipitation exclusively in a marine phreatic environment [105–107]. In contrast to more humid coastal beachrock [31], there is no evidence at Al-Mejarma for cementation within the meteoric environment [105,106,108]. Magnesium calcite cement favors precipitation from seawater, and lithification either occurs at depth after percolation or seawater evaporation [109]. Similar isopachous rims of HMC blade cement have been documented from the beachrocks of the Arabian Gulf [110]. The rare occurrence of meniscus cement suggests a subtidal environment [111].

Carbon and oxygen stable isotopes, in combination with petrographic characteristics, have been widely utilized to trace the source of carbonate cement and to interpret the diagenetic environment and processes of beachrock formation [15,18,107,112–114]. The $\delta^{13}C$ values from the beachrock cement are consistent with the values of marine carbonates [115]. The water from which the cement was derived has relatively homogeneous temperature and uniform composition as shown by observed isotopic homogeneity of the studied samples [18,108]. The $\delta^{18}O$ isotopes of carbonates normally reflect the temperature, salinity, and isotopic composition of the fluid from which they precipitate, thus freshwater carbonates possess negative $\delta^{18}O$ values whilst marine cement shows positive values [104]. The positive values correlate well with cement precipitated under the seawater environment [104]. A low average value of $\delta^{18}O$ (0.5‰), similar to the cement data obtained in the present study, was reported by Holail and Rashed [18] in the Mediterranean and the Red Sea. The Z values for the Al-Mejarma beachrocks range from 134.3 to 135.6, suggesting that the cement was generated predominantly from seawater [76]. This is similar to the average Z value of 135 from the Holocene beachrock in northeastern Brazil that also indicates dominantly marine cementation [23].

The mild modifications of the initial cement fabrics by etching and dissolution features observed in the upper part of the beachrock sequence (Figure 5e–f) are possibly related to rare rainstorms or desert dew. The beachrocks were not exposed to pervasive meteoric diagenesis. However, calcite cements showing signs of dissolution and depletion of their Mg content in the upper part of the beachrock profile have been attributed to the effect of infiltration of meteoric water and subaerial exposure [12,16,17,24].

The Al-Mejarma coastal evolution is in agreement with previous barrier island systems [88]. The results from this study add greater detail to the development of a beach barrier system in arid regions. Though the shortcoming of the present study is the lack of precise dating of the beachrocks, their formation is probably associated with the mid to late Holocene sea-level and climate changes. The initiation of many modern coastal barrier-lagoon systems worldwide is linked to the deceleration of sea-level rise and initial fall between 7000 and 4000 BP [116]. The cementation of Al-Mejarma beachrock may be correlated with the nearby Al-Shuaiba beachrocks (about 40 km to the north of Al-Mejarma). The beachrocks of Al-Shuaiba also occur at an elevation of 1.2 m above mean sea level and dated back as 3160 ± 110 years BP [117].

7. Conclusions

The Al-Mejarma beachrock geometry, orientation, and lithofacies characteristics pertain to an upper shoreface to foreshore barrier system formed by onshore and longshore migration of sand. The framework composition includes mixed siliciclastic and calcareous biogenic grains of variable grain sizes, ranging from medium sand to granule, cemented by high Mg-calcite. The cement occurs mainly as an isopachous rim around grains, pore-filling with scalenohedral and blade-shaped crystals, and rarely forms a meniscus bridge between grains. The cement developed mainly under marine phreatic conditions and the uniform $\delta^{18}O$ isotopic and $\delta^{13}C$ values of the cement suggest that the cement was derived mainly from seawaters of constant composition and temperature. The scalenohedral crystals in the upper part of the beachrock exhibit dissolution features possibly related to the rare rainstorms and desert dew.

The coastal evolution and the formation of the Al-Mejarma beachrock reflect the effect of the Holocene sea-level and climate changes. During the early-middle Holocene humid interval, huge amounts of siliciclastic sediments were transported through wadis from the Arabian Shield and stored offshore. These siliciclastic sediments along with shallow marine calcareous skeletal remains migrated shoreward to the east, SSE, and NNE by wind-generated onshore waves and longshore drift to form a barrier island/spit in the Al-Mejarma area during the middle and late Holocene sea-level highstand and fall. The beachrock sediments were preserved by rapid marine cementation. The late Holocene sea-level fall, and a progressive increase in aridity, led to the emergence of the shoreface-beach barrier. In addition, the Al-Mejarma Paleolagoon was gradually infilled, becoming shallower and closed, while a new lagoon formed seaward of the beachrock.

Author Contributions: Conceptualization, I.M.G., H.A.A.-W. and A.A.K.; methodology, I.M.G., A.A.M. and M.H.A.; software, I.M.G.; validation, I.M.G., H.A.A.-W. and A.A.K.; formal analysis, I.M.G.; investigation, I.M.G., A.A.M., M.H.A.; resources, I.M.G.; data curation, I.M.G. and B.G.J.; writing—original draft preparation, I.M.G., H.A.A.-W., A.A.K.; writing—review and editing, I.M.G. and B.G.J.; supervision, H.A.A.-W.; project administration, I.M.G.; funding acquisition, I.M.G. All authors have read and agreed to the published version of the manuscript.

Funding: This project was funded by the Deanship of Scientific Research (DSR) at King Abdulaziz University, Jeddah, grant number G-312-150-37.

Institutional Review Board Statement: Not applicable.

Informed Consent Statement: Not applicable.

Data Availability Statement: The study did not report any data.

Acknowledgments: This project was funded by the Deanship of Scientific Research (DSR) at King Abdulaziz University, Jeddah, under grant number G-312-150-37. The authors thank the DSR for this technical and financial support. The authors are grateful to Maurice Tucker (Univ. of Bristol) for the fruitful discussion about the cement and depositional setting. We extend our thanks to Essam Aboud, Aaid Al-Zubeiri, and Bandar Al-Zahrani (King Abdulziz University) for their assistance in field work. We are very grateful to the editor and the reviewers for their constructive comments and editorial handling.

Conflicts of Interest: The authors declare no conflict of interest.

References

1. Anthony, E.J. Wave influence in the construction, shaping and destruction of river deltas: A review. *Mar. Geol.* **2015**, *361*, 53–78. [CrossRef]
2. Forde, T.C.; Nedimović, M.R.; Gibling, M.R.; Forbes, D.L. Coastal evolution over the past 3000 years at Conrads Beach, Nova Scotia: The influence of local sediment supply on a paraglacial transgressive system. *Estuaries Coasts* **2016**, *39*, 363–384. [CrossRef]
3. Falkenroth, M.; Adolphs, S.; Cahnbley, M.; Bagci, H.; Kázmér, M.; Mechernich, S.; Hoffmann, G. Biological indicators reveal small-scale sea-level variability during MIS 5e (Sur, Sultanate of Oman). *Open Quat.* **2020**, *6*, 1–20. [CrossRef]
4. Rivers, J.; Engel, M.; Dalrymple, R.; Yousif, R.; Strohmenger, C.J.; Al-Shaikh, I. Are carbonate barrier islands mobile? Insights from a mid to late-Holocene system, Al Ruwais, northern Qatar. *Sedimentology* **2020**, *67*, 534–558. [CrossRef]
5. Mulhern, J.S.; Johnson, C.L.; Green, A.N. When is a barrier island not an island? When it is preserved in the rock record. *Front. Earth Sci.* **2021**, *8*, 560437. [CrossRef]
6. Hein, C.J.; Fitzgerald, D.M.; Cleary, W.J.; Albernaz, M.B.; de Menezes, J.T.; Klein, A.H.D.F. Evidence for a transgressive barrier within a regressive strandplain system: Implications for complex coastal response to environmental change. *Sedimentology* **2013**, *60*, 469–502. [CrossRef]
7. Cooper, J.A. Beachrock formation in low latitudes: Implications for coastal evolutionary models. *Mar. Geol.* **1991**, *98*, 145–154. [CrossRef]
8. Dickinson, W.R. Paleoshoreline record of relative Holocene sea levels on Pacific islands. *Earth-Sci. Rev.* **2001**, *55*, 191–234. [CrossRef]

9. Harvey, N. Holocene coastal evolution: Barriers, beach ridges, and tidal flats of South Australia. *J. Coast. Res.* **2006**, *22*, 90–99. [CrossRef]
10. Kelly, C.S.; Green, A.N.; Cooper, J.A.G.; Wiles, E. Beachrock facies variability and sea level implications: Preliminary study. *J. Coast. Res.* **2014**, *70*, 736–742. [CrossRef]
11. Duong, N.; Lieu, N.T.; Cúc, N.T.; Saito, Y.; Hương, N.T.; Phượng, N.T.; Thuy, A.H. Holocene paleoshoreline changes of the Red River Delta, Vietnam. *Rev. Palaeobot. Palynol.* **2020**, *278*, 104235. [CrossRef]
12. Erginal, A.E. Beachrock as evidence of sea-level lowstand during the Clasical period, Parion antique city, Marmara Sea, Turkey. *Geodin. Acta* **2012**, *25*, 96–103. [CrossRef]
13. Stattegger, K.; Tjallingii, R.; Saito, Y.; Michelli, M.; Thanh, N.T.; Wetzel, A. Mid to late Holocene sea level reconstruction of southeast Vietnam using beachrock and beach-ridge deposits. *Glob. Planet. Chang.* **2013**, *110*, 214–222. [CrossRef]
14. Öztürk, M.Z.; Erginal, A.E.; Kiyak, N.G.; Öztürk, T. Cement fabrics and optical luminescence ages of beachrock, North Cyprus: Implications for Holocene sea-level changes. *Quat. Int.* **2016**, *401*, 132–140. [CrossRef]
15. Spurgeon, D.; Davis, J.R.A.; Shinn, E.A. Formation of 'beachrock' at Siesta Key, Florida, and its influence on barrier island development. *Mar. Geol.* **2003**, *200*, 19–29. [CrossRef]
16. Erginal, A.E.; Ekinci, Y.L.; Demirci, A.; Bozcu, M.; Ozturk, M.Z.; Avcioglu, M.; Oztura, M.Z. First record of beachrock on Black Sea coast of Turkey: Implications for late Holocene sea-level fluctuations. *Sediment. Geol.* **2013**, *294*, 294–302. [CrossRef]
17. Mauz, B.; Vacchi, M.; Green, A.; Hoffmann, G.; Cooper, A. Beachrock: A tool for reconstructing relative sea level in the far-field. *Mar. Geol.* **2015**, *362*, 1–16. [CrossRef]
18. Holail, H.; Rashed, M. Stable isotopic composition of carbonate cemented recent beachrock along the Mediterranean and the Red Sea coasts of Egypt. *Mar. Geol.* **1992**, *106*, 141–148. [CrossRef]
19. Kelletat, D. Beachrock as sea level indicator? Remarks from a geomorphological point of view. *J. Coast. Res.* **2006**, *22*, 1555–1564. [CrossRef]
20. Vousdoukas, M.I.; Velegrakis, A.F.; Plomaritis, T.A. Beachrock occurrence, characteristics, formation mechanisms and impacts. *Earth-Sci. Rev.* **2007**, *85*, 23–46. [CrossRef]
21. Erginal, A.E.; Kıyak, N.G.; Öztürk, B. Investigation of beachrock using microanalyses and OSL dating: A case study from Bozcaada Island, Turkey. *J. Coast. Res.* **2010**, *26*, 350–358. [CrossRef]
22. Danjo, T.; Kawasaki, S. Formation mechanisms of beachrocks in Okinawa and Ishikawa, Japan, with a focus on cements. *Mater. Trans.* **2014**, *55*, 493–500. [CrossRef]
23. Júnior, A.V.; Paes, B.C.; Vieira, M.M.; Sial, A.N.; Neumann, V.H. Diagenesis of Holocene beachrock in northeastern Brazil: Petrology, isotopic evidence and age. *Quat. Environ. Geosci.* **2018**, *2*, 26–35.
24. Nayanthara, P.G.N.; Dassanayake, A.B.N.; Nakashima, K.; Kawasaki, S. Distribution and cementation characteristics of beachrocks along southern, southwestern and western coast of Sri Lanka. *J. Sediment. Environ.* **2021**, *6*, 93–106. [CrossRef]
25. Bathurst, R.G.C. *Carbonate Sediments and Their Diagenesis. Developments in Sedimentology 12*; Elsevier: Amsterdam, The Netherlands, 1975; p. 658.
26. Milliman, J.D. *Marine Carbonates. Part 1, Recent Sedimentary Carbonates*; Springer: New York, NY, USA, 1974; p. 375.
27. Moore, C.H.; Billings, G.K. Preliminary model of beachrock cementation, Grand Cayman Island, B.W.I. In *Carbonate Cements*; Bricker, O.P., Ed.; Johns Hopkins University Press: Baltimore, MD, USA, 1971; pp. 41–43.
28. Hanor, J.S. Precipitation of beachrock cements: Mixing of marine and meteoric waters vs. CO_2-degassing. *J. Sediment. Petrol.* **1978**, *48*, 489–501.
29. Neumeier, U. Experimental modelling of beachrock cementation under microbial influence. *Sediment. Geol.* **1999**, *126*, 35–46. [CrossRef]
30. Webb, G.E.; Jell, J.S.; Baker, J.C. Cryptic intertidal microbialites in beachrock, Heron Island, Great Barrier Reef: Implications for the origin of microcrystalline beachrock cement. *Sediment. Geol.* **1999**, *126*, 317–334. [CrossRef]
31. Wiles, E.; Green, A.N.; Cooper, J.A.G. Rapid beachrock cementation on a South African beach: Linking morphodynamics and cement style. *Sediment. Geol.* **2018**, *378*, 13–18. [CrossRef]
32. Al-Ramadan, K. Diagenesis of Holocene beachrocks: A comparative study between the Arabian Gulf and the Gulf of Aqaba, Saudi Arabia. *Arab. J. Geosci.* **2014**, *7*, 4933–4942. [CrossRef]
33. Ghandour, I.M.; Al-Washmi, H.A.; Bantan, R.A.; Gadallah, M.M. Petrographical and petrophysical characteristics of asynchronous beachrocks along Al-Shoaiba Coast, Red Sea, Saudi Arabia. *Arab. J. Geosci.* **2014**, *7*, 355–365. [CrossRef]
34. Koeshidayatullah, A.; Al-Ramadan, K. Unraveling cementation environment and patterns of Holocene beachrocks in the Arabian Gulf and the Gulf of Aqaba: Stable isotope approach. *Geol. Q.* **2014**, *58*, 207–216. [CrossRef]
35. Haredy, R.A.; Ghandour, I.M.; Erginal, A.E.; Bozcu, M. Beachrock cementation patterns along the Gulf of Aqaba coast, Saudi Arabia. *Arab. J. Sci. Eng.* **2019**, *44*, 479–487. [CrossRef]
36. Mannaa, A.A.; Haredy, R.A.; Ghandour, I.M. Beachrock as a Paleoshoreline Indicator: Example from Wadi Al-Hamd, South Al-Wajh, Saudi Arabia. *J. Mar. Sci. Eng.* **2021**, *9*, 984. [CrossRef]
37. DeMenocal, P.; Ortiz, J.; Guilderson, T.; Sarnthein, M. Abrupt onset and termination of the African humid period: Rapid climate responses to gradual insolation forcing. *Quat. Sci. Rev.* **2000**, *19*, 347–361. [CrossRef]
38. Mitrovica, J.X.; Milne, G.A. On the origin of late Holocene sea level highstands within equatorial ocean basins. *Quat. Sci. Rev.* **2002**, *21*, 2179–2190. [CrossRef]

39. Jado, A.R.; Zötl, J.G. *Quaternary Period in Saudi Arabia*; Springer: Berlin/Heidelberg, Germany, 1984; Volume 2, p. 360.
40. Ritchie, J.C.; Eyles, C.H.; Haynes, C.V. Sediment and pollen evidence for an early to mid-Holocene humid period in the eastern Sahara. *Nature* **1985**, *314*, 352–355. [CrossRef]
41. Bailey, G. The Red Sea, coastal landscapes, and hominin dispersals. In *The Evolution of Human Populations in Arabia*; Petraglia, M.D., Rose, J.I., Eds.; Springer: Berlin/Heidelberg, Germany, 2010; pp. 15–37.
42. Engel, M.; Matter, A.; Parker, A.G.; Parton, A.; Petraglia, M.D.; Preston, G.; Preusser, F. Lakes or wetlands? A comment on "The middle Holocene climatic records from Arabia: Reassessing lacustrine environments, shift of ITCZ in Arabian Sea, and impacts of the southwest Indian and African monsoons" by Enzel et al. *Glob. Planet. Chang.* **2017**, *148*, 258–267. [CrossRef]
43. Biton, E.; Gildor, H.; Trommer, G.; Siccha, M.; Kucera, M.; van der Meer, M.T.; Schouten, S. Sensitivity of Red Sea circulation to monsoonal variability during the Holocene: An integrated data and modeling study. *Paleoceanography* **2010**, *25*, PA4209. [CrossRef]
44. Arz, H.W.; Lamy, F.; Pätzold, J.; Müller, P.J.; Prins, M. Mediterranean moisture source for an early-Holocene humid period in the northern Red Sea. *Science* **2003**, *300*, 118–121. [CrossRef]
45. Lézine, A.M.; Robert, C.; Cleuziou, S.; Inizan, M.L.; Braemer, F.; Saliège, J.F.; Sylvestre, F.; Tiercelin, J.J.; Crassard, R.; Méry, S.; et al. Climate change and human occupation in the southern Arabian lowlands during the last deglaciation and the Holocene. *Glob. Planet. Chang.* **2010**, *72*, 412–428. [CrossRef]
46. Morcos, S.A. Physical and chemical oceanography of the Red Sea. *Oceanogr. Mar. Biol. Annu. Rev.* **1970**, *8*, 73–202.
47. Sofianos, S.S.; Johns, W.E. An oceanic general circulation model (OGCM) investigation of the Red Sea circulation, 1. Exchange between the Red Sea and the Indian Ocean. *J. Geophys. Res. Ocean.* **2002**, *107*, 17-1–17-11. [CrossRef]
48. Manaa, A.A.; Jones, B.G.; McGregor, H.V.; Zhao, J.X.; Price, D.M. Dating Quaternary raised coral terraces along the Saudi Arabian Red Sea coast. *Mar. Geol.* **2016**, *374*, 59–72. [CrossRef]
49. Fouda, M.M.; Gerges, M.A. *Implication of Climate Change in the Red Sea and Gulf of Aden Region: An Overview*; United Nations Environment Programme, Regional Seas Reports and Studies: Nairobi, Kenya, 1994.
50. Shanas, P.R.; Aboobacker, V.M.; Albarakati, A.M.A.; Zubier, K.M. Superimposed wind-waves in the Red Sea. *Ocean. Eng.* **2017**, *138*, 9–22. [CrossRef]
51. Langodan, S.; Cavaleri, L.; Pomaro, A.; Portilla, J.; Abualnaja, Y.; Hoteit, I. Unraveling climatic wind and wave trends in the Red Sea using wave spectra partitioning. *J. Clim.* **2018**, *31*, 1881–1895. [CrossRef]
52. Langodan, S.; Cavaleri, L.; Viswanadhapalli, Y.; Hoteit, I. The Red Sea: A natural laboratory for wind and wave modeling. *J. Phys. Oceanogr.* **2014**, *44*, 3139–3159. [CrossRef]
53. Jiang, H.; Farrar, J.T.; Beardsley, R.C.; Chen, R.; Chen, C. Zonal surface wind jets across the Red Sea due to mountain gap forcing along both sides of the Red Sea. *Geophys. Res. Lett.* **2009**, *36*, L19605. [CrossRef]
54. Kemp, A.C.; Horton, B.; Donnelly, J.P.; Mann, M.E.; Vermeer, M.; Rahmstorf, S. Climate related sea-level variations over the past two millennia. *Proc. Natl. Acad. Sci. USA* **2011**, *108*, 11017–11022. [CrossRef] [PubMed]
55. Woodroffe, C.D.; Webster, J.M. Coral reefs and sea-level change. *Mar. Geol.* **2014**, *352*, 248–267. [CrossRef]
56. Lambeck, K.; Rouby, H.; Purcell, A.; Sun, Y.; Sambridge, M. Sea level and global ice volumes from the last glacial maximum to the Holocene. *Proc. Natl. Acad. Sci. USA* **2014**, *111*, 15296–15303. [CrossRef] [PubMed]
57. Brooke, B.P.; Huang, Z.; Nicholas, W.A.; Oliver, T.S.; Tamura, T.; Woodroffe, C.D.; Nichol, S.L. Relative sea-level records preserved in Holocene beach-ridge strandplains—An example from tropical northeastern Australia. *Mar. Geol.* **2019**, *411*, 107–118. [CrossRef]
58. Dougherty, A.J.; Thomas, Z.A.; Fogwill, C.; Hogg, A.; Palmer, J.; Rainsley, E.; Williams, A.N.; Ulm, S.; Rogers, K.; Jones, B.G.; et al. Redating the earliest evidence of the mid-Holocene relative sea-level highstand in Australia and implications for global sea-level rise. *PLoS ONE* **2019**, *14*, e0218430. [CrossRef] [PubMed]
59. Rushby, G.T.; Richards, G.T.; Gehrels, W.R.; Anderson, W.P.; Bateman, M.D.; Blake, W.H. Testing the mid-Holocene relative sea-level highstand hypothesis in north Wales, United Kingdom. *Holocene* **2019**, *29*, 1491–1502. [CrossRef]
60. Shaked, Y.; Marco, S.; Lazar, B.; Stein, M.; Cohen, C.; Sass, E.; Agnon, A. Late Holocene shorelines at the Gulf of Aqaba: Migrating shorelines under conditions of tectonic and sea level stability. *Eur. Geosci. Union Stephan Mueller Spec. Publ. Ser.* **2002**, *2*, 1–7. [CrossRef]
61. Hein, C.J.; FitzGerald, D.M.; Milne, G.A.; Bard, K.; Fattovich, R. Evolution of a Pharaonic harbor on the Red Sea: Implications for coastal response to changes in sea level and climate. *Geology* **2011**, *39*, 687–690. [CrossRef]
62. Al-Mikhlafi, A.S.; Hibbert, F.D.; Edwards, L.R.; Cheng, H. Holocene relative sea-level changes and coastal evolution along the coastlines of Kamaran Island and As-Salif Peninsula, Yemen, southern Red Sea. *Quat. Sci. Rev.* **2021**, *252*, 106719. [CrossRef]
63. Behairy, A.K.A.; Durgaprasada Rao, N.V.N.; El-Shater, A. A siliciclastic coastal sabkha, Red Sea coast, Saudi Arabia. *J. King Abdulaziz Univ. Mar. Sci.* **1991**, *2*, 65–77. [CrossRef]
64. Ghandour, I.M.; Haredy, R.A. Facies analysis and sequence stratigraphy of Al-Kharrar Lagoon coastal sediments, Rabigh Area, Saudi Arabia: Impact of sea level and climate changes on coastal evolution. *Arab. J. Sci. Eng.* **2019**, *44*, 505–520. [CrossRef]
65. Ghandour, I.M.; Majeed, J.; Al-Zubieri, A.G.; Mannaa, A.A.; Aljahdali, M.H.; Bantan, R.A. Late Holocene Red Sea coastal evolution: Evidence from shallow subsurface sedimentary facies, north Al-Wajh, Saudi Arabia. *Thalass. Int. J. Mar. Sci.* **2021**, *37*, 1–12.

66. Ghandour, I.M.; Al-Zubieri, A.G.; Basaham, A.S.; Mannaa, A.A.; Al-Dubai, T.A.; Jones, B.G. Mid-late Holocene paleoenvironmental and sea level reconstruction on the Al Lith Red Sea coast, Saudi Arabia. *Front. Mar. Sci.* **2021**, *8*, 677010. [CrossRef]
67. Abu-Zied, R.H.; Bantan, R.A. Palaeoenvironment, palaeoclimate and sea-level changes in the Shuaiba Lagoon during the late Holocene (last 3.6 ka), eastern Red Sea coast, Saudi Arabia. *Holocene* **2015**, *25*, 1301–1312. [CrossRef]
68. Bantan, R.A.; Abu-Zied, R.H.; Al-Dubai, T.A. Late Holocene environmental changes in a sediment core from Al-Kharrar Lagoon, eastern Red Sea Coast, Saudi Arabia. *Arab. J. Sci. Eng.* **2019**, *44*, 6557–6570. [CrossRef]
69. Basaham, A.S.; El Sayed, M.A.; Ghandour, I.M.; Masuda, H. Geochemical background for the Saudi Red Sea coastal systems and its implication for future environmental monitoring and assessment. *Environ. Earth Sci.* **2015**, *74*, 4561–4570. [CrossRef]
70. Basaham, A.S.; Ghandour, I.M.; Haredy, R. Controlling factors on the geochemistry of Al-Shuaiba and Al-Mejarma coastal lagoons, Red Sea, Saudi Arabia. *Open Geosci.* **2019**, *11*, 426–439. [CrossRef]
71. Lisitzin, E. *Sea Level Changes*; Elsevier: Amsterdam, The Netherlands, 1974.
72. Johnson, P.R. *Explanatory Notes to the Map of Proterozoic Geology of Western Saudi Arabia*; Technical Report; Saudi Geological Survey: Jeddah, Saudi Arabia, 2006.
73. Greenwood, W.R.; Stoeser, D.B.; Fleck, R.J.; Stacey, J.S. *Late Proterozoic Island-Arc Complexes and Tectonic Belts in the Southern Part of the Arabian Shield, Kingdom of Saudi Arabia*; Open-File Report, 1982-83-296; Ministry of Petroleum and Mineral Resources, Deputy Ministry for Mineral Resources: Jeddah, Saudi Arabia, 1982.
74. Goldsmith, J.R.; Graf, D.L.; Heard, H.C. Lattice constants of the calcium-magnesium carbonates. *Am. Mineral.* **1961**, *46*, 453–457.
75. Craig, H. Isotopic standards for carbon and oxygen correction factors for mass spectrometric analysis of carbon dioxide. *Geochim. Cosmochim. Acta* **1957**, *12*, 133–149. [CrossRef]
76. Keith, M.L.; Weber, J.N. Carbon and oxygen isotopic composition of selected limestones and fossils. *Geochim. Cosmochim. Acta* **1964**, *28*, 1787–1816. [CrossRef]
77. Reading, H.G. Clastic coast. In *Sedimentary Environments: Processes, Facies and Stratigraphy*; Reading, H.G., Ed.; Blackwell Science: Oxford, UK, 1996; pp. 154–231.
78. Lessa, G.C.; Angulo, R.J.; Giannini, P.C.; Arau´jo, A.D. Stratigraphy and Holocene evolution of a regressive barrier in south Brazil. *Mar. Geol.* **2000**, *165*, 87–108. [CrossRef]
79. González-Villanueva, R.; Pérez-Arlucea, M.; Costas, S.; Bao, R.; Otero, X.L.; Goble, R. 8000 years of environmental evolution of barrier-lagoon systems emplaced in coastal embayments (NW Iberia). *Holocene* **2015**, *25*, 1786–1801. [CrossRef]
80. Bezzi, A.; Casagrande, G.; Martinucci, D.; Pillon, S.; del Grande, C.; Fontolan, G. Modern sedimentary facies in a progradational barrier-spit system: Goro lagoon, Po delta, Italy. *Estuar. Coast. Shelf Sci.* **2019**, *227*, 106323. [CrossRef]
81. McCubbin, D.G. Barrier-island and strand-plain facies. In *Sandstone Depositional Environments*; Scholle, P.A., Darwin Spearing, D., Eds.; AAPG Memoir: Tulsa, OK, USA, 1982; Volume 31, pp. 247–279.
82. Clifton, H.E. A re-examination of facies models for clastic shorelines. In *Facies Models Revisited*; Posamentier, H.G., Walker, R.G., Eds.; SEPM Special Publication: Tulsa, OK, USA, 2006; Volume 84, pp. 293–337.
83. Lindhorst, S.; Betzler, C.; Hass, H.C. The sedimentary architecture of a Holocene barrier spit (Sylt, German Bight): Swash-bar accretion and storm erosion. *Sediment. Geol.* **2008**, *206*, 1–16. [CrossRef]
84. Van Heteren, S. Barrier systems. In *Coastal Environments and Global Change*; Masselink, G., Gehrels, R., Eds.; John Wiley and Sons: Hoboken, NJ, USA, 2015; pp. 194–226.
85. Sousa, C.; Boski, T.; Pereira, L. Holocene evolution of a barrier island system, Ria Formosa, south Portugal. *Holocene* **2019**, *29*, 64–76. [CrossRef]
86. FitzGerald, D.M.; Rosen, P.S. *Glaciated Coasts*; Academic Press: London, UK, 1987.
87. FitzGerald, D.; van Heteren, S. Classification of paraglacial barrier systems: Coastal New England, U.S.A. *Sedimentology* **1999**, *46*, 1083–1108. [CrossRef]
88. Sloss, C.R.; Jones, B.G.; Murray-Wallace, C.V.; McClennen, C.E. Holocene sea level fluctuations and the sedimentary evolution of a barrier estuary: Lake Illawarra, New South Wales, Australia. *J. Coast. Res.* **2005**, *21*, 943–959. [CrossRef]
89. Joury, M.R.F.; James, N.P.; James, C. Nearshore cool-water carbonate sedimentation and provenance of Holocene calcareous strandline dunes, southeastern Australia. *Aust. J. Earth Sci.* **2018**, *65*, 221–242. [CrossRef]
90. Durgaprasada Rao, N.V.N.; Behairy, A.K.A. Intertidal conglomerate south of Yanbu—An episodic clastic deposit in the eastern Red Sea. *Palaeogeogr. Palaeoclimatol. Palaeoecol.* **1987**, *58*, 221–228. [CrossRef]
91. Fruergaard, M.; Tessier, B.; Poirier, C.; Mouazé, D.; Weill, P.; Noël, S. Depositional controls on a hypertidal barrier-spit system architecture and evolution, Pointe du Banc spit, north-western France. *Sedimentology* **2020**, *67*, 502–533. [CrossRef]
92. Flemming, B.W. Siliciclastic back-barrier tidal flats. In *Principles of Tidal Sedimentology*; Davis, R.A., Dalrymple, R.W., Eds.; Springer: Berlin/Heidelberg, Germany, 2012; pp. 231–267.
93. Storms, J.E.A. Event-based stratigraphic simulation of wave-dominated shallow-marine environments. *Mar. Geol.* **2003**, *199*, 83–100. [CrossRef]
94. Aagaard, T.; Davidson-Arnott, R.; Greenwood, B.; Nielsen, J. Sediment supply from shoreface to dunes: Linking sediment transport measurements and long-term morphological evolution. *Geomorphology* **2004**, *60*, 205–224. [CrossRef]
95. Snedden, J.W.; Nummedal, D. Coherence of surf zone and shelf current flow on the Texas (USA) coastal margin: Implications for interpretation of paleo-current measurements in ancient coastal sequences. *Sediment. Geol.* **1990**, *67*, 221–236. [CrossRef]

96. Reineck, H.E.; Singh, I.B. *Depositional Sedimentary Environments with Reference to Terrigenous Clastics*; Springer: Berlin/Heidelberg, Germany, 1980.
97. Klein, D. *Sandstone Depositional Models for Exploration for Fossil Fuels*; International Human Resources Development Corporation: Boston, MA, USA, 1982.
98. Fox, W.T.; Davis, R.J. Seasonal variation in beach erosion and sedimentation on the Oregon Coast. *Geol. Soc. Am. Bull.* **1978**, *89*, 1541–1549. [CrossRef]
99. Peltier, W.R. On eustatic sea level history: Last Glacial Maximum to Holocene. *Quat. Sci. Rev.* **2002**, *21*, 377–396. [CrossRef]
100. Clemmensen, L.B.; Andreasen, F.; Nielsen, S.T.; Sten, E. The late Holocene coastal dunefield at Vejers, Denmark: Characteristics, sand budget and depositional dynamics. *Geomorphology* **1996**, *17*, 79–98. [CrossRef]
101. Psuty, N.P.; Eugénia, M.; Moreira, S.A. Holocene sedimentation and sea level rise in the Sado Estuary, Portugal. *J. Coast. Res.* **2000**, *16*, 125–138.
102. Walker, R.G.; James, N.P. *Facies Models Response to Sea Level Change*, 2nd ed.; Geological Association of Canada: St. John's, NS, Canada, 1992; p. 409.
103. De Falco, G.; Antonioli, F.; Fontolan, G.; lo Presti, V.; Simeone, S.; Tonielli, R. Early cementation and accommodation space dictate the evolution of an overstepping barrier system during the Holocene. *Mar. Geol.* **2015**, *369*, 52–66. [CrossRef]
104. Moore, C.H. *Carbonate Diagenesis and Porosity*; Elsevier: Amsterdam, The Netherlands, 1989.
105. Longman, M.W. Carbonate diagenetic textures from near surface diagenetic environments. *Am. Assoc. Pet. Geol. Bull.* **1980**, *64*, 461–487.
106. James, N.P.; Choquette, P.W. Limestone: The sea floor diagenetic environment. In *Diagenesis*; McIlreath, I., Morrow, D., Eds.; Canada Geological Association of Canada Reprint Series: St John's, ND, Canada, 1990; pp. 13–34.
107. Scholle, P.A.; Ulmer-Scholle, D.S. A color guide to the petrography of carbonate rocks: Grains, textures, porosity, diagenesis. *Am. Assoc. Pet. Geol. Mem.* **2003**, *77*, 1–474.
108. Vieira, M.M.; Sial, A.N.; de Ros, L.F.; Morad, S. Origin of Holocene beachrock cements in northeastern Brazil: Evidence from carbon and oxygen isotopes. *J. South Am. Earth Sci.* **2017**, *79*, 401–408. [CrossRef]
109. Hopley, D. Beachrock as sea level indicator. In *Sea Level Research*; van de Plassche, O., Ed.; Great Yarmouth: Galliard Printers, UK, 1986; pp. 157–173.
110. Holail, H.M.; Shaaban, M.N.; Mansour, A.S. Cementation of Holocene beachrock in the Aqaba and the Arabian Gulfs: Comparative study. *Carbonates Evaporites* **2004**, *19*, 142–150. [CrossRef]
111. Hillgärtner, H.; Dupraz, C.; Hug, W. Microbially induced cementation of carbonate sands: Are micritic meniscus cements good indicators of vadose diagenesis? *Sedimentology* **2001**, *48*, 117–131. [CrossRef]
112. Calvet, F.; Cabrera, M.C.; Carracedo, J.C.; Mangas, J.; Pérez-Torrado, F.J.; Recio, C.; Travé, A. Beachrocks from the island of La Palma (Canary Islands, Spain). *Mar. Geol.* **2003**, *197*, 75–93. [CrossRef]
113. Friedman, G.M. Climatic significance of Holocene beachrock sites along shorelines of the Red Sea. *Am. Assoc. Pet. Geol. Bull.* **2005**, *89*, 849–852. [CrossRef]
114. Vieira, M.M.; de Ros, L.F. Cementation patterns and genetic implications of Holocene beachrocks from northeastern Brazil. *Sediment. Geol.* **2006**, *192*, 207–230. [CrossRef]
115. Hoefs, J. *Stable Isotope Geochemistry*; Springer: Berlin/Heidelberg, Germany, 1997.
116. Woodroffe, C.D. *Coasts: Form, Process, and Evolution*; Cambridge University Press: Cambridge, UK, 2003.
117. Dullo, W.C.; Montaggioni, L. Modern Red Sea coral reefs: A review of their morphologies and zonation. In *Sedimentary and Tectonic Evolution of Rift Basins–The Red Sea-Gulf of Aden*; Purser, B.H., Bosence, D.W.J., Eds.; Springer: Berlin/Heidelberg, Germany, 1998; pp. 583–594.

Journal of
Marine Science and Engineering

MDPI

Article

Beachrock as a Paleoshoreline Indicator: Example from Wadi Al-Hamd, South Al-Wajh, Saudi Arabia

Ammar A. Mannaa [1], Rabea A. Haredy [1] and Ibrahim M. Ghandour [1,2,*]

1 Marine Geology Department, Faculty of Marine Science, King Abdulaziz University, P.O. Box 80200, Jeddah 21589, Saudi Arabia; amanna@kau.edu.sa (A.A.M.); rhardy@kau.edu.sa (R.A.H.)
2 Geology Department, Faculty of Science, Tanta University, Tanta 31527, Egypt
* Correspondence: ighandour@kau.edu.sa

Abstract: The present study concerns the Holocene inland beachrocks that are exposed in the Red Sea coastal plain at the mouth of Wadi Al-Hamd, South Al-Wajh City, Saudi Arabia, and their utility as an indicator for Holocene climate and sea level changes. In addition, the framework composition, and carbon and oxygen isotopic data, are employed to interpret the origin of their cement. The beachrock consists mainly of gravel and coarse-grained terrigenous sediments dominated by lithic fragments of volcanic rocks, cherts and rare limestones along with quartz, feldspars and traces of amphiboles and heavy minerals. In addition, rare skeletal remains dominated by coralline algae, benthic foraminifera and mollusca remains are recognized. The allochems are cemented by high Mg-calcite (HMC) formed mainly in the intertidal zone under active marine phreatic conditions. The cement takes the form of isopachous to anisopachous rinds of bladed crystals, micritic rim non-selectively surrounding siliciclastic and skeletal remains, and pore-filling micrite. Pore-filling micrite cement occasionally displays a meniscus fabric, suggesting a vadose environment. The $\delta^{18}O$ and $\delta^{13}C$ values of carbonate cement range from −0.35‰ to 1‰ (mean 0.25‰) and −0.09‰ to 3.03‰ (mean 1.85‰), respectively, which are compatible with precipitation from marine waters. The slight depletion in $\delta^{18}O$ and $\delta^{13}C$ values in the proximal sample may suggest a slight meteoric contribution.

Keywords: beachrocks; paleoshoreline; Holocene sea-level and climate changes; Wadi Al-Hamd; Red Sea

Citation: Mannaa, A.A.; Haredy, R.A.; Ghandour, I.M. Beachrock as a Paleoshoreline Indicator: Example from Wadi Al-Hamd, South Al-Wajh, Saudi Arabia. *J. Mar. Sci. Eng.* **2021**, *9*, 984. https://doi.org/10.3390/jmse9090984

Academic Editor: Gemma Aiello

Received: 13 August 2021
Accepted: 5 September 2021
Published: 8 September 2021

Publisher's Note: MDPI stays neutral with regard to jurisdictional claims in published maps and institutional affiliations.

Copyright: © 2021 by the authors. Licensee MDPI, Basel, Switzerland. This article is an open access article distributed under the terms and conditions of the Creative Commons Attribution (CC BY) license (https://creativecommons.org/licenses/by/4.0/).

1. Introduction

The development of the Red Sea coastal zone is an important axis in the new economic vision of Saudi Arabia. Development plans and the forecasting of future changes along the Red Sea coastal area should take into account Holocene sea level and climate changes. The impact of Holocene sea level and climate changes, and their influence on the depositional and erosional processes of the coastal zone, is a pertinent topic in geological and archaeological research and coastal zone development plans [1]. The Holocene sea level rose rapidly from the Last Glacial Maximum (LGM) until 6–7 ka ago, when most of the continental ice sheets melted [2]. Correspondingly, the climate was more humid during that time, and turned extremely arid starting from about 5 ka ago [3]. During the humid early and middle Holocene, floods may have been more frequent than they are today, and perhaps more capable of carrying a larger coarse sediment load.

Several proxies have been utilized worldwide to infer Holocene climate and sea level changes. These proxies include shoreline shifts and subsequent vertical variations in sedimentary facies [4–7], temporal variations of microfossils [8,9], beachrocks, marine notches and raised coral reefs [10–14]. Recent descriptions of coastal sedimentology of the tropics and warm temperate zones render beachrocks as potentially useful indicators of sea level and climatic changes and former paleoshoreline positions [13,15,16]. Beachrocks are an outstanding sea level proxy since they hold information on not only the vertical but also the lateral evolution of the shoreline [17,18].

Beachrocks are carbonate cemented littoral sediments of different compositions (terrigenous, biogenic and anthropogenic) and grain sizes (sand to gravel) typically found within the intertidal zones and, rarely, in supratidal and lower subtidal zones along low latitude coastlines ([15,19–22]. The cementation process is accomplished under various climatic conditions by supersaturation of interstitial water, intensive evaporation, mixing of marine and meteoric waters [23], CO_2 degassing of intertidal groundwater [24] and by the aid of biological activities [25]. The texture, microfabrics and mineralogy of carbonate cements provide information about the formation mechanisms and environmental conditions of cementation.

Beachrocks of different composition and grain size are distributed intermittently along the Saudi Red Sea and the Gulf of Aqaba coasts. They consist mainly of gravel- and sand-sized siliciclastic grains and/or calcareous skeletal remains, cemented by aragonite and/or high Mg-calcite (HMC) and, rarely, low Mg-calcite [22,26–29]. The beachrocks along the Saudi Red Sea and the Gulf of Aqaba coasts have received little attention compared to the Egyptian and Jordanian coasts ([19,25,30–32]. An interesting, not previously documented beachrock crops out far inland, 1.1 km from the recent shoreline at the mouth of Wadi Al-Hamd, 50 km south of Al-Wajh City, at latitude N 25° 57.737′ and longitude E 36° 42.721′ (Figure 1). This implies an in-situ cementation at the intertidal level within the sedimentary pile after deposition. The present study aims to interpret the depositional setting of these beachrocks and their relationship to Holocene climate and sea level changes. The study will further determine the cement's mineralogical composition and microfabric in an attempt to gain information about the source of cement.

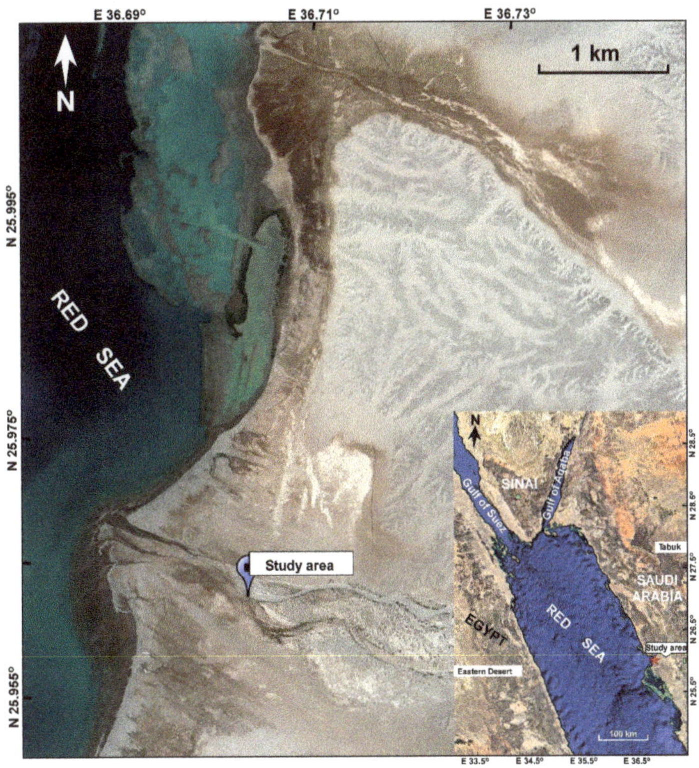

Figure 1. Landsat image shows the area of study and the location of the beachrock exposure at the mouth of Wadi Al-Hamd.

2. Area of Study

The area of study is located at the mouth of Wadi Al-Hamd, northern Red Sea, about 55 km south of Al Wajh City, Saudi Arabia, at latitude N 25° 57.737′ and longitude E 036° 42.721′ (Figure 1). Wadi Al-Hamd is one of the largest wadis shedding the Arabian Shield in northern Saudi Arabia. It extends from mountain scarp near Al-Medina City, flows inland to NW and discharges sediments and freshwaters into the Red Sea. For most of the year, Wadi Al-Hamd is extremely dry; however, in the winter (October and March), it temporarily activates during episodic major flood seasons in the form of short-duration showers generally associated with thunderstorms. The climate in the area is identified by its arid desert; dry, hot summers with maximum daily temperatures that range from 20 °C to 35 °C in January and up to 48 °C in July. The precipitation rate ranges from 0.5 to 116 mm/year, whereas the evaporation rate is high, up to 2 m^3/year or slightly more [33]. The prevailing winds are from the NNW to SSE over the entire year with speed varying from 2 to 10 m/s [33]. Wave heights of up to 2.5 m and, rarely, up to 4 m are generated by the strong onshore-directed winds. The area has a semidiurnal microtidal regime, with a spring and neap tidal range of 0.7 m and 0.5 m, respectively. Although relatively weak, the strength of flood tidal currents is stronger than the almost insignificant ebb currents. From an aerial view, the clastic-dominated coast of Wadi Al-Hamd is linear to slightly arched, projecting itself seaward at the mouth of the wadi. It is an asymmetric lobate delta, flanked to the north by a narrow strandplain with a relatively steep foreshore profile and a wide, low gradient intertidal flat to the south, separated by a low relief beach ridge [5,34]. Sabkhas and coastal sand dunes occupy the flood plain, while the channel is barred seaward by a low relief beach ridge with back-barrier swale separating the channel from the sea. The strandplain to the north contains a series of sand bars and a runnel system running parallel to slightly slanted to the shoreline [5]. A fringing reef and mangrove tickets occur to the north and away from the influence of wadi sedimentation. The nearshore sediment composition is dominated by detrital siliciclastic minerals, which are dominated by plagioclase with subordinate feldspars, amphiboles and traces of heavy minerals [35].

The coastal plain of the area south Al-Wajh is covered by Quaternary siliciclastic sediments (Figure 2). The sediments are derived mainly from the Precambrian Basement Complex of the Arabian shield and the Cenozoic volcanic and sedimentary rocks. These rocks crop out in the surrounding area and along the axis of the wadi from the catchment area near Al-Medina City to the downstream area. The downstream area of Wadi Al Hamd is covered by Quaternary alluvial terraces in the lower part consisting of consolidated, very coarse sediments of Pleistocene age, overlain by Recent unconsolidated gravel and sand [35–37].

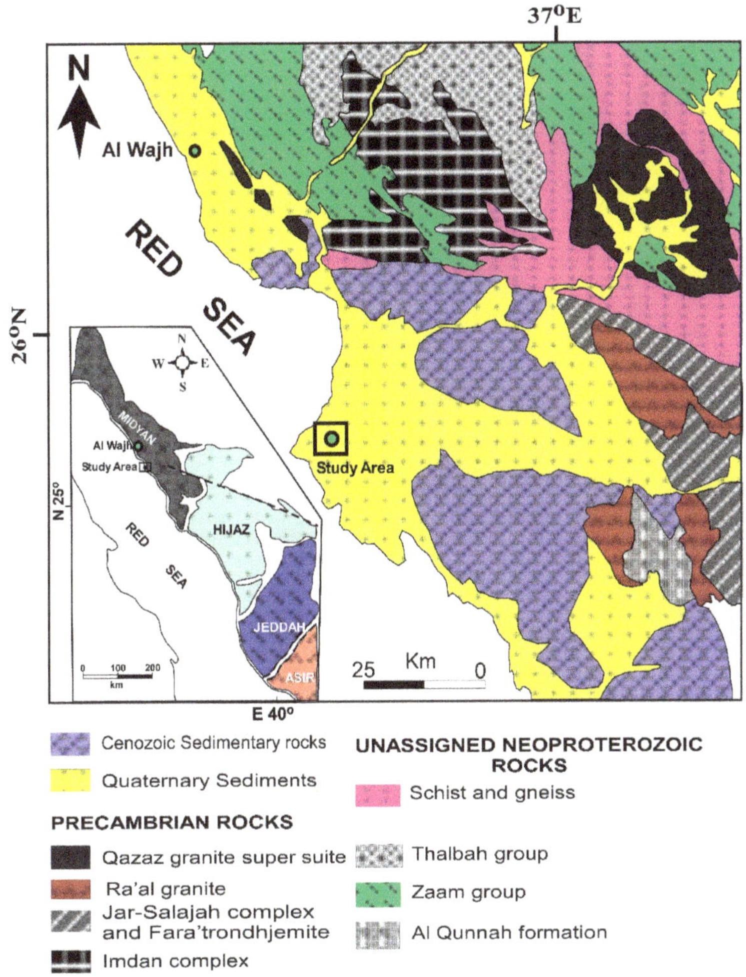

Figure 2. Simplified geologic map of the area of study (adapted from [37]).

3. Materials and Methods

Five samples were collected from the beachrock exposure at the mouth of Wadi Al-Hamd (Figure 3). Thin sections were prepared following traditional methods at the Petrology Lab, Faculty of Sciences, Cairo University. The thin sections were investigated by a petrographic microscope, in order to determine textural attributes, pores, framework grains, cement morphology and micro-structures. The morphology and micro-structures of the cement were examined using SEM. The mineral composition was determined using powder X-ray diffraction (XRD). The $CaCO_3$ content was determined using pressure-calcimeter [38]. Stable isotopic values of oxygen and carbon ($\delta^{18}O$ and $\delta^{13}C$) of the carbonate cement were estimated for five samples using an automated carbonate preparation device (KIEL-III) coupled to a gas-ratio mass spectrometer (Finnigan MAT 252) at the Environmental Isotope Laboratory, Geosciences Department, University of Arizona. The cement was carefully separated using a micro-driller to avoid contamination. The cement was then reacted with dehydrated phosphoric acid under vacuum at 70 °C. The isotope ratio measurement is calibrated based on repeated measurements of NBS-19 and NBS-18, and precision is ±0.10 ‰ for $\delta^{18}O$ and ±0.08‰ for $\delta^{13}C$ (1 sigma).

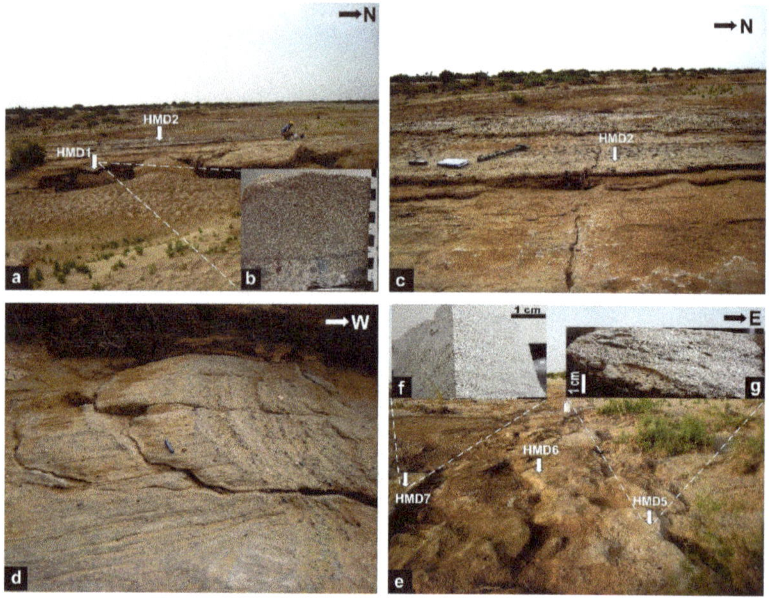

Figure 3. Field Photographs showing the characteristic features and locations of the samples of the beachrock at the mouth of Wadi Al-Hamd. (**a–c**) The near horizontal to gently westward inclined gravel-rich proximal strata. (**b**) Close-up view of sample HMD1, showing normal grading from well-rounded gravel to massive sands. (**d**) Westward gently inclined graded bedded strata. (**e**) Fracture system in the distal beds. (**f,g**) Close-up view showing internal features of the poorly-sorted gravelly sand (HMD5) and the low angle cross bedded with foresets displaying normal grading in sample HMD 7.

4. Results

Inland beachrocks are exposed at the mouth of Wadi Al-Hamd, South Al-Wajh, at about 1.1 km to the east of the recent shoreline, formed by an accumulation of coarse clastic sediments with a westward decrease in the grain size (Figure 3a–g). They occur at about a meter above the current sea-level and extend for about 60 m in N-S direction and up to 40 m in the E-W direction. They consist of 10 to 20 cm thick tabular beds of well-cemented terrigenous very coarse sands and gravels gently inclined (<10°) to the west (seaward) (Figure 3d). They vary with downdip in the grain size and internal sedimentary structures. Beds in the proximal part are sharp-based, mantled with poorly-sorted, well-rounded pebble-grade spherical to elongated, gravel-grade sand, and vertical to well-sorted sand (Figure 3b). In the distal location, beds vary from massive, parallel horizontal laminated and planar cross-stratified, with foresets defined by gravel concentration and displaying normal grading (Figure 3f). They are fractured orthogonally to the bedding plane (Figure 3e). Body and trace fossils are not recognized in the exposed rocks.

4.1. Framework Composition and Cementation

The beachrocks of Wadi Al-Hamd cover a range of grain sizes, varying from generally poorly- to moderately-sorted sandy conglomerate to gravelly sandstones. Pebbles are rounded to subrounded and consist of lithic fragments, dominantly basalt and chert. Thin-section investigations showed that the framework composition consists mainly of siliciclastic grains, dominated by mono- and polycrystalline quartz, feldspars, lithic fragments of quartzite, basalts, chert and rare limestones, amphiboles and rare heavy minerals rutile and tourmaline (Figure 4a–h). The siliciclastic terrigenous grains are derived from rocks of the Hijaz and Midyan terranes and from the Quaternary limestones in the hinterland.

Skeletal remains (Figure 4b,d,f) are very rare (<2%) and dominated by algae, bivalves, echinoids, gastropods and larger benthic foraminifera. The skeletal remains were possibly transported by a longshore drift from offshore. The original porosity of the beachrocks was intergranular (and very rarely intragranular) and was totally to partially occluded by cementation and, locally, by internal sediments. The $CaCO_3$ content is generally low to moderate, varying from 28 to 53% with an average of 25% (Table 1).

Figure 4. Photomicrographs showing the framework composition and the cement morphology in the Wadi Al-Hamd beachrocks. (**a–c**,**f**) The beachrock consists mainly of siliciclastic grains surrounded by isopachous Mg-calcite cement, a blade shape with a micritic base (red arrows). Pore-filling micrite taking a meniscus fabric is shown by yellow arrow in a. (**a**,**b**): sample HMD 1, (**c**): sample HMD 6, (**d**): sample HMD 7. (**e**,**g**,**h**) Samples HMD 2 and 5, showing the isopachous HMC cement that takes a blade shape without micritic base. (The bars in (**a–f**) = 200 μm, and in (**g**) and (**h**) = 40 μm).

Table 1. The oxygen and carbon stable isotope, Z values and $CaCO_3$ % content of the Wadi Al-Hamd beachrocks.

Sample	$\delta\ ^{13}C_{VPDB}$‰	$\delta\ ^{18}O_{VPDB}$‰	Z Values	$CaCO_3$%
HMD 1	−0.09	−0.22	127	28
HMD 2	3.03	1.00	134	29
HMD 5	2.02	0.74	131.8	27
HMD 6	2.41	0.08	132.3	29
HMD 7	1.89	−0.35	131	53

The cement takes the form of isopachous and anisopachous bladed crystals up to 18.4 μm surrounding grains, micrite envelope and pore filling microcrystalline cement (Figure 4a–h). Isopachous rims around the grains are non-selective, covering both siliciclastic and carbonate grains (Figure 4b–d). The cement lines grains or completely fills the intergranular pores (Figure 3g,h). Some grains are surrounded directly by isopachous blade crystals, and some display micritic base as a first generation of cement, followed by the bladed crystals. The crystals of the rims are arranged sub-perpendicularly to chaotically on the surface of siliciclastic grains (Figure 5a–h). SEM analysis shows that radial aggregate encrusting a previous, bladed cement is observed (Figure 5e). In addition, the micritic cement is associated in sample HMD 5 with filaments (Figure 5c,d). The cement consists mainly of high-Mg calcite, as revealed by XRD, petrographic and SEM investigations.

4.2. Carbon and Oxygen Stable Isotopes

Carbon and oxygen stable isotopes are widely used to determine the origin of the water from which the cement is precipitated [39–41]. The overall $\delta^{13}C_{VPDB}$ values of the cement vary between −1.09 ‰ to 3.03 ‰ (mean 1.85‰), and the $\delta^{18}O_{VPDB}$ values range from −0.35 ‰ to 1.00 ‰ (mean 0.25‰) (Table 1). Only the cement of the most proximal sample (HMD1) is depleted in oxygen and carbon isotopic values (−0.09 and −0.22‰, respectively). However, the average $\delta^{13}C_{VPDB}$ and $\delta^{18}O_{VPDB}$ values are higher than the limit established for modern marine carbonates ($\delta^{13}C_{VPDB}$; −2.00‰) and modern seawater ($\delta^{18}O_{VPDB}$; −4.00‰) [42].

The $\delta^{18}O$ values of the cements of the Wadi Al-Hamd beachrocks are slightly lighter isotopically than those recorded from the Egyptian Red Sea and Mediterranean beachrocks [19]. The $\delta^{18}O$ and $\delta^{13}C$ values of the Mediterranean beachrock cements range between −0.40 and +1.20‰ (mean of + 0.50‰) and +1.00 to +2.10, respectively. On the other hand, the $\delta^{18}O$ and $\delta^{13}C$ values of the Red Sea beachrock cements range between −0.10 and +1.20 (mean of +0.50‰) and from +2.10 to 4.50%, respectively [19]. Therefore, it appears that the mean values of oxygen and carbon isotopic compositions are relatively identical for the cements of both localities (Figure 6).

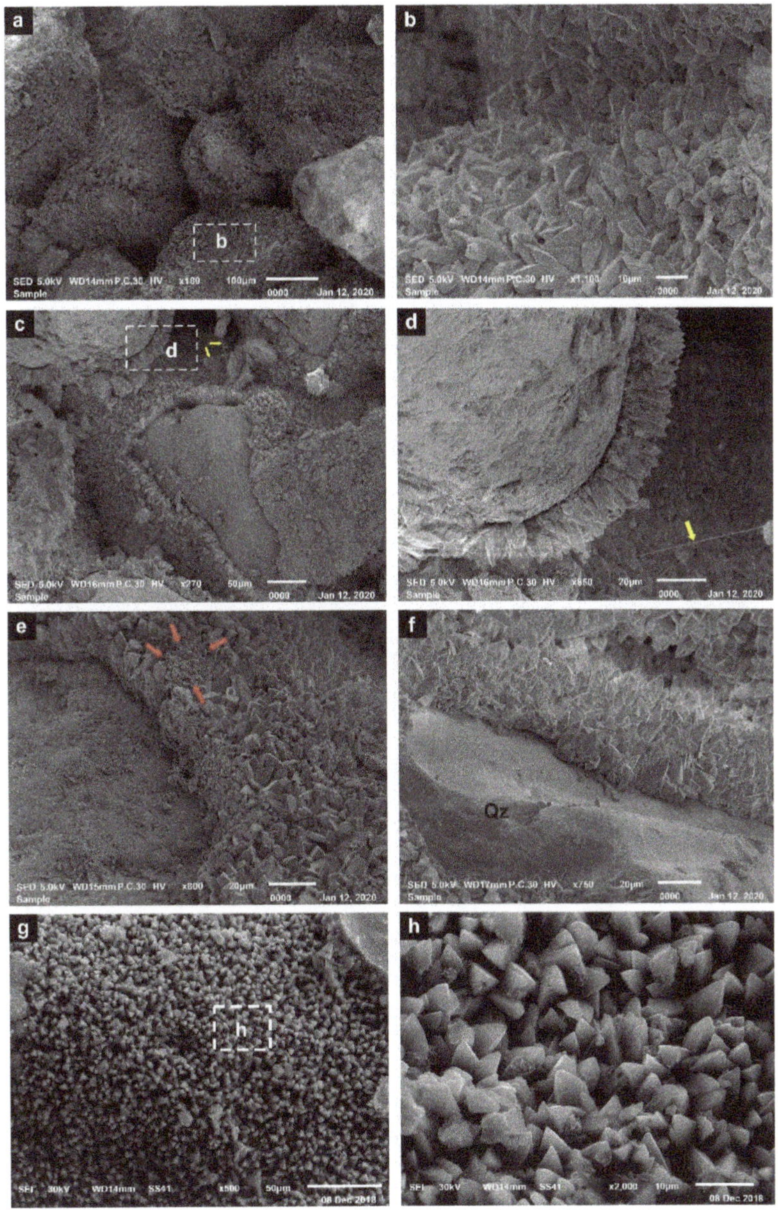

Figure 5. SEM photo micrographs showing the characteristic features of the carbonate cement in the Wadi Al-Hamd beachrocks. (**a**,**b**) HMC blades surrounding grains and the crystals show random orientation (sample HMD 1). (**c**–**e**) Isopachous HMC calcite blades grow perpendicular to the grain surface (sample HMD 5). The length of the blades is up to 18.3 μm; note the organic filaments (yellow arrows) that support the contribution of algal activity in cement formation. Small radial aggregate encrusting the bladed cement is shown by red arrows. (**f**–**h**) Pore-filling HMC calcite cements taking a well-developed bladed shape (sample HMD 7) surrounding quartz (Qz).

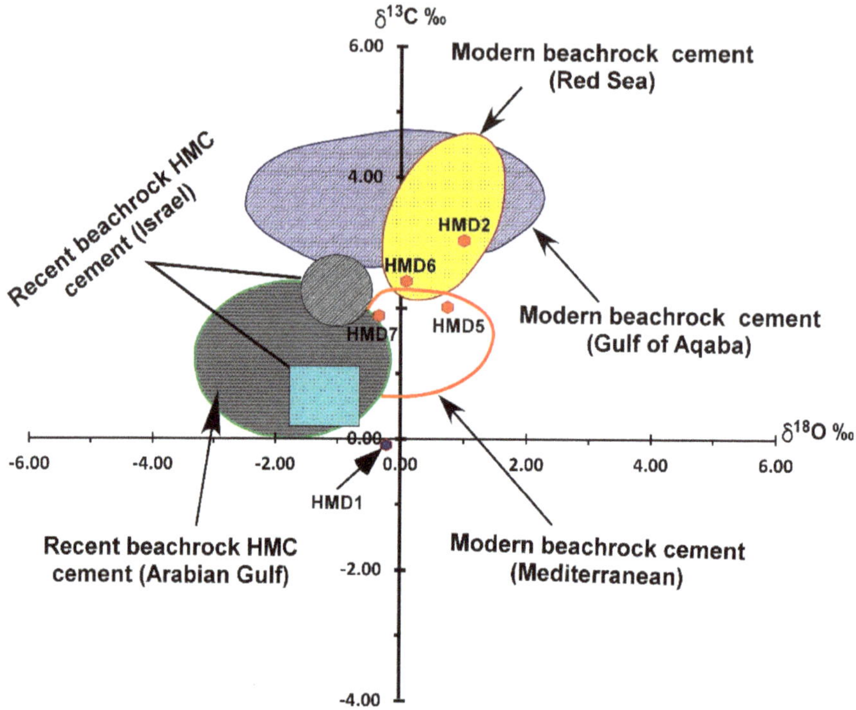

Figure 6. Bivariate plot between carbon and oxygen stable isotope composition of the Wadi Al-Hamd beachrocks and other beachrocks in the Arabian and Aqaba gulfs [32], Bahamian beachrocks [43], Red Sea and Mediterranean Sea [19], Mediterranean beachrocks of Israel [42].

The Z value [44] is determined to distinguish the type of porewater from which the carbonate cement is precipitated. The Z value is calculated as follows:

$$Z = a(\delta^{13}C + 50) + b(\delta^{18}O + 50)$$

where a and b are constants values equal 2.048 and 0.498, respectively. Z values > 120 indicate a cement of marine-water origin, whereas values < 120 suggest a freshwater origin. The Z values obtained for the studied samples, ranging from 127.003 to 134.006‰ (average; 131.22‰), suggest that the cement was generally generated by a dominantly marine origin.

5. Discussion

The grain size, sedimentary structures, framework composition and cement composition and fabric of the Wadi Al-Hamd beachrocks provide information of the climate and sea level changes surrounding it. The data indicate that the climate was more humid during the deposition of the sediments constituting the beachrock, and the sea level was higher than its present position. The prevalence of siliciclastic coarse sediments in the beachrocks suggests a period of active wadi discharge. The low content of $CaCO_3$ reflects the dominance of siliciclastic terrigenous constituents. The $CaCO_3$ content is attributed to the carbonate cement and rare calcareous skeletal remains. Active wadi discharge is related to a humid climate with relatively high rates of precipitation. In general, coarse clastic sediments are remarkably absent or constitute only a minor proportion of the modern Red Sea nearshore sediments [45]. Most wadis (the main conduit of terrigenous sediments) disappear halfway into the Red Sea coast, and very little fine-grained terrigenous sediments arrive to the nearshore zone because of the extreme aridity. The modern unconsolidated beach and nearshore sediments consist mainly of fine- to medium-grained sands [46]. Similarly, the shallow subsurface sediments of the wadi mouth and nearshore environments are dominantly medium- and, rarely, coarse-grained sands [47], finer than that of beachrocks. The gravel-rich beachrocks at the mouth of Wadi Al-Hamd were transported, therefore, and deposited by a greater runoff, under a much more humid climate than what prevails today. This humid interval possibly coincides with the middle Holocene pluvial period [45]. Several examples have been reported from inland basins and Red Sea coastal zones in western Saudi Arabia, suggesting a middle Holocene long-term climatic humid climate ([6,9,45,48]. The carbon and oxygen isotopic data of the Gulf of Aqaba fossil corals suggest that the climate of the area was humid prior to 4.9 ka ago, and that after that, the climate turned extremely arid [49]. Evidence for regional aridity starting about 5 ka has been reported [3]. The gravel and medium- to coarse-grained sands containing rare skeletal remains suggest deposition in a relatively higher energy environment. The occurrence of skeletal remains suggests a marine-influenced environment. The horizontal to gently seaward inclined beds, displaying horizontal to planar cross lamination, suggest deposition under variable energy conditions within the beach (upper foreshore) environment [50].

The area of study is situated in a tectonically stable zone [51], which extends from the north of Al-Wajh City (~N 26°) to the north of the Al-Quattan (~N 21°). The occurrence of the beachrocks 1.1km inland suggests that the sea level must have maintained such a high position for some time to allow deposition of the coarse sediments. Therefore, during the deposition of beachrocks the Red Sea coastal plain at the mouth of Wadi Al-Hamd was flooded during a period of relative sea level rise, and the shoreline migrated farther inland. The beachrocks under investigation are permanently emergent above the present high tide level. This may indicate that the sea level during deposition of the beachrock sediments was 1.3 m, or slightly higher than the current sea level. Along the Red Sea coastline, a mid-Holocene sea level highstand with high spatial and temporal variability is observed [51,52]. A sea level up to a maximum of 1–2 m higher than the present was recorded for the mid-Holocene [51,53].

Cementation Patterns

Petrographic examination and the isotopic values of the cement suggest that cementation took place in the intertidal environment under marine conditions, and there is no evidence for cementation within the meteoric environment. The beach setting is ideal for the precipitation of abiotic marine cements. High energy conditions, including wave and tidal activity, and relatively coarse, highly porous and permeable sediments, ensure that adequate volumes of supersaturated marine water are able to move through the pore system to enable cementation with the aid of hot weather. Interestingly, the samples had similar cements, indicating that the same processes of cementation occurred.

Clasts cemented with HMC bladed cement indicate cementation in an intertidal zone under active marine phreatic conditions [54–56]. Similar cement types were reported from the Mediterranean and Bermuda beachrocks, and were interpreted as a product of marine cementation [57]. The micrite HMC cement, with local occurrence of organic filaments in the studied beachrocks, may be related to biological activity. The role of micro-organisms in the formation of micrite cement has been reported by several authors [25,58]. Micritic pore-fillings suggest the involvement of biological activity [55]. In cases of no relict cellular or laminar structures, micrite cement indicates rapid supersaturation of pore waters and multiple instances of nucleation [55]. The grain-to-grain meniscal bridges following micrite encrustations imply vadose conditions [58,59].

Based on the isotopic data, it is apparent that cementation took place within the marine beach environment. The isotopic values of the cement show mostly positive carbon and oxygen values. The negative $\delta^{13}C_{VPDB}$ and $\delta^{18}O_{VPDB}$ values from the most proximal sample (HMD1) may indicate the possible contribution of fresh water in the cement precipitation. The depletion in $\delta^{18}O$ and $\delta^{13}C$ isotopic values of beachrock carbonate cements from Brazil and the Canary Islands (Spain) were attributed to the slight modification by meteoric water at relatively low salinity and at relatively high temperatures [60–62]. The limited isotopic heterogeneity among samples suggests a stability of water composition and temperature during the cement formation [19,62]. Isotopic values of most samples plot within the modern marine beachrock cement field of the Red Sea, Mediterranean Sea, Gulf of Aqaba and straddle Arabian Gulf field [19,32,42]. The high-Mg calcites both the Arabian and the Aqaba Gulfs possess relatively depleted $\delta^{13}C$ values than those of the cement of the Wadi Al-Hamd beachrocks. High-Mg calcite precipitated in the present day under warm shallow seawater is relatively enriched in oxygen (+3.00 ‰) and carbon (+3.5 ± 1.5 ‰) isotopic values [23,63]. The $\delta^{18}O$ (−0.35 to +1.00 ‰) and the $\delta^{13}C$ (−0.09 to +3.03 ‰) values of the high-Mg calcite cement of the Wadi Al-Hamd beachrocks indicate precipitation in equilibrium with present-day seawater. A value of 0.54 ‰ $\delta^{18}O$ indicates cementation in the intertidal zone [64].

6. Conclusions

An interesting beachrock is exposed 1.1 km inland at the Red Sea coastal plain at the mouth of Wadi Al-Hamd. It bears criteria that can be used to interpret Holocene climate and sea level changes. The following concluding remarks are:

1. The grain size, sedimentary structures and framework compositions indicate that the sediments were derived during humid intervals mainly from a terrigenous source, with little contribution of marine derived calcareous skeletal remains. They were deposited in a high-energy beach environment.
2. Deposition in the beach environment suggests that the shoreline backstepped more landward during a period of relative sea level rise. The beachrock at the mouth of Wadi Al-Hamd is therefore a paleoshoreline indicator.
3. The suggested humid climate and relative sea level highstand may be correlated with the middle Holocene humid interval and sea level highstand.
4. The dominant isopachous bladed and micritic Mg-calcite cement suggests that cement growth in the intertidal zone under active marine phreatic conditions. This is supported by the isotopic data and Z values that indicate that the cement generally originated from marine water. The meniscus fabric and the depletion of oxygen and carbon isotope values in the most proximal sample may suggest a slight contribution from meteoric water in the vadose environment.

Author Contributions: All authors contributed to every stage of the work. Conceptualization, fieldwork and data collection were done by all authors. The microscopic investigations, data analysis and first draft of the manuscript was written by I.M.G., A.A.M. and R.A.H. Reviewed and commented on early versions of the manuscript written by I.M.G. Project administration and funding acquisition were under the supervision of A.A.M. All authors have read and agreed to the published version of the manuscript.

Funding: This project was funded by the Deanship of Scientific Research (DSR) at King Abdulaziz University, Jeddah, Grant No. G22-150-1439.

Institutional Review Board Statement: Not applicable.

Informed Consent Statement: Not applicable.

Data Availability Statement: Not applicable.

Acknowledgments: This project was funded by the Deanship of Scientific Research (DSR) at King Abdulaziz University, Jeddah, under Grant No. G22-150-1439. The authors thank the DSR for this technical and financial support. The authors thank Aaid Al-Zubeiri and Bandar El-Zahrani (KAU) for their assistance during field studies and sample collection. We are very grateful to the editor and the reviewers for their constructive comments and editorial handling.

Conflicts of Interest: The authors declare no conflict of interest.

References

1. Vacchi, M.; Marriner, N.; Morhange, C.; Spada, G.; Fontana, A.; Rovere, A. Multiproxy assessment of Holocene relative sea-level changes in the western Mediterranean: Sea-level variability and improvements in the definition of the isostatic signal. *Earth-Sci. Rev.* **2016**, *155*, 172–197. [CrossRef]
2. Nakada, M.; Kimura, R.; Okuno, J.; Moriwaki, K.; Miura, H.; Maemoku, H. Late Pleistocene and Holocene melting history of the Antarctic ice sheet derived from sea-level variations. *Mar. Geol.* **2000**, *167*, 85–103. [CrossRef]
3. Demenocal, P.; Ortiz, J.; Guilderson, T.; Adkins, J.; Sarnthein, M.; Baker, L.; Yarusinsky, M. Abrupt onset and termination of the African Humid Period: Rapid climate responses to gradual insolation forcing. *Quat. Sci. Rev.* **2000**, *19*, 347–361. [CrossRef]
4. Hein, C.J.; FitzGerald, D.M.; Milne, G.A.; Bard, K.; Fattovich, R. Evolution of a Pharaonic harbor on the Red Sea: Implications for coastal response to changes in sea level and cli-mate. *Geology* **2011**, *39*, 687–690. [CrossRef]
5. Ghandour, İ.M.; Al-Washmi, H.A.; Haredy, R.A.; Al-Zubieri, A.G. Facies evolution and depositional model of an arid microtidal coast: Example from the coastal plain at the mouth of Wadi Al-Hamd, Red Sea, Saudi Arabia. *Turk. J. Earth Sci.* **2016**, *25*, 256–273. [CrossRef]
6. Ghandour, I.M.; Al-Zubieri, A.G.; Basaham, A.S.; Mannaa, A.A.; Al-Dubai, T.A.; Jones, B.G. Mid-Late Holocene Paleoenvironmental and Sea Level Reconstruction on the Al Lith Red Sea Coast, Saudi Arabia. *Front. Mar. Sci.* **2021**. [CrossRef]
7. Ghandour, I.M.; Haredy, R.A. Facies analysis and sequence stratigraphy of Al-Kharrar lagoon coastal sediments, Rabigh Area, Saudi Arabia: Impact of sea-level and climate changes on coastal evolution. *Arab. J. Sci. Eng.* **2019**, *44*, 505–520. [CrossRef]
8. Abu-Zied, R.H.; Bantan, R.A. Palaeoenvironment, palaeoclimate and sea-level changes in the Shuaiba Lagoon during the late Holocene (last 3.6 ka), eastern Red Sea coast, Saudi Arabia. *Holocene* **2015**, *25*, 1301–1312. [CrossRef]
9. Bantan, R.A.; Abu-Zied, R.H.; Al-Dubai, T.A. Late Holocene environmental changes in a sediment core from Al-Kharrar Lagoon, eastern Red Sea Coast, Saudi Arabia. *Arab. J. Sci. Eng.* **2019**, *44*, 6557–6570. [CrossRef]
10. Dickinson, W.R. Paleoshoreline record of relative Holocene sea levels on Pacific islands. *Earth-Sci. Rev.* **2001**, *55*, 191–234. [CrossRef]
11. Shaked, Y.; Marco, S.; Lazar, B.; Stein, M.; Cohen, C.; Sass, E.; Agnon, A. Late Holocene shorelines at the Gulf of Aqaba: Migrating shorelines under conditions of tectonic and sea level stability. *EGU Stephan Mueller Spec. Publ. Ser.* **2002**, *2*, 105–111. [CrossRef]
12. de Oliveira Caldas, L.H.; Stattegger, K.; Vital, H. Holocene sea-level history: Evidence from coastal sediments of the northern Rio Grande do Norte coast, NE Brazil. *Mar. Geol.* **2006**, *228*, 39–53. [CrossRef]
13. Mauz, B.; Vacchi, M.; Green, A.; Hoffmann, G.; Cooper, A. Beachrock: A tool for re-constructing relative sea level in the far-field. *Mar. Geol.* **2015**, *362*, 1–6. [CrossRef]
14. Manaa, A.A.; Jones, B.G.; McGregor, H.V.; Zhao, J.X.; Price, D.M. Dating Quaternary raised coral terraces along the Saudi Arabian Red Sea coast. *Mar. Geol.* **2016**, *374*, 59–72. [CrossRef]
15. Kelletat, D. Beachrock as sea-level indicator? Remarks from a geomorphological point of view. *J. Coast. Res.* **2006**, *22*, 1558–1564. [CrossRef]
16. Ozturk, M.Z.; Erginal, A.E.; Kiyak, N.G.; Ozturk, T. Cement fabrics and optical luminescence ages of beachrock, North Cyprus: Implications for Holocene sea-level changes. *Quat. Int.* **2016**, *401*, 132–140. [CrossRef]
17. Karkani, A.; Evelpidou, N.; Vacchi, M.; Morhange, C.; Tsukamoto, S.; Frechen, M.; Maroukian, H. Tracking shoreline evolution in central Cyclades (Greece) using beachrocks. *Mar. Geol.* **2017**, *388*, 25–37. [CrossRef]
18. Falkenroth, M.; Schneider, B.; Hoffmann, G. Beachrock as sea-level indicator–a case study at the coastline of Oman (Indian Ocean). *Quat. Sci. Rev.* **2019**, *206*, 81–98. [CrossRef]
19. Holail, H.; Rashed, M. Stable isotopic composition of carbonate-cemented recent beachrock along the Mediterranean and the Red Sea coasts of Egypt. *Mar. Geol.* **1992**, *106*, 141–148. [CrossRef]
20. Vousdoukas, M.I.; Velegrakis, A.F.; Plomaritis, T.A. Beachrock occurrence, characteris-tics, formation mechanisms and impacts. *Earth-Sci. Rev.* **2007**, *85*, 23–46. [CrossRef]
21. Erginal, A.E.; Güneç, N.K.; Bozcu, M.; Ertek, A.; Güngüneş, H.; Sungur, A.; Türker, G. On the origin and age of the Arıburnu beachrock, Gelibolu Peninsula, Turkey. *Turk. J. Earth Sci.* **2008**, *17*, 803–819.
22. Ghandour, I.M.; Al-Washmi, H.A.; Bantan, R.A.; Gadallah, M.M. Petrographical and petrophysical characteristics of asynchronous beachrocks along Al-Shoaiba Coast, Red Sea, Saudi Arabia. *Arab. J. Geosci.* **2014**, *7*, 355–365. [CrossRef]
23. Milliman, J.D. *Recent Sedimentary Carbonates: Part 1 Marine Carbonates*; Springer: New York, NY, USA, 1974; p. 375.
24. Hanor, J.S. Precipitation of beachrock cements; mixing of marine and meteoric waters vs. CO_2-degassing. *J. Sediment. Res.* **1978**, *48*, 489–501.
25. Neumeier, U. Experimental modelling of beachrock cementation under microbial influence. *Sediment. Geol.* **1999**, *126*, 35–46. [CrossRef]
26. El-Shater, A.; Abu Ouf, M.A. Beachrock in South Jeddah, The Red Sea Coast of Saudi Arabia. *J King Abdulaziz Univ. Mar. Sci.* **1995**, *6*, 53–65. [CrossRef]
27. Al-Ramadan, K. Diagenesis of Holocene beachrocks: A comparative study between the Arabian Gulf and the Gulf of Aqaba, Saudi Arabia. *Arab. J. Geosci.* **2014**, *7*, 4933–4942. [CrossRef]
28. Koeshidayatullah, A.; Al-Ramadan, K. Unraveling cementation environment and patterns of Holocene beachrocks in the Arabian Gulf and the Gulf of Aqaba: Stable isotope ap-proach. *Geol. Q.* **2014**, *58*, 207–216. [CrossRef]
29. Haredy, R.A.; Ghandour, I.M.; Erginal, A.E.; Bozcu, M. Beachrock Cementation Patterns Along the Gulf of Aqaba Coast, Saudi Arabia. *Arab. J. Sci. Eng.* **2019**, *44*, 479–487. [CrossRef]

30. Strasser, A.; Strohmenger, C.; Davaud, E.; Bach, A. Sequential evolution and diagenesis of Pleistocene coral reefs (South Sinai, Egypt). *Sediment. Geol.* **1992**, *78*, 59–79. [CrossRef]
31. Strasser, A.; Strohmenger, C. Early diagenesis in Pleistocene coral reefs, southern Sinai, Egypt: Response to tectonics, sea-level and climate. *Sedimentology* **1997**, *44*, 537–558. [CrossRef]
32. Holail, H.M.; Shaaban, M.N.; Mansour, A.S. Cementation of Holocene beachrock in the Aqaba and the Arabian Gulfs: Comparative study. *Carbonates Evaporites* **2004**, *19*, 142–150. [CrossRef]
33. Morcos, S.A. Physical and chemical oceanography of the Red Sea. *Oceanogr. Mar. Biol. Annu. Rev.* **1970**, *8*, 73–202.
34. Ghandour, I.M.; Al-Washmi, H.A.; Haredy, R.A. Gravel-sized mud clasts on an arid microtidal sandy beach: Example from the northeastern Red Sea, South Al-Wajh, Saudi Arabia. *J. Coast. Res.* **2013**, *29*, 110–117. [CrossRef]
35. Bantan, R.A.; Ghandour, I.M.; Al-Zubieri, A.G. Mineralogical and geochemical composition of the subsurface sediments at the mouth of Wadi Al-Hamd, Red Sea coast, Saudi Arabia: Implication for provenance and climate. *Environ. Earth Sci.* **2020**, *79*, 1–20. [CrossRef]
36. Moufti, A.M. Mineralogy and mineral chemistry of auriferous stream sediments from Al Wajh area, NW Saudi Arabia. *Arab. J. Geosci.* **2009**, *2*, 1–17. [CrossRef]
37. Davies, F.B. *Explanatory Notes on the Geologic Map of Al Wajh Quadrangle, Sheet 26 B*; Ministry of Petroleum and Mineral Resources, Deputy Ministry for Mineral Resources: Jeddah, Saudi Arabia, 1985; p. 27.
38. Sherrod, L.A.; Dunn, G.; Peterson, G.A.; Kolberg, R.L. Inorganic carbon analysis by modified pressure-calcimeter method. *Soil Sci. Soc. Am. J.* **2002**, *66*, 299–305. [CrossRef]
39. Allan, J.R.; Matthews, R.K. Isotope signatures associated with early meteoric diagenesis. *Sedimentology* **1982**, *29*, 797–817. [CrossRef]
40. Spurgeon, D.; Davis, R.A., Jr.; Shinnu, E.A. Formation of Beach Rock at Siesta Key, Florida and its influence on barrier island development. *Mar. Geol.* **2003**, *200*, 19–29. [CrossRef]
41. Armstrong-Altrin, J.S.; Lee, Y.I.; Verma, S.P.; Worden, R.H. Carbon, oxygen, and strontium isotope geochemistry of carbonate rocks of the upper Miocene Kudankulam Formation, southern India: Implications for paleoenvironment and diagenesis. *Geochemistry* **2009**, *69*, 45–60. [CrossRef]
42. Magaritz, M.; Gavish, E.; Bakler, N.; Kafri, U. Carbon and oxygen isotope composition-indicators of cementation environment in Recent, Holocene and Pleistocene sediments along the coast of Israel. *J. Sediment. Petrol.* **1979**, *49*, 401–411.
43. Beier, J.A. Diagenesis of Quaternary Bahamian beachrock; petrographic and isotopic evidence. *J. Sediment. Res.* **1985**, *55*, 755–761. [CrossRef]
44. Keith, M.L.; Weber, J.N. Carbon and oxygen isotopic composition of selected limestones and fossils. *Geochim. et Cosmochim. Acta* **1964**, *28*, 1787–1816. [CrossRef]
45. Rao, N.D.; Behairy, A.K. Intertidal conglomerate south of Yanbu-An episodic clastic deposit in the eastern Red Sea. *Palaeogeogr. Palaeoclimatol. Palaeoecol.* **1987**, *58*, 221–228.
46. Ghandour, I.M. Controlling factors on the textural characteristics of the coastal sediments around Wadi Al-Hamd mouth Delta, NE Red Sea, Saudi Arabia. *J. King Abdulaziz Univ. Mar. Sci.* **2014**, *25*, 61–78. [CrossRef]
47. Al-Zubieri, A.G.; Ghandour, I.M.; Haredy, R.A. Controlling factors on the grain size distribution of shallow subsurface coastal sediments at the mouth of Wadi Al-Hamd, northeastern Red Sea, Saudi Arabia. *J. King Abdulaziz Univ. Mar. Sci.* **2018**, *27*, 1–17.
48. Jado, A.R.; Zötl, J.G. *Quaternary period in Saudi Arabia*; Springer: Wien, Austria, 1984; p. 360.
49. Moustafa, Y.A.; Pätzold, J.; Loya, Y.; Wefer, G. Mid-Holocene stable isotope record of corals from the northern Red Sea. *Int. J. Earth Sci.* **2000**, *88*, 742–751. [CrossRef]
50. Clifton, H.E. Progradational sequences in Miocene shoreline deposits, southeastern Caliente Range, California. *J. Sediment. Res.* **1981**, *51*, 165–184.
51. Khanna, P.; Petrovic, A.; Ramdani, A.I.; Homewood, P.; Mettraux, M.; Vahrenkamp, V. Mid-Holocene to present circum-Arabian sea level database: Investigating future coastal ocean inundation risk along the Arabian plate shorelines. *Quat. Sci. Rev.* **2021**, *261*, 106959. [CrossRef]
52. Al-Mikhlafi, A.S.; Hibbert, F.D.; Edwards, L.R.; Cheng, H. Holocene relative sea-level changes and coastal evolution along the coastlines of Kamaran Island and As-Salif Pen-insula, Yemen, southern Red Sea. *Quat. Sci. Rev.* **2021**, *252*, 106719. [CrossRef]
53. Murray-Wallace, C.V.; Woodroffe, C.D. *Quaternary Sea-Level Changes: A Global Perspective*; Cambridge University Press: New York, NY, USA, 2014; p. 484.
54. James, N.P.; Choquette, P.W. Limestone: The sea floor diagenetic environment. In *Diagenesis*; McIlreath, I., Morrow, D., Eds.; Geological Association of Canada Reprint Series: St John's, ND, Canada, 1990; pp. 13–34.
55. Vieira, M.M.; De Ros, L.F. Cementation patterns and genetic implications of Holocene beachrocks from northeastern Brazil. *Sediment. Geol.* **2006**, *192*, 207–230. [CrossRef]
56. Desruelles, S.; Fouache, É.; Ciner, A.; Dalongeville, R.; Pavlopoulos, K.; Kosun, E.; Coquinot, Y.; Potdevin, J.L. Beachrocks and sea level changes since Middle Holocene: Comparison between the insular group of Mykonos–Delos–Rhenia (Cyclades, Greece) and the southern coast of Turkey. *Glob. Planet. Chang.* **2009**, *66*, 19–33. [CrossRef]
57. Alexanderson, T. Mediterranean beachrock cementation: Marine precipitation of Mg-calcite. In *The Mediterranean Sea: A Natural Sedimentation Laboratory*; Stanley, D.J., Ed.; Dowden, Hutchinson and Ross Publishers: Stroudsburg, PA, USA, 1972; pp. 203–223.
58. Strasser, A.; Davaud, E.; Jedoui, Y. Carbonate cements in Holocene beachrock: Example from Bahiret et Biban, southeastern Tunisia. *Sediment. Geol.* **1989**, *62*, 89–100. [CrossRef]

59. Meyers, J.H. Marine vadose beachrock cementation by cryptocrystalline magnesian cal-cite, Maui, Hawaii. *J. Sediment. Res.* **1987**, *57*, 558–570.
60. Chaves, N.S.; Sial, A.N. Mixed oceanic and freshwater depositional conditions for beachrocks of Northeast Brazil: Evidence from carbon and oxygen isotopes. *Int. Geol. Rev.* **1998**, *40*, 748–754. [CrossRef]
61. Calvet, F.; Cabrera, M.C.; Carracedo, J.C.; Mangas, J.; Pérez-Torrado, F.J.; Recio, C.; Travé, A. Beachrocks from the island of La Palma (Canary Islands, Spain). *Mar. Geol.* **2003**, *197*, 75–93. [CrossRef]
62. Vieira, M.M.; Sial, A.N.; De Ros, L.F.; Morad, S. Origin of Holocene beachrock cements in northeastern Brazil: Evidence from carbon and oxygen isotopes. *J. South Am. Earth Sci.* **2017**, *79*, 401–408. [CrossRef]
63. Hudson, J.D. Stable isotopes of limestone lithification. *J. Geol. Soc. Lond.* **1977**, *133*, 637–660. [CrossRef]
64. Coudray, J.; Montaggioni, L. The diagenetic products of marine carbonate as sea level indicators. In *Sea-Level Research: A Manual for the Collection and Evaluation of Data*; Plassche, O.V.D., Ed.; Geo Books: Norwich, UK, 1986; pp. 311–360.

Journal of
Marine Science and Engineering

Review

Marine Geological Studies of the Bay of Naples (Southern Tyrrhenian Sea, Italy): Revised Applications of the Seismo-Stratigraphic Concepts and Evolving Technologies to a Late Quaternary Volcanic Area

Gemma Aiello

Istituto di Scienze Marine (ISMAR), Consiglio Nazionale delle Ricerche (CNR), Sezione Secondaria di Napoli, 80133 Naples, Italy; gemma.aiello@cnr.it

Abstract: Marine geological studies of Naples Bay are discussed and reviewed, focusing on the application of the seismo-stratigraphic concepts to a Late Quaternary volcanic area. The Naples Bay represents an active volcanic area in which the interactions between volcanic and sedimentary processes controlled a complex stratigraphic architecture during the Late Quaternary period. While the volcanic processes took place in correspondence with the activity of the Somma–Vesuvius, Campi Flegrei Ischia, and Procida volcanic complexes, the sedimentary processes were controlled by the fluvial processes in the Sarno-Sebeto coastal plain and by the tectonic uplift in correspondence with the Sorrento Peninsula's structural high Key geophysical and stratigraphic studies of the three active volcanic complexes are revised and discussed. The seismo-stratigraphic concepts applied in the geological interpretation of seismic profiles of Naples Bay are reviewed and discussed: here, the classical concepts of seismic and sequence stratigraphy have been successfully applied, but only partly, due to the occurrence of several buried volcanoes and volcanic seismic units and tephra layers, calibrated by gravity cores.

Keywords: marine geology; seismic stratigraphy; sequence stratigraphy; Naples Bay; Somma–Vesuvius; Campi Flegrei; Ischia; bradyseism; earthquakes; landslides

Citation: Aiello, G. Marine Geological Studies of the Bay of Naples (Southern Tyrrhenian Sea, Italy): Revised Applications of the Seismo-Stratigraphic Concepts and Evolving Technologies to a Late Quaternary Volcanic Area. *J. Mar. Sci. Eng.* **2024**, *12*, 416. https://doi.org/10.3390/jmse12030416

Academic Editor: Michael Lazar

Received: 15 January 2024
Revised: 17 February 2024
Accepted: 19 February 2024
Published: 26 February 2024

Copyright: © 2024 by the author. Licensee MDPI, Basel, Switzerland. This article is an open access article distributed under the terms and conditions of the Creative Commons Attribution (CC BY) license (https://creativecommons.org/licenses/by/4.0/).

1. Introduction

The seismic stratigraphy of Naples Bay is herein revised and discussed, starting from the old (1980–2000) seismo-stratigraphic data and proceeding towards the new (2000–2023) seismo-stratigraphic data. Focusing on case histories located in the Somma–Vesuvius, Campi Flegrei, and Ischia and Procida offshore, the technological content of the seismo-stratigraphic data, if applied to the Naples Bay, is also taken into account, reviewing the basic seismic systems used in the acquisition of single-channel, multichannel, and magnetic data.

This introduction assesses reflection seismic overlays, diffraction, seismic sources, receivers, filters, move out, migration, and deconvolution processes, reporting which systems were deployed and which data were collected with them [1–4].

The Sparker is among the most important seismic sources for marine surveys and has been used extensively in Naples Bay, providing high-resolution seismic profiles useful in geological interpretation. It consists of an electrode array powered by a capacitor bank of a few hundred joules and has been widely employed as a source for studies of the continental shelf [1–4].

From 1998 to 2000, a large database of single-channel reflection seismic measurements has been recorded in the Naples and Pozzuoli Bays by the Parthenope University of Naples, using both the Multispot Extended Array Sparker (MEAS) [5] and the Surfboom. Both these systems are suitable for surveying the continental shelf and slope of Naples Bay [5].

A further development of marine seismic technology in Naples Bay is the Sparker source SAM 96, employed in Antarctica expeditions from 2001 to the present (Figure 1) [6], which is characterized by a varying number of electrodes that can be arranged as "dual-in-line".

Figure 1. SAM 96: Sparker Array Multi-tip (96 electrodes), developed in the framework of the PNRA (Programma Nazionale Ricerche Antartide) [6].

Boomers (EG&G Uniboom; Figure 2) with pulse lengths of 0.1–0.2 ms have been developed for reflection profiling in water depths of a few meters [7,8]. The Uniboom triggers a seismic wave capable of penetrating to significant depths below the seabed. The Uniboom is composed of an electrical coil magnetically coupled to a plate. It has only been subordinately used in Naples Bay, where the use of Sparker and Subbottom Chirp seismic sources has been preferred.

Figure 2. The Uniboom seismic source.

The post-collapse evolution of the Phlegrean caldera is characterized by unrest and eruptions. The most recent evolution of the Phlegrean caldera has been characterized by short-term ground deformation (or bradyseism), culminating in unrest between 1969 and 1972 and between 1982 and 1984. After these bradyseismic crises, the Institute of Oceanology at the Istituto Universitario Navale (IUN) and the National Research Council (CNR ISMAR) in Naples, Italy, conducted high-resolution reflection seismic surveys in the Pozzuoli area [9–14]. These surveys were primarily single-channel, but occasionally multi-channel. Ranieri and Mirabile [9] studied the deep geological structure of the Campi Flegrei volcanic complex with the aim of improving knowledge on seismic technologies and seismic sources more appropriate for the study of volcanic margins. Additional geological and geophysical findings have been presented by Mirabile et al. [15] for the underwater volcanic region in Naples Bay.

D'Argenio et al. [10] applied Airgun technologies to the geological study of Naples Bay, providing interpreted seismic sections of the bay. Some structures are volcanic and their stratigraphic architecture has been reconstructed in detail [16–22]. Recently, Giordano et al. [23] have developed a new type of Sparker source that has been tested along seismic lines located in the Gulf of Pozzuoli and in Naples harbor in order to provide specific seismic characteristics, which has included its use in shallow water depths and the recognition of the first sedimentary layers. The graphical interfaces have been provided by the D Seismic software (version 1), developed in the frame of the PNRA project [24–26]. This software allows for a real-time analysis of the Fast Fourier Transform (FFT) spectrum of the recorded seismic signals [23].

In this paper, a review of significant seismo-stratigraphic studies performed in Naples Bay is provided, referring to the key areas of the Somma–Vesuvius offshore, of Naples Bay, including the Gulf of Pozzuoli, and of the Ischia and Procida offshore, in order to construct significant stratigraphic correlations in the bay.

2. Geological Setting

The Campania–Latium Tyrrhenian margin is distinguished from a continental slope, having a complex architecture. This slope bounds the Vavilov basin and is characterized by a southwards widening, proceeding from the Pontine Islands (20 km wide) to the Palinuro volcanic complex [27] (150 km wide; Figure 3b). A tectonic sketch map of the Southern Apennines shows the regional geological setting of the area [28] (Figure 3a). The Southern Apennines–Calabrian Arc system resulted from the convergence from the Upper Cretaceous to the Quaternary between the Corsica–Sardinia–Calabria block, of European origin, and the Apulian–Adriatic block, of African affinity [28] (Figure 3a). This process involved the subduction of the Neotethys oceanic lithosphere, originally interposed between the two paleo-continental margins.

Transverse systems along the Tyrrhenian extensional margin of central and southern Italy have strongly controlled the volcanism [29] (Figure 4). Extensional processes have affected the Tyrrhenian margin of the Italian Peninsula since the Late Miocene period [30]. The relaxation occurred behind the eastward-migrating Apennine orogen and was largely controlled by the subduction of the oceanic lithosphere beneath the Calabrian Arc [30–33] (Figure 4). Along the Tyrrhenian margin of central Italy, active extensional tectonics migrated eastwards from the Tyrrhenian area (mainly during the Miocene–Pliocene) towards the Apennine divide, where the limit between the extensional and compressional domains is located [29–33] (Figure 4).

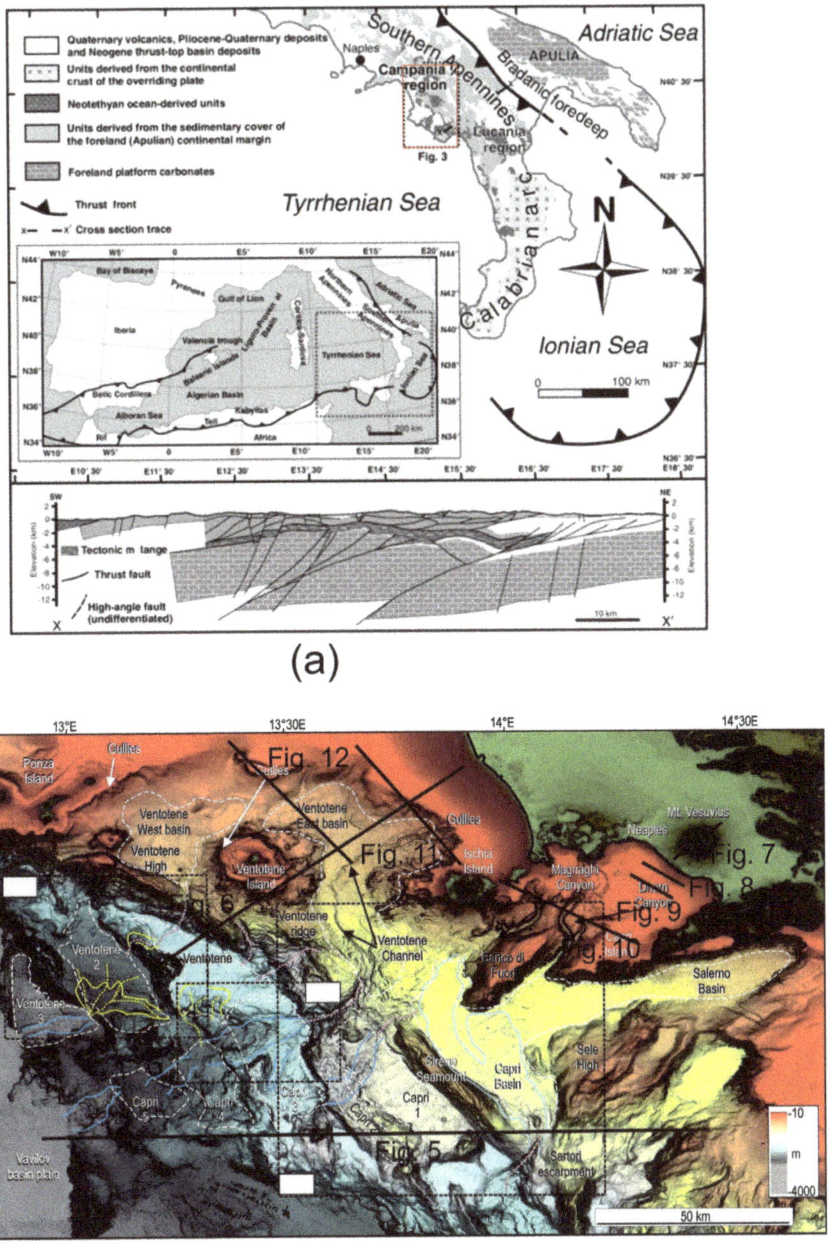

Figure 3. (**a**) Tectonic sketch map of Southern Italy (modified after Ciarcia et al. [28]). (**b**) Shaded relief bathymetric map of the Latium–Campania continental margin (modified after Gamberi et al. [27]). The location of seismic profiles shown in this paper has also been reported. Blanked areas correspond to detailed figures not reported herein.

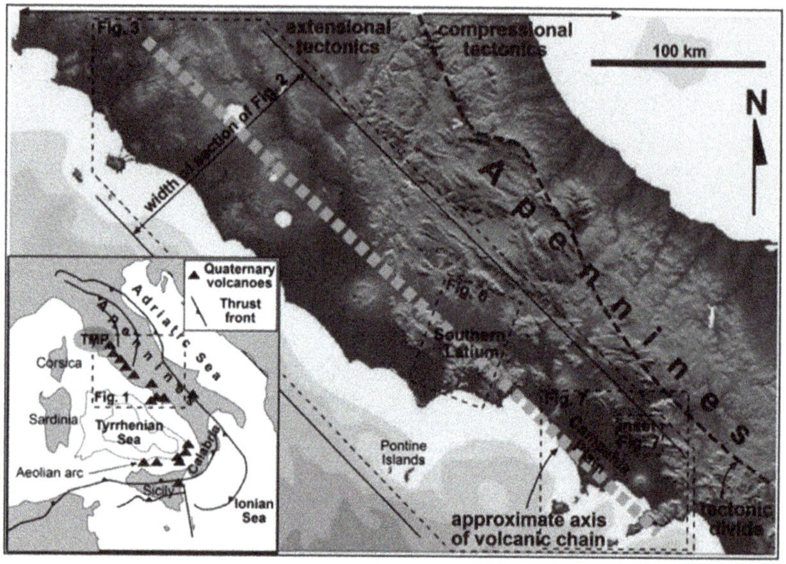

Figure 4. Transverse system along the extensional margin of central and southern Italy (modified after Acocella and Funiciello [29]).

A key seismo-stratigraphic and regional geological study on the Campania Latium continental margin is that of Sartori et al. [34] (Figure 5). Figure 5 shows seismo-stratigraphic sections of the relationships between the Campania continental margin and the Marsili and Vavilov basins, where both continental and oceanic crust have been represented. The oceanic crust has been calibrated through ODP site 651. This sector is made up of horizontal, continuous Plio-Quaternary successions and covers an area with an asymmetrical basement. The sequences are composed of pelagic oozes and turbiditic deposits, the former becoming more and more prevalent (ODP Sites 650, 651). The Plio–Quaternary depocenters (more than 1.4 s twt) occur close to the Campania margin (Figure 5), hosted into half grabens and bounded by faults trending from N–S to NW–SE.

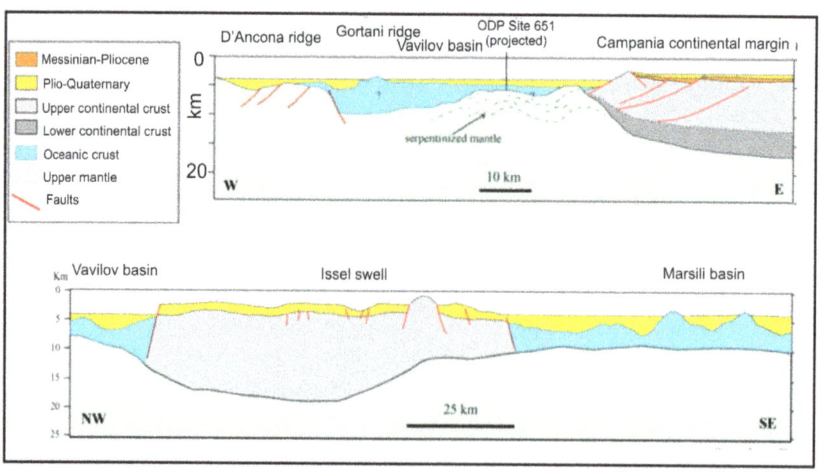

Figure 5. Seismo-stratigraphic sections of the Campania continental margin, showing the relationships with the Vavilov and Marsili basins (modified after Sartori et al. [34]).

Another important seismo-stratigraphic study is that of Conti et al. [35], focusing on the tectono-stratigraphic evolution of the Campania continental margin (Figure 6). The seismic profile TIR10-16, trending from Ventotene to Ischia, displays the extension occurring on the Campania–Latium continental margin, as shown by the systems of normal faults trending NE–SW [35] (Figure 6).

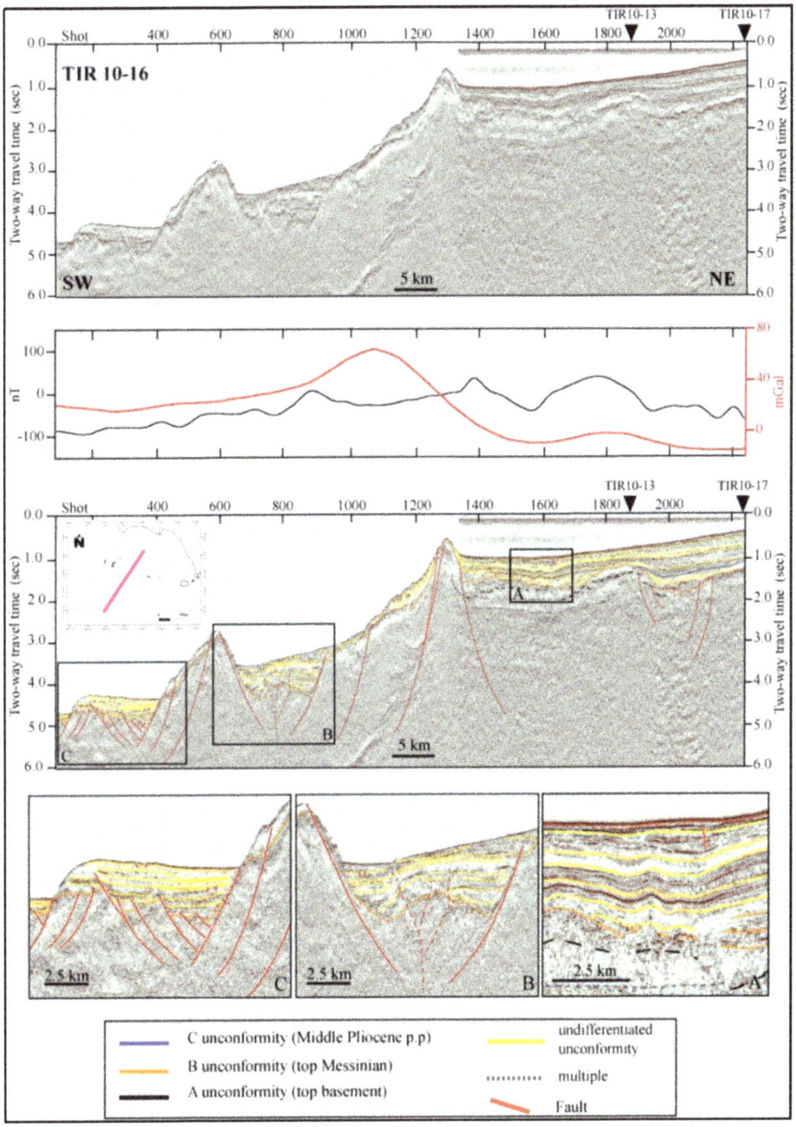

Figure 6. Seismic section TIR 10–16 and corresponding geological interpretation (modified after Conti et al. [35]).

3. Applications of Seismic Stratigraphy

Seismo-stratigraphic concepts and methods have been applied to volcanic sequences [20,36–40]. The buried volcanoes of Ischia have been studied in detail based on seismo-stratigraphic methodologies integrated with the Kingdom software (version 2) [20].

Three serial geological sections, constructed based on seismic interpretation, have shown the geometric characteristics and the stratigraphic relationships between the volcanic and the sedimentary units in the northern Ischia offshore. The seismic units have been correlated with the volcanic units' outcropping in the adjacent northern coastal belt of Ischia [20]. In particular, the V1 seismo-stratigraphic unit is genetically related to a buried volcano next to the Castello d'Ischia volcanic edifice, located in the northeastern Ischia offshore. The V2 seismo-stratigraphic unit is genetically related to a lava dome offshore of Mt. Vico, located in the northwestern Ischia offshore. The V3 seismo-stratigraphic unit is correlated with the Campanian Ignimbrite (39 ky B.P.) based on seismic facies, stratigraphic relationships, and correlation with the onshore geological units of Procida and Vivara [21]. Here, a transition from siliciclastic to bioclastic deposits has been highlighted [21]. The V4 seismo-stratigraphic unit has been correlated with the Sant'Anna lavas (22 ky B.P.). The V5 seismo-stratigraphic unit is genetically related to the Spiaggia degli Inglesi lava flows [22]. The V6 seismo-stratigraphic unit has been interpreted as a lava and scoria volcano, genetically related to the S. Pietro lavas, located eastwards of the Ischia harbor, where the S. Pietro lavas crop out. The V7 seismo-stratigraphic unit is a shallow buried volcano resembling a dome or a dyke, uplifting and deforming the surrounding sediments, and located offshore of Punta della Scrofa (northern Ischia).

Planke et al. [36] have highlighted that the seismic units of volcano stratigraphy consist of landward flows, lava deltas, inner flows, inner SDR (Seaward Dipping Reflectors), outer high, and outer SDR. A five-stage volcano stratigraphic model has been proposed for a rifted continental margin, including initial explosive volcanism with the formation of basaltic complexes poorly imaged on seismic data, effusive subaerial volcanism forming the inner flows, lava deltas and landward flows' seismic units, continuing effusive subaerial volcanism forming the inner SDR at the narrowing rift, explosive shallow marine volcanism including the outer high, and deep marine volcanism forming the outer SDR [36]. Jerram et al. [37] have provided a volcano stratigraphic framework identifying a set of lava sequences, including tabular simple flows, compound-braided flows, and subaqueously deposited hyaloclastite facies. Abdelmalak et al. [38] have applied the volcano stratigraphic model of Planke et al. [36], recording the early encroachment of flood basalt into the basin and the buildup of a lava delta system. Bischoff et al. [39] proposed the classification of buried volcanoes into three main geomorphic categories, including the small-volume clusters, craters, and cones, the wide, composite, shield, and caldera volcanoes, and the voluminous lava fields. Martì et al. [40] have reviewed the volcano-stratigraphic concepts, highlighting that different types of volcanic unconformities exist. The first type of volcanic unconformities ('minor unconformities') appear between successive eruptions that are separated by phases of quiescence, or even between different pulses of the same eruption, and are not associated with varying dynamics of the volcanic system. The second type of volcanic unconformities ("moderate unconformities") are formed during the erosional processes of previous volcanic deposits, during the deposition of reworked materials and the formation of paleosoil separating different volcanic deposits. The third type of volcanic unconformities ("major unconformities") individuates during the construction of a stratovolcano and involves sector failures, caldera collapses, and changes in the clustering of the volcanic cones.

4. Results

4.1. *Seismic Stratigraphy of the Somma–Vesuvius Offshore*

The seismic stratigraphy offshore of the Somma–Vesuvius volcanic complex, as studied by several authors, is herein revised [41–49]. Gravimetric data have been used to complement seismo-stratigraphic knowledge [50]. Bruno et al. [49] have shown the seismo-stratigraphic setting of the Mesozoic carbonate basement onshore of the Somma–Vesuvius volcanic complex to be based on multichannel seismic profiles. Three important seismic reflectors have been recognized, corresponding with the top of the Meso-Cenozoic carbonate sequence, the top of Pliocene deposits, and a strong unconformity within the Quaternary deposits, respectively (Figure 7) [49].

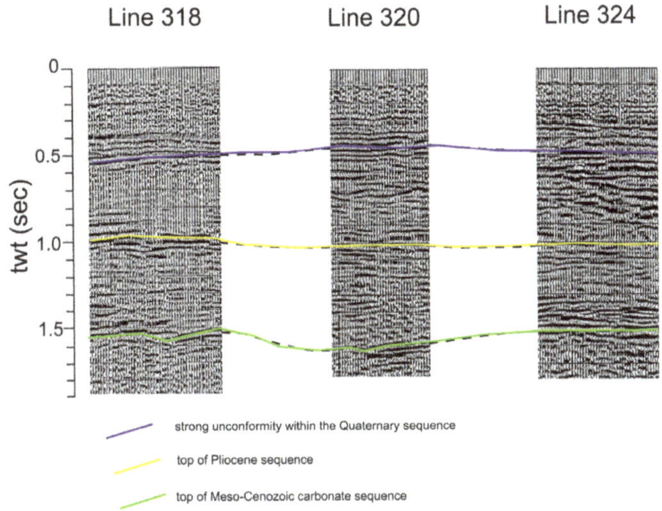

Figure 7. Sketch diagram showing significant seismic horizons onshore the Somma–Vesuvius volcanic complex (modified after Bruno et al. [49]).

Three representative examples of the reflective subsurface image from the area are shown in Figure 7 at 0.5 s, 1 s, and 1.5 s, for two-way time. The three horizons can be traced within the two other sequences, as well. The Trecase 1 well has calibrated the third horizon at 1.5 s, concurrent with the top of the carbonate determined down the well. The low seismic data quality on line 316 does not permit a certain reciprocity of the other two upper reflectors with the post-Mesozoic stratigraphy. The reflector at around 1 s, with two-way time, is relevant to the top of the Pliocene sequence beneath the Quaternary sediments. The uppermost horizon, at about 0.5 s, stands for a strong unconformity within the Quaternary deposits. The general structure of the carbonate basement is more composite than a monoclinalic structure, dipping towards the northwest. In general, the carbonates deepen towards Naples Bay and towards the center of the Campania Plain. They are downthrown by regional faults, controlling a horst and graben-like structure, and reach a maximum depth of approximately 2600 m in the Acerra graben, located in the Campania Plain [49].

Bruno and Rapolla [43] have integrated their previous results through a seismic profile recorded within the Avellino caldera, at the southwestern edge of the Great Cone of Vesuvius. An important seismic horizon suggests an SW lateral collapse was undergone by the volcano, probably between 35 and 11 ky B.P. Buried craters, dykes, and pyroclastic deposits were discovered on the eastern side of the volcanic complex. The existence of NW–SE striking faults that cut across Vesuvius has been substantiated, and probably played a major role in the genesis of the volcano. Ring faults were found mainly in the northeastern and southeastern sectors of the volcano. These faults appeared on the seismic data as a low signal-to-noise area qualified by high lateral heterogeneity and the existence of numerous fractures due to the stress field created by the volcano collapse [43].

Aiello [41] has recently proposed an up-to-date seismo-stratigraphic setting offshore of the Somma–Vesuvius volcano. Figure 8 shows the main seismo-stratigraphic results. The seismic profiles are located on the continental shelf of the Bay of Naples offshore of the Somma–Vesuvius volcanic complex and have been recorded through a Watergun seismic source.

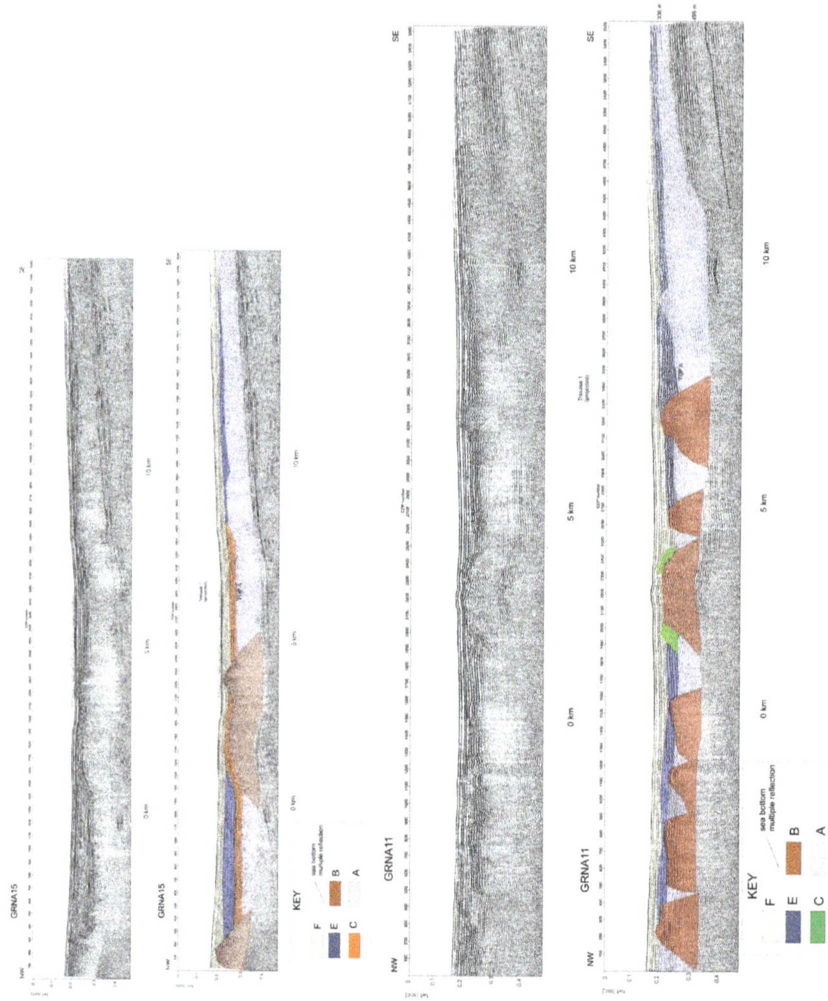

Figure 8. Seismo-stratigraphic setting of the Somma–Vesuvius offshore (modified after Aiello [41] and reinterpreted here as part of the current study). G: PDC older than the AD 79 deposits (detected on other seismic lines and shown in Aiello [41]). F: Highstand System Tract. E: Debris flow genetically related to Somma–Vesuvius. C: Lowstand System Tract. D: NYT (Neapolitan Yellow Tuff deposits; detected on other seismic lines as shown in Aiello [41]). B: Buried tuff rings (cryptodomes in the previous literature). A: Campanian Ignimbrite.

The seismic units previously recognized offshore of Somma–Vesuvius have been reviewed and compared with the seismic units recognized in this paper in a sketch table (Table 1). Previous studies on the submarine structure of the Somma–Vesuvius volcano have reported on the occurrence of buried volcanic structures and thick volcanic seismic units, genetically related with the eruptive events of the Somma–Vesuvius volcano (Table 1) [42,46,47,51].

Table 1. Revision of the seismic units offshore of the Somma–Vesuvius volcanic complex.

Authors	Letter Designation	Seismic Facies	Description	Geological Interpretation
Milia et al. [42]	S	Chaotic	Slump	Debris flow associated with the 1631 eruption
	H	Parallel reflectors, good continuity, high amplitude	Highest unit	Holocene marine deposits
	NYT	Chaotic	Wedge-shaped unit in the northwestern sector of Naples Bay	Undersea component of the Neapolitan Yellow Tuff erupted from Campi Flegrei 15 ky ago
"	VSP	Reflection free with diffractions at the top	The unit thins southwards of Vesuvius volcano	Major volcanic surge and pyroclastic deposits related to the Mt. Somma collapse 17 ky ago
	V	Reflection free	Dome-shaped unit warping the unit L	Cryptodomes coeval with the Mt. Somma collapse 17 ky ago
"	L	Parallel reflectors, good continuity, high amplitude	Constant thickness across Naples Bay	Lowstand marine deposits
	I	Chaotic	Wedge-shaped seismic unit	Campanian Ignimbrite (37 ky B.P.)
	P	High amplitude, low continuity seismic reflectors	The unit dips northwestwards and onlaps on the MC unit	Marine deposits
	MC	Chaotic	Lowest unit	Meso-Cenozoic carbonates
Aiello et al. [46]	None	Parallel and continuous seismic reflectors	Highest unit	Late Pleistocene–Holocene marine and coastal deposits
	None	Chaotic seismic reflectors	Mound-shaped buried volcanic structures	Submerged and/or buried parasitic vents genetically related to the Somma–Vesuvius volcanic complex
"	None	Discontinuous to chaotic seismic reflectors	Tabular seismic unit	Campanian Ignimbrite
"	None	Parallel and discontinuous seismic reflectors	Thin drape onlapping underlying seismic units	Marine deposits of a lowstand phase older than 35 ky B.P.
"	None	Inclined to parallel seismic reflectors	Relict prograding wedge	Middle-Late Pleistocene prograding wedge
Aiello et al. [51]	E	Parallel seismic reflectors	Wedge-shaped unit cropping out at the sea bottom and overlying volcanic structures	Holocene mud wedge
"	D	Parallel and discontinuous seismic reflectors	Seismic unit filling depressed palaeo-morphologies and onlapping the top of volcanic domes (unit B)	Late Pleistocene marine and coastal deposits

Table 1. *Cont.*

Authors	Letter Designation	Seismic Facies	Description	Geological Interpretation
"	BV	Chaotic	Mound-shaped volcanic seismic units	Submerged and/or buried volcanic vents genetically related to the Somma–Vesuvius volcanic complex
"	CI	Chaotic	Tabular	Campanian Ignimbrite
Passaro et al. [47]	PGLM	Parallel seismic reflectors	Outcropping at the sea bottom	Transgressive and Highstand System Tracts
"	FV			Fluid vents
"	V1-V6	Chaotic	Mound-shaped	Cryptodomes genetically related with the Somma–Vesuvius volcanic complex

In particular, Milia et al. [42] have recognized different seismic units in the Somma–Vesuvius offshore based on the interpretation of Sparker seismic profiles (Table 1). These authors have considered two possible explanations for the buried mounds. The first hypothesis is that these buried mounds are hummocks in a debris flow, produced during the Mt. Somma collapse, creating the breached crater. The second account is that they are cryptodomes, rising upwards and warping the horizontal layers of the lowstand marine deposits. The cryptodomes are represented by intrusions triggering the uprising of the overlying sediments, whose comprehension is important for the volcanic hazard assessment.

A three-dimensional reconstruction of the volcanic structures located offshore of the Torre del Greco town (Vesuvius area) has been carried out [51]. In particular, the seismic units include the Holocene mud wedge, overlying the underlying volcanic structures; the Late Pleistocene marine and coastal deposits, characterized by parallel to sub-parallel seismic reflectors, with lateral terminations of onlap on the flanks of the volcanic domes; the BV unit, a wide volcanic vent located offshore of the volcano; the B unit, corresponding with the mound-shaped, isolated, and buried volcanic structures located near the top of the CI seismic unit; and the Campanian Ignimbrite (CI) seismic unit, showing a complex palaeo topography, with some palaeo-terraced surfaces located at its top, controlled by normal faults with a little throw (Table 1).

Several seismic units have been reported on the submarine slopes of the Somma–Vesuvius volcanic complex (Table 1) [47], mainly represented by the highstand and transgressive deposits (PGLM) overlying a regional unconformity, corresponding with the Last Glacial Maximum (LGM; 19 ky B.P.). Fluid vents (FV) and mound-shaped cryptodomes (V1-V6) have also been distinguished (Table 1). Four volcanic vents, located offshore of the Somma–Vesuvius volcanic complex, have been emplaced after the LGM, while a fifth vent was emplaced during more recent times, suggesting the reactivation of volcanic activity during historical times.

4.2. Seismic Stratigraphy of Naples Bay

One of the key papers on the seismic stratigraphy of Naples Bay is that of Fusi et al. [52], whose seismic sections have been recently revised by Aiello and Caccavale [46]. Figure 9 shows one of the most significant seismic sections of the Sparker database of Fusi et al. [52], as reassessed by Aiello and Caccavale [46], which is located in the central sector of Naples Bay, corresponding to the Dohrn canyon (Dohrn western branch and Dohrn eastern branch). In particular, this seismic profile shows that the Dohrn eastern branch was completely incised in the Late Pleistocene prograding wedge. The upper part of the wedge on the canyon's southeastern flank is being eroded away. This implies that the origin of this canyon has been controlled by both the sedimentary processes and by the volcanic ones (Figure 9).

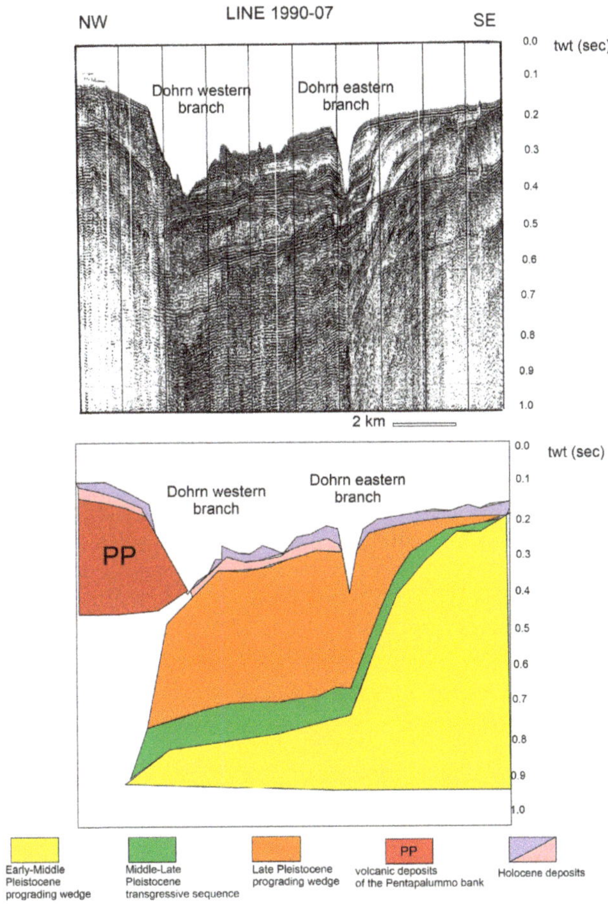

Figure 9. Sparker profile 1990-07 and corresponding geological interpretation (modified after Fusi et al. [52] and Aiello and Caccavale [46] and reinterpreted here as part of the current study).

The key seismic section showing the stratigraphy of Naples Bay has been reinterpreted, aimed at showing a complete seismo-stratigraphic setting of the area [10] (Figure 10). The seismic profile of Figure 10 is located in Naples Bay from offshore of the Ischia and Procida Islands and offshore of the Sorrento Peninsula [10]. The carbonate acoustic basement is characterized by a NW-dipping monoclinalic structure and is genetically related to the Meso-Cenozoic carbonate unit outcropping in the adjacent coastal belt in correspondence with the Sorrento Peninsula and Capri Island. The deepest seismic unit is represented by a NW-dipping prograding wedge, with eroded topsets and preserved clinoforms, Early–Middle Pleistocene in age (Figure 10). The B unconformity bounds the seismic unit 1 in its upper part and represents an erosional truncation extending from the continental shelf to the upper slope. On the northwestern flank of the basin, the unconformity is down-thrown by normal faults with a little vertical throw, while on the southeastern one it rises along the NW-dipping monoclinalic structure composed of the carbonate basement and seismic unit 1 (Figure 10).

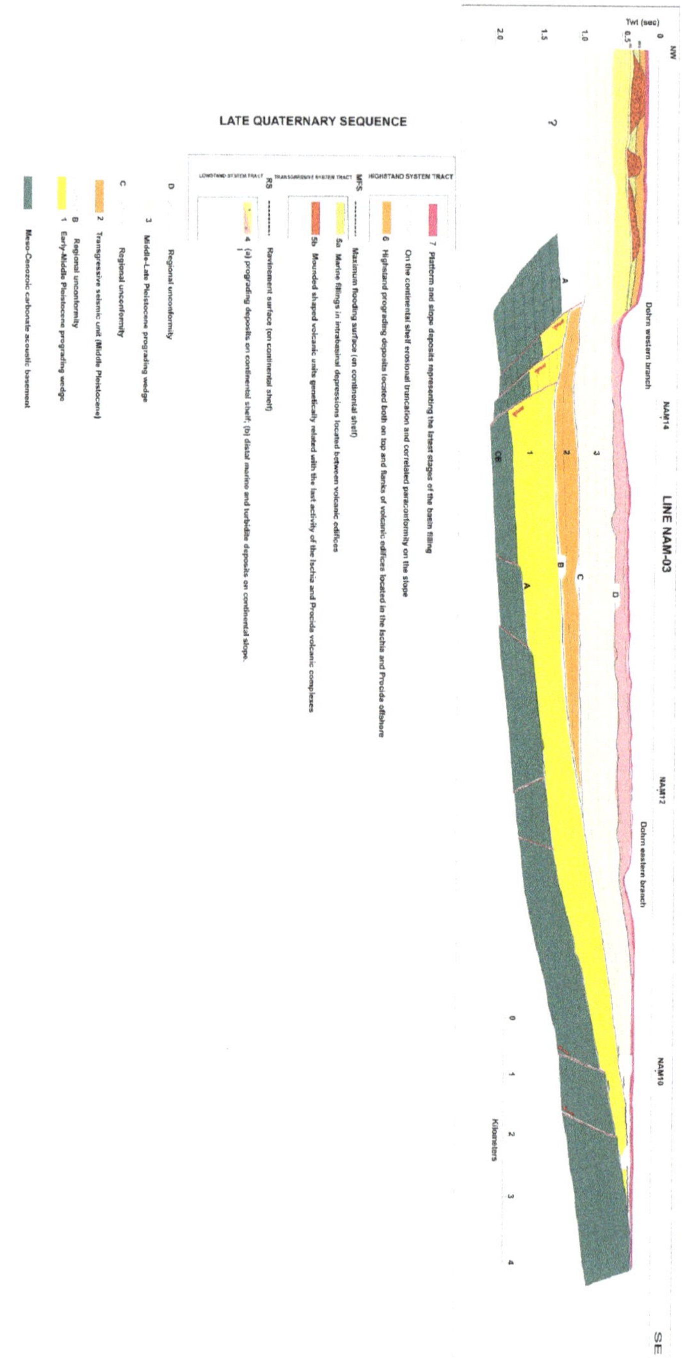

Figure 10. Geological interpretation of the NAM3 multichannel profile (modified after D'Argenio et al. [10] and reinterpreted here as part of the current study).

Seismic unit 2 is wedge-shaped and shows a progressive onlap of the seismic reflectors on the B unconformity. The stratigraphic relationships with the adjacent seismic units and the overall seawards shifting of the coastal and marine facies suggest its deposition during a transgressive phase younger than the B unconformity. The wedging of seismic unit 2 highlights its synsedimentary nature and that it was deposited during active tectonic phases of the adjacent mainland (Figure 10). The C unconformity bounds seismic unit 2 in its upper part and is characterized by an erosional to non-depositional hiatus ranging in age from the Late Pleistocene to the Holocene.

Seismic unit 3 is a shelf and slope sequence characterized by sigmoidal/oblique to parallel reflectors (Figure 10). It is widespread in the western sector of the bay and is older than the emplacement of the Campanian Ignimbrite seismic unit (39 ky B.P.). The entry points of the prograding wedge indicate feeding sources located in correspondence with the Sarno river mouth. The Dohrn and Magnaghi canyons are carved deeply into the unit, displaying relict morphologies located at the Dohrn eastern head and suggesting its deposition during lowstand phases of the Middle–Late Pleistocene era. The D unconformity is located at the top of seismic unit 3 and at the base of the Late Quaternary depositional sequence. It corresponds to an erosional unconformity, probably subaerial, as shown by the occurrence of palaeochannels (Figure 10).

The Late Quaternary depositional sequence is composed of lowstand, transgressive, and highstand system tracts (Figure 10).

The lowstand system tract is composed of two kinds of deposits, namely the prograding deposits on the continental shelf (4a in Figure 10) and the distal marine and turbidite deposits on the continental slope (4b in Figure 10). Prograding deposits are cut by the seismic section perpendicular to their direction of progradation, and they may appear as sub-horizontal reflectors. On the slope, distal marine facies and gravitational reworked sediments alternating with chaotic and acoustically transparent intervals lie in paraconformity on the underlying seismic units. In its upper part, the lowstand system tract is bounded by a ravinement surface (Figure 10).

The transgressive system tract constitutes volcanic bodies, mound-shaped (5a in Figure 10), and marine fillings deposited in intrabasinal depressions located between the volcanic edifices (5b in Figure 10). It is bounded upwards by the maximum flooding surface.

The highstand system tract is represented by highstand prograding deposits located both on top and on the flanks of volcanic edifices and by gravitational reworked sediments and turbidite deposits, representing the latest stages of basin filling (Figure 10).

4.3. Seismic Stratigraphy of the Ischia Offshore

Key seismo-stratigraphic studies on the Ischia volcanic complex are those of Bruno et al. [53], Aiello et al. [54], Aiello et al. [55], Aiello [56], Aiello and Caccavale [21], and Milia et al. [20]. All these studies have highlighted that the occurrence of isolated volcanic bodies renders the sequence stratigraphic approach particularly challenging for the geological interpretation of seismic profiles. The volcanic bodies of Ischia have been successfully mapped using the software Kingdom (version 2) [20]. Bruno et al. [53] have reported E–W and NE–SW volcanic ridges following the regional faults, NE–SW trending. Submerged volcanoes are cut by fault scarps that balance, with a strong vertical component, Holocene marine and volcaniclastic deposits. Aiello et al. [54] have highlighted submarine gravitational instabilities occurring in the northwestern off-shore of the island as resulting from widespread debris avalanches, often interlayered in the stratigraphic succession. Aiello et al. [55] have established the general seismo-stratigraphic setting of the Ischia off-shore, correlating onshore volcanic stratigraphy with the seismic stratigraphy of the offshore sectors of Punta Imperatore, Grotta del Mavone, Punta del Chiarito, S. Angelo, Barano, and Maronti. Figure 11 shows the seismo-stratigraphic setting of the Ischia structural high in the regional framework of the Bay of Naples.

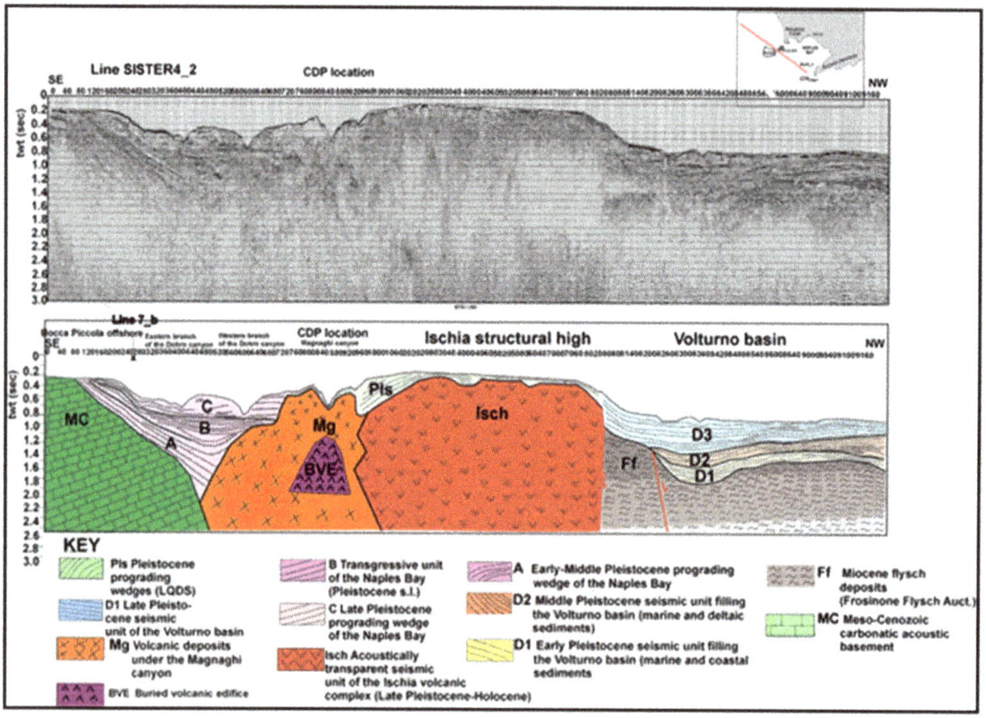

Figure 11. Geological interpretation of the SISTER4_2 multichannel profile (modified after Aiello et al. [13] and reinterpreted here as part of the current study). Volcanic structures: Mg (Magnaghi canyon), BVE (Buried Volcanic Edifice), Isch (Ischia structural high).

5. Discussion and Conclusions

Marine geological studies on Naples Bay have been revised, focusing on the application of seismo-stratigraphic concepts to a Late Quaternary volcanic area, and in particular, to the Somma–Vesuvius, central Naples Bay and Ischia offshore areas.

The seismo-stratigraphic method has been fully applied in central Naples Bay, where four regional unconformities (A, B, C, and D in Figure 10), separating incomplete depositional sequences in the Naples basin filling, have been recognized (Figure 10). The only complete depositional sequence is the Late Quaternary one.

In the Somma–Vesuvius offshore area, the seismic profiles have recorded a low seismic unit (CI), corresponding to the 39 ky old Campanian Ignimbrite and another seismic unit (B), located seawards, that is located below the LST deposits and is overlain by the HST deposits (Figure 8) [41]. B is associated with cryptodomes. If lowstand deposits lie on the B unit, the B unit should have emerged at the moment of LST deposition. A cryptodome is, by definition, a subsurface structure.

In proximal regions, the lowstand deposits are overlaid by a chaotic seismic unit, explained as pyroclastic flow deposits (PF) [41]. The lowstand deposits, which have parallel and continuous seismic reflectors, lie above the Campanian Ignimbrite. Due to its seismic facies and stratigraphic relationships, the lowstand unit has been explained as the sediments that were put in the falling and as lowstand of the sea level. A local uplift has occurred in recent times, as commanded by the cryptodome intrusion, and can be observed in the seafloor deformation. These events have been troubled by the eruption of the NYT unit (15 ky B.P.) in the Naples offshore area, representing a main volcano–tectonic case in the Naples area.

A close linkage of the Somma–Vesuvius seismo-stratigraphic data with the Quaternary volcanism of the Campania continental margin is suggested. Offshore of the Vesuvius volcano, seven seismic units are known, in particular the G, the F, the E, the D, the C, the B, and the A units (Figure 8) [41]. These units, corresponding to PDC older than the 79 A.D. eruption (G), to the Highstand System Tract (F), to a debris flow genetically related to Somma–Vesuvius (E), to the Lowstand System Tract (C) D: to the NYT deposits (D), to the buried tuff rings (B, cryptodomes in the previous literature), and to the Campanian Ignimbrite (A), respectively, are associated with the Quaternary volcanism of the Campania continental margin [11,12,30,48].

The transverse extensional systems occurring on the Tyrrhenian extensional margin of central and southern Italy have strongly influenced the distribution of volcanism [28,29]. This is demonstrated by the structural maps showing the main fault systems and the Plio-Quaternary volcanoes and confirmed by the location of volcanic edifices along the margin (Vulsini, Sabatini, Colli Albani, Roccamonfina, and volcanoes of the Neapolitan district) [29]. Moreover, this has been confirmed by regional seismo-stratigraphic evidence on the continental margin. The seismic profiles TIR 10–16 (Figure 6) and TIR10-14 have shown the extension on the Campania–Latium margin as controlled by NE–SW trending normal faults [35], and a strike-slip component, locally occurring, suggests transverse systems.

Important buried volcanic structures occur on the Campania continental margin. Apart the buried volcanoes shown in the Somma–Vesuvius offshore area (Figure 8), central Naples Bay (Figures 9 and 10), and Ischia (Figure 11), another important volcanic edifice, namely the Pithecusa volcano, has been detected at the Volturno river mouth based on seismic interpretation (Figure 12). Its emplacement is genetically related to the eruptive phases of the Neapolitan volcanoes (Figure 12). In its upper part, the volcanic edifice is eroded by a regional unconformity and its northeastern slope is down thrown by a normal fault, marking the passage to a seismo-stratigraphic unit correlated to the Frosinone Flysch (Figure 12). This unit appears clearly on the northeastern side of the buried volcanic edifice, where it is down-thrown by normal faults. The western side of the volcano, as well as the seismic unit genetically related to the flysch deposits, are overlain by a thick seismic unit, genetically related to Early Pleistocene marine and coastal deposits (Figure 12).

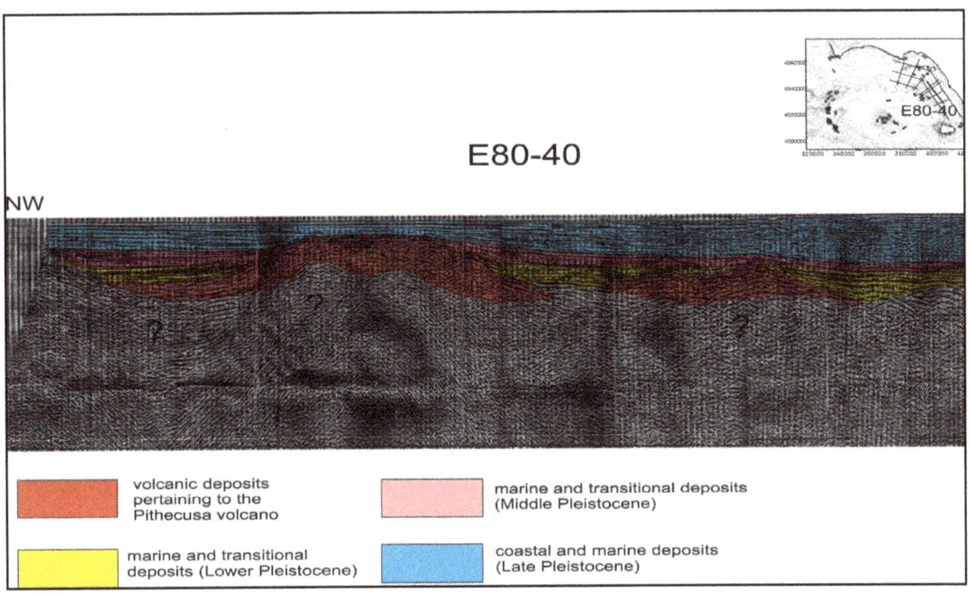

Figure 12. Seismo-stratigraphic setting of the Pithecusa volcano on the Campania continental margin (Volturno river mouth).

Funding: This research received no external funding.

Institutional Review Board Statement: Not applicable.

Informed Consent Statement: Not applicable.

Data Availability Statement: Not applicable.

Conflicts of Interest: The author declares no conflicts of interest.

References

1. Jones, E.J.W. *Marine Geophysics*, 1st ed.; John Wiley & Sons Inc: Hoboken, NJ, USA, 1999; pp. 1–474.
2. Robinson, E.A. *Seismic Velocity Analysis and the Convolutional Model*, 1st ed.; Springer: Berlin/Heidelberg, Germany, 1982; pp. 1–290.
3. Cole, R.H. *Underwater Explosions*; Dover Publications: New York, NY, USA, 1965; pp. 1–490.
4. Knott, S.T.; Hersey, J.B. Interpretation of high-resolution echo-sounding techniques and their use in bathymetry, marine geophysics and biology. *Deep Sea Res.* **1956**, *4*, 36–44. [CrossRef]
5. Mirabile, L.; Fevola, F.; Galeotti, F.; Ranieri, G.; Tangaro, G. Sismica monocanale ad alta risoluzione con sorgente multi spot di tipo sparker: Applicazione ai dati di tecniche di deconvoluzione. In Proceedings of the 10th GNGTS Congress, Rome, Italy, 18–21 October 1991.
6. Corradi, N.; Ferrari, M.; Giordano, F.; Giordano, R.; Ivaldi, R.; Sbrana, A. SAM-source and D-seismic system: The Use in Marine Geological Mapping CARG and PNRA Projects. In Proceedings of the 27th IAS Meeting of Sedimentologists, Alghero, Italy, 18–20 September 1990.
7. Edgerton, H.E.; Hayward, G.G. The boomer sonar source for seismic profiling. *J. Geophys. Res.* **1964**, *68*, 3033–3042. [CrossRef]
8. Simpkin, P.G.; Davis, A. For seismic profiling in very shallow water, a novel receiver. *Sea Technol.* **1993**, *34*, 9.
9. Ranieri, G.; Mirabile, L. Ricerca ed applicazione di metodi geofisici al rilievo sperimentale della struttura medio-profonda dell'area flegrea con uso di sorgenti sismiche watergun. *Ann. Ist. Univ. Navale Napoli* **1991**, *57*, 149–186.
10. D'Argenio, B.; Aiello, G.; de Alteriis, G.; Milia, A.; Sacchi, M.; Tonielli, R.; Angelino, A.; Budillon, F.; Chiocci, F.L.; Conforti, A.; et al. *Digital Elevation Model of the Naples Bay and Adjacent Area (Eastern Tyrrhenian Sea, Italy)*; Atlante di Cartografia Geologica, "Mapping Geology in Italy"; APAT: Rome, Italy, 2004.
11. Aiello, G.; Angelino, A.; D'Argenio, B.; Marsella, E.; Ruggieri, S.; Siniscalchi, A. Buried volcanic structures in the Gulf of Naples (Southern Tyrrhenian Sea, Italy) resulting from high resolution magnetic survey and seismic profiling. *Ann. Geophys.* **2005**, *48*, 1–15.
12. Aiello, G.; Cicchella, A.G.; Di Fiore, V.; Marsella, E. New seismo-stratigraphic data of the Volturno Basin (Northern Campania, Tyrrhenian margin, Southern Italy): Implications for tectono-stratigraphy of the Campania and Latium sedimentary basins. *Ann. Geophys.* **2011**, *54*, 265–283.
13. Aiello, G.; Marsella, E.; Cicchella, A.G.; Di Fiore, V. New insights on morpho-structures and seismic stratigraphy of the Campania continental margin based on deep multichannel profiles. *Rend. Lince-Sci. Fis. Nat.* **2011**, *22*, 349–373. [CrossRef]
14. Aiello, G.; Cicchella, A.G. Dati sismostratigrafici sul margine continentale della Campania tra Ischia, Capri ed il bacino del Volturno (Tirreno meridionale, Italia) in base al processing sismico ed all'interpretazione geologica di profili sismici a riflessione multicanale. *Quad. Geofis.* **2019**, *149*, 1–52.
15. Mirabile, L.; De Marinis, E.; Frattini, M. The Phlegrean fields beneath the sea: The underwater volcanic district of Naples, Italy. *Boll. Geof. Teor. Appl.* **2000**, *41*, 159–186.
16. Aiello, G.; Angelino, A.; Marsella, E.; Ruggieri, S.; Siniscalchi, A. Carta magnetica di alta risoluzione del Golfo di Napoli (Tirreno meridionale). *Boll. Soc. Geol. Ital.* **2004**, *123*, 333–342.
17. Secomandi, M.; Paoletti, V.; Aiello, G.; Fedi, M.; Marsella, E.; Ruggieri, S.; D'Argenio, B.; Rapolla, A. Analysis of the magnetic anomaly field of the volcanic district of the Bay of Naples, Italy. *Mar. Geophys. Res.* **2003**, *24*, 207–221. [CrossRef]
18. Berrino, G.; Cerutti, G.; Corrado, G.; De Maria, P.; Riccardi, U. Gravity studies on active Italian volcanoes: A comparison between absolute and relative gravimetry. *Boll. Geof. Teor. Appl.* **1998**, *40*, 497–510.
19. Fedi, M.; Quarta, T. Wavelet analysis for the regional-residual and local separation of potential field anomalies. *Geophys. Prospect.* **1998**, *46*, 50–57. [CrossRef]
20. Milia, A.; Aiello, G.; Iannace, P.; Torrente, M.M. Complex stratigraphic relationships between volcanic features and sedimentary deposits in a submarine environment: The northern offshore Holocene Ischia volcanic field (Italy). *J. Volcanol. Geotherm. Res.* **2021**, *419*, 107379. [CrossRef]
21. Aiello, G.; Caccavale, M. From Siliciclastic to Bioclastic Deposits in the Gulf of Naples: New Highlights from Offshore Ischia and Procida–Pozzuoli Based on Sedimentological and Seismo-Stratigraphic Data. *Quaternary* **2021**, *4*, 44. [CrossRef]
22. Aiello, G. New insights on the late Quaternary geologic evolution of the Ischia Island coastal belt based on high-resolution seismic profiles. *Ital. J. Geosci.* **2018**, *137*, 87–106. [CrossRef]
23. Giordano, A.; De Luca, L.; Giordano, P. Design and implementation of a marine seismic source by electrical discharge. *Boll. Geof. Teor. Appl.* **2020**, *61*, 589–606.

24. Giordano, F.; Giordano, R.; Corradi, N. D-Seismic: A very flexible low cost hardware/software system for acquisition, real time and post processing of seismic data of Ross Sea (Antartica 2002 expedition). In Proceedings of the Forum Acusticum Seville 2002, Seville, Spain, 16–20 September 2002.
25. Giordano, F.; Giordano, R.; Corradi, N.; Nicotra, G.; Ortosecco, I.; Pittà, A. Improving "S/N" in reflection seismic marine records by mean SAM96 (Sparker array multitip) and D-Seismic (hardware—Software system for seismic data acquisition and processing). In Proceedings of the 5th European Conference on Noise Control, EURONOISE 2003, Naples, Italy, 19–21 May 2003.
26. Corradi, N.; Giordano, F.; Giordano, R. The application of a very high resolution hardware and software (D-Seismic) system for the loss of seismic data for the study of the Ross Sea sedimentary. In Proceedings of the AIOL, Genova, Italy, 14–18 September 2004; Volume 17, pp. 115–124.
27. Gamberi, F.; Della Valle, G.; Marani, M.P.; Mercorella, A.; Distefano, S.; Di Stefano, A. Tectonic controls on sedimentary system along the continental slope of the central and southeastern Tyrrhenian Sea. *Ital. J. Geosci.* **2019**, *138*, 317–332. [CrossRef]
28. Ciarcia, S.; Vitale, S.; Di Staso, A.; Iannace, A.; Mazzoli, S.; Torre, M. Stratigraphy and tectonics of an Internal Unit of the southern Apennines: Implications for the geodynamic evolution of the peri-Tyrrhenian mountain belt. *Terra Nova* **2012**, *21*, 88–96. [CrossRef]
29. Acocella, V.; Funiciello, R. Transverse systems along the extensional Tyrrhenian margin of central Italy and their influence on volcanism. *Tectonics* **2006**, *25*, TC2003. [CrossRef]
30. Jolivet, L.; Faccenna, C.; Goff, B.; Mattei, M.; Rossetti, F.; Brunet, C.; Storti, F.; Funiciello, R.; Cadet, J.P.; d'Agostino, N.; et al. Midcrustal shear zones in postorogenic extension: Example from the northern Tyrrhenian Sea. *J. Geophys. Res. Solid Earth* **1998**, *103*, 12123–12160. [CrossRef]
31. Malinverno, A.; Ryan, W.B.F. Extension in the Tyrrhenian Sea and shortening in the Apennines as result of arc migration driven by sinking of the lithosphere. *Tectonics* **1986**, *5*, 227–245. [CrossRef]
32. Royden, L.; Patacca, E.; Scandone, P. Segmentation and configuration of subducted lithosphere in Italy: An important control on thrust-belt and foredeep-basin evolution. *Geology* **1987**, *15*, 714–717. [CrossRef]
33. Patacca, E.; Sartori, R.; Scandone, P. Tyrrhenian basin and Apenninic arcs: Kinematic relations since Late Tortonian times. *Mem. Soc. Geol. Ital.* **1990**, *45*, 425–451.
34. Sartori, R.; Torelli, L.; Zitellini, N.; Carrara, G.; Magaldi, M.; Mussoni, P. Crustal features along a W–E Tyrrhenian transect from Sardinia to Campania margins (Central Mediterranean). *Tectonophysics* **2004**, *383*, 171–192. [CrossRef]
35. Conti, A.; Bigi, S.; Cuffaro, M.; Doglioni, C.; Scrocca, D.; Muccini, F.; Cocchi, L.; Ligi, M.; Bortoluzzi, G. Transfer zones in an oblique back-arc basin setting: Insights from the Latium-Campania segmented margin (Tyrrhenian Sea). *Tectonics* **2017**, *36*, 78–107. [CrossRef]
36. Planke, S.P.A.; Symonds, E.; Avelstad, J.; Skogseid, J. Seismic volcanostratigraphy of large-volume basaltic extrusive complexes on rifted margins. *J. Geophys. Res. Solid Earth* **2000**, *105*, 333–351. [CrossRef]
37. Jerram, D.A.; Single, R.T.; Hobbs, R.W.; Nelson, C.E. Understanding the offshore flood basalt sequence using onshore volcanic facies analogues: An example from the Faroe–Shetland basin. *Geol. Mag.* **2009**, *146*, 353–367. [CrossRef]
38. Abdelmalak, M.M.; Planke, S.; Faleide, J.I.; Jerram, D.A.; Zastrozhnov, D.; Eide, S.; Myklebust, R. The development of volcanic sequences at rifted margins: New insights from the structure and morphology of the Vøring Escarpment, mid-Norwegian Margin. *J. Geophys. Res. Solid Earth* **2016**, *121*, 5212–5236. [CrossRef]
39. Bischoff, A.; Planke, S.; Holford, S.; Nicol, A. Seismic Geomorphology, Architecture and Stratigraphy of Volcanoes Buried in Sedimentary Basins. In *Updates in Volcanology—Transdisciplinary Nature of Volcano Science*; IntechOpen: Rijeka, Croatia, 2021. [CrossRef]
40. Martí, J.; Groppelli, G.; Brum da Silveira, A. Volcanic stratigraphy: A review. *J. Volcanol. Geotherm. Res.* **2018**, *357*, 68–91. [CrossRef]
41. Aiello, G. Submarine Stratigraphy of the Eastern Bay of Naples: New Seismo-Stratigraphic Data and Implications for the Somma-Vesuvius and Campi Flegrei Volcanic Activity. *J. Mar. Sci. Eng.* **2022**, *10*, 1520. [CrossRef]
42. Milia, A.; Mirabile, L.; Torrente, M.M.; Dvorak, J.J. Volcanism offshore of Vesuvius volcano in Naples Bay. *Bull. Volcanol.* **1998**, *59*, 404–413. [CrossRef]
43. Bruno, P.P.G.; Rapolla, A. Study of sub-surface structure of Somma-Vesuvius (Italy) by seismic reflection data: Implications for Campanian volcanism. *J. Volcanol. Geoth. Res.* **1999**, *92*, 373–387. [CrossRef]
44. Milia, A.; Torrente, M.M.; Bellucci, F. A possible link between faulting, cryptodomes and lateral collapses at Vesuvius Volcano (Italy). *Glob. Planet Change* **2012**, *90–91*, 121–134. [CrossRef]
45. Sacchi, M.; Insinga, D.D.; Milia, A.; Molisso, F.; Raspini, A.; Torrente, M.M.; Conforti, A. Stratigraphic signature of the Vesuvius 79 AD event off the Sarno prodelta system, Naples Bay. *Mar. Geol.* **2005**, *222–223*, 443–469. [CrossRef]
46. Aiello, G.; Caccavale, M. The Coastal Areas of the Bay of Naples: The Sedimentary Dynamics and Geological Evolution of the Naples Canyons. *Geosciences* **2023**, *13*, 226. [CrossRef]
47. Passaro, S.; Sacchi, M.; Tamburrino, S.; Ventura, G. Fluid Vents, Flank Instability, and Seafloor Processes along the Submarine Slopes of the Somma-Vesuvius Volcano, Eastern Tyrrhenian Margin. *Geosciences* **2018**, *8*, 60. [CrossRef]
48. Milia, A.; Torrente, M.M. Space-time evolution of an active volcanic field in an extensional region: The example of the Campania margin (eastern Tyrrhenian Sea). In *Vesuvius, Campi Flegrei, and Campanian Volcanism*, 1st ed.; De Vivo, B., Belkin, H., Rolandi, G., Eds.; Elsevier: New York, NY, USA, 2020; pp. 297–321.

49. Bruno, P.P.; Cippitelli, G.; Rapolla, A. Seismic study of the Mesozoic carbonate basement around Mt. Somma—Vesuvius, Italy. *J. Volcanol. Geotherm. Res.* **1998**, *84*, 311–322. [CrossRef]
50. Linde, N.; Ricci, T.; Baron, L.; Shakas, A.; Berrino, G. The 3-D structure of the Somma-Vesuvius volcanic complex (Italy) inferred from new and historic gravimetric data. *Sci. Rep.* **2017**, *7*, 8434. [CrossRef]
51. Aiello, G.; Marsella, E.; Ruggieri, S. Three-dimensional magneto-seismic reconstruction of the 'Torre del Greco' submerged volcanic structure (Naples Bay, Southern Tyrrhenian Sea, Italy): Implications for Vesuvius's marine geophysics and volcanology. *Near Surf. Geophys.* **2010**, *8*, 17–32. [CrossRef]
52. Fusi, N.; Mirabile, L.; Camerlenghi, A.; Ranieri, G. Marine geophysical survey of the Gulf of Naples (Italy): Relationship between submarine volcanic activity and sedimentation. *Mem. Soc. Geol. Ital.* **1991**, *47*, 95–114.
53. Bruno, P.P.G.; de Alteriis, G.; Florio, G. The western undersea section of the Ischia volcanic complex (Italy, Tyrrhenian sea). *Geophys. Res. Lett.* **2002**, *29*, 57-1–57-4. [CrossRef]
54. Aiello, G.; Marsella, E.; Passaro, S. Submarine instability processes on the continental slopes off the Campania region (Southern Tyrrhenian sea, Italy): The case history of Ischia island (Naples Bay). *Boll. Geof. Teor. Appl.* **2009**, *50*, 193–207.
55. Aiello, G.; Marsella, E.; Passaro, S. Stratigraphic and structural setting of the Ischia volcanic complex (Naples Bay, Southern Italy) revealed by submarine seismic reflection data. *Rend. Lince-Sci. Fis. Nat.* **2012**, *23*, 387–408. [CrossRef]
56. Aiello, G. New sedimentological and coastal and marine geological data on the Quaternary marine deposits of the Ischia Island (Gulf of Naples, Southern Tyrrhenian Sea, Italy). *Geomarine Lett.* **2020**, *40*, 593–618. [CrossRef]

Disclaimer/Publisher's Note: The statements, opinions and data contained in all publications are solely those of the individual author(s) and contributor(s) and not of MDPI and/or the editor(s). MDPI and/or the editor(s) disclaim responsibility for any injury to people or property resulting from any ideas, methods, instructions or products referred to in the content.

MDPI
St. Alban-Anlage 66
4052 Basel
Switzerland
www.mdpi.com

Journal of Marine Science and Engineering Editorial Office
E-mail: jmse@mdpi.com
www.mdpi.com/journal/jmse

Disclaimer/Publisher's Note: The statements, opinions and data contained in all publications are solely those of the individual author(s) and contributor(s) and not of MDPI and/or the editor(s). MDPI and/or the editor(s) disclaim responsibility for any injury to people or property resulting from any ideas, methods, instructions or products referred to in the content.

www.ingramcontent.com/pod-product-compliance
Lightning Source LLC
Chambersburg PA
CBHW040801250926
51850CB00007BA/1054
* 9 7 8 3 7 2 5 8 1 0 8 7 1 *